Plantation Production and Political Power

Plantation Production and Political Power

Plantation Development in South-west India in a Long-Term Historical Perspective, 1743–1963

Paul Erik Baak

DELHI
OXFORD UNIVERSITY PRESS
CALCUTTA CHENNAI MUMBAI
1997

Oxford University Press, Walton Street, Oxford OX2 6DP

Oxford New York
Athens Auckland Bangkok Calcutta
Cape Town Chennai Dar es Salaam Delhi
Florence Hong Kong Istanbul Karachi
Kuala Lumpur Madrid Melbourne Mexico City
Mumbai Nairobi Paris Singapore
Taipei Tokyo Toronto
and associates in
Berlin Ibadan

ISBN 0 19 564103 5

Typeset by Resodyn, New Delhi 110070
Printed in India at Rekha Printers Pvt. Ltd., New Delhi 110020
and published by Manzar Khan, Oxford University Press
YMCA Library Building, Jai Singh Road, New Delhi 110001

To Joseph

Preface

In my experience, working on a Ph.D. thesis in the Social Sciences is an exciting enterprise. Probably to a much greater extent than those working in any of the other scientific disciplines, one is confronted with the relation between subject and object, between 'self' and 'other', between inner motives and the outside world.

My study grew out of my absolute fascination with tropical areas, particularly in combination with social issues. Though I do not exactly know why, I had always dreamed of living in the tropics. Perhaps my father's birthplace, the *Sumatrastraat* in Amsterdam, or my mother's interest in the Netherlands East Indies/Indonesia attracted my attention to (European) life in (exotic) Asia, though the cheerless Dutch climate was most probably the main reason for my daydreams. My family background played a much more serious role in my social awareness: my grandparents, having enjoyed only very limited or no education, started working at a very early age, toiling long hours for a pittance. Probably for that reason, a strong sense of social justice constituted *our* political orientation. My interest in North-South, East-West relations was further stimulated by lively discussions with friends, and various journeys which I undertook. I think these are the reasons I specialized in Non-Western History and, when the opportunity occurred, I left for Kerala to begin my research.

Inspired by works dealing with European expansion and capitalist development (Wallerstein 1974, 1980, 1989; Stavrianos 1981; Wolf 1982) but without much knowledge about planting or India, I started my work as an almost complete outsider. The information obtained from various institutions, associations, companies, and individuals gradually enabled me to become acquainted with the historical ins and outs of plantation development in south-west India. During my three stays in Kerala (1989, 1991–2 and 1994), I consulted archival material and books from the following government, planters and labour organizations:

Centre for Development Studies, Kerala University, Kerala State Archives, Legislative Library, Rubber Research Institute of India; Association of Planters of Kerala, Central Travancore Planters' Association, Kannan Devan Planters' Association, Mundakayam Planters' Association, Tata Tea, United Planters' Association of Southern India; and the Communist Party of India. I am extremely thankful for the openness and practical assistance they offered.

Paradoxically enough, my being a complete outsider was my greatest asset in acquiring inside information. Had I been an official, a planter, or a delegate of a leftist political party, I would never have been able to discuss matters as openly as I did with representatives of all interest groups involved in plantation development. Former representatives of the Kerala government informed me about their relative lack of authority in the plantation districts; the influential planters' lobby; the encroachments on government lands by planters, smallholders and landless labourers; the powerful labour unions. In addition, planters taught me about the daily stress with which they had to cope, being situated between the Board of Directors/Visiting Agents on the one hand and staff/field officers/workers on the other, facing influential labour leaders and corrupt officers, and confronted with social isolation. And the workers told me their often cruel present and past realities; hard labour; low wages; poor housing; and lack of sufficient education, in their opinion probably the cause nowadays of widespread alcoholism and the large number of suicides in former times. In addition, being a 'naïve' Dutchman, I was allowed to ask freely about the implications of 'caste' and 'community' in plantation life, perhaps to a greater extent than an Indian scholar could discuss these politically sensitive issues. And lastly, being a European gave also certain advantages; most informants were also interested in me, my life, and Western society at large.

During my research, I had the pleasure of working with four highly-inspiring supervisors. Dr Dick Kooiman (Vrije Universiteit, Amsterdam) introduced me to the region, its history, historiography, and historical sources. In addition, he encouraged me to continue my initial work in the form of a Ph.D. project, especially since I passed my *doctorandus* examinations with distinction. Dr P.K. Michael Tharakan (Centre for

Development Studies) was my local guide. His knowledge of Kerala history, which he was always willing to share with me, is impressive. Professor Jan Breman (Centre for Asian Studies, Amsterdam), whom I happened to meet for the first time in Kerala, was also a source of inspiration. He emphasized the wider context of plantation development in colonial Asia. His *Taming the Cooly Beast* greatly influenced me as well. Professor Heather Sutherland (Vrije Universiteit and Centre for Asian Studies Amsterdam) is admired for both her unorthodox method of supervision – providing me with sufficient freedom and self-confidence during the stage of writing the thesis – and her refreshing attitude towards historiography and theory: unbiased and open-minded towards new concepts and ideas, in that supported by her excellent analytical skills. With regard to all four of my supervisors: they gently guided me, never heartlessly turning me inside out.

In addition, I could not have achieved much without the help of the following institutions and individuals. Besides the archives and libraries in Kerala, I consulted the sources available in the India Office Library and Records (now renamed the Oriental and India Office Collections) of the British Library, the School of Oriental and African Studies (both in London), the Centre for South Asian Studies (Cambridge), the Royal Tropical Institute, the University of Amsterdam and the Vrije Universiteit (all situated in Amsterdam). Furthermore, I had the opportunity to pay a short visit to the estate regions of Sri Lanka (greatly helped by the Satyodaya Centre for Social Research and Encounter, Kandy), Sumatra and Java. I thank the Foundation for Scientific Research in Developing Countries (WSO) for financing my first trip to India, the Centre for Asian Studies Amsterdam (CASA) for giving me full support during the period 1989–94. Also, CASA, often in collaboration with the Dutch Organization for Scientific Research (NWO), provided me with various opportunities to present my findings to wider audiences by attending conferences on Capitalist Plantations in Colonial Asia (Amsterdam, 1990), South Asia (Amsterdam, 1990), European Expansion (Lisbon, 1992), Social History (Baltimore, 1993), Kerala Studies (Trivandrum, 1994) and Free and Unfree Labour (Amsterdam, 1995). Furthermore, I would like to thank those scholars who gave constructive criticism on earlier drafts of (parts of) this book, including Dr Ron Aminzade, Professor Peter Boomgaard, Pascale Chaput, Dr James L. Chiriyankandath, Dr V. Haridasan, Professor Frans Hüsken, Professor Dirk Kolff, Dr K.K. Kusuman, Dr Raman Mahadevan, Professor Raul Pertierra, Ravi Raman MA, Dr Tharian George, Professor Charles Tilly, Professor S. Uma Devi, Professor David Washbrook, and Dr Peter van der Werff. In my efforts to present a user-friendly dissertation, I was aided by Rosemary Robson (language), Menne-Chris Kosian (maps) and Sjoukje Rienks (layout). Lastly, I should acknowledge that certain aspects of my study have been published in *Abstracts International Congress on Kerala Studies, Archipel, CASA Nova, Economic and Political Weekly, Itinerario, Journal of Kerala Studies, and South Asia Research*, though mostly in a somewhat different form.

Of course, without the support of fellow students, friends and family, these years would have been less pleasant. At CASA, I was most familiar with 'CASA Novas' like Arjan de Haan, Jos Mooij and Christiaan Heersink. Among my 'Amsterdam' friends, Gert and Corine, Jan-Marten, Jur and Ingrid, Peter, Peter, Pieter and Selma in particular have to be mentioned. My 'Kerala' companions include Christien, Jean-François and Heide, Koji, Motti, Nannu, Paulette, Sylvie, Joël and Maïva, and Victor, the latter also functioning as my interpreter during field trips. My stimulating parents, 'managerial' brother, optimistic grandmother and 'Indian' family-in-law always did their utmost to support me. Annelies, successively my girlfriend, fiancee and wife in the course of my study, made my dreams come more than true, living in the tropics *together*.

Contents

List of Maps, Figures and Tables

Maps

Figures

Tables

Abbreviations

AITUC	All India Trade Union Congress
APK	Association of Planters of Kerala (Ernakulam, Kerala)
APT	*Acts and Proclamations of Travancore*
CDS	Centre for Development Studies (Ulloor, Kerala)
CMS	Church Missionary Society
CPI	Communist Party of India (Trivandrum, Kerala)
CPI(ML)	Communist Party of India Marxist-Leninist
CSAS	Centre for South Asian Studies (Cambridge, England)
CSI	Church of South India
CSF	Cover System File (Kerala State Archives)
CTPA	Central Travancore Planters' Association (Vandiperiyar, Kerala)
FUA	Free University (Amsterdam, The Netherlands)
HDKU	History Department, Kerala University (Kariavattam, Kerala)
INC	Indian National Congress
INTUC	Indian National Trade Union Congress
IOL&R	India Office Library and Records (London, England)
KAR	*Kerala Administration Report*
KARB	Kerala Agrarian Relations Bill
KDPA	Kannan Devan Planters' Association (Munnar, Kerala)
KLA	Kerala Legislative Assembly
KSA	Kerala State Archives (Trivandrum, Kerala)
LL	Legislative Library (Trivandrum, Kerala)
LMS	London Missionary Society
LS	Lok Sabha
LSE	London School of Economics and Political Sciences (London, England)
MA	Madras Archives (Madras, Tamil Nadu)
MPA	Mundakayam Planters' Association (Mundakayam, Kerala)
MRPA	Mundakayam Rubber Planters' Association
NSS	Nayar Service Society

OBLR	*Opinions regarding the Bills relating to Land Reforms*
PTSMC	*Papers regarding the Ten Square Miles Concession*
PWD	Public Works Department
R/../..	Crown Representative's Records (India Office Library and Records)
RCL	Royal Commission on Labour
RLSM	Report on the enquiry into matters arising from the Labour Strike Munnar
RPT	*Regulations and Proclamations of Travancore*
RSIPEC	*Report of the South of India Planters' Enquiry Committee*
RSP	Revolutionary Socialist Party
RTI	Royal Tropical Institute (Amsterdam, The Netherlands)
RRII	Rubber Research Institute of India (Kottayam, Kerala)
SARBUPRI	Sarekat Buruh Perkebuban Republik Indonesia (Union of Indonesian Plantation Workers)
SIA	South Indian Association
SNDPY	Sree Narayana Dharma Paripalana Yogam
SOAS	School of Oriental and African Studies (London, England)
TAD	*Travancore Almanac and Directory*
TAR	*Travancore Administration Report*
TCAR	*Travancore-Cochin Administration Report*
TCLA	Travancore-Cochin Legislative Assembly
TLA	Travancore Legislative Assembly
TLRM	*Travancore Land Revenue Manual*
TSCSC	Travancore Sri Chitra State Council
TSMA	Travancore Sri Mulam Assembly
TT	Tata Tea (Munnar, Kerala)
UA	University of Amsterdam (Amsterdam, The Netherlands)
ULKU	University Library, Kerala University (Trivandrum, Kerala)
UPASI	United Planters' Association of Southern India (Coonoor, Tamil Nadu)
UTUC	United Trade Union Congress

Glossary

Aikya Kerala	United Kerala
acre	0.4047 hectare
anna	see British rupee
a-varna	untouchables, those not belonging to a varna
brahmins	the priestly varna
British rupee (Rs)	= 16 annas (as) = 192 pice (p)
British pound (£)	= 20 shilling = 240 pence
British Resident	the diplomatic representative of the British government (referred to as Agent to the Governor General, Madras States after October 1923 and as Resident for the Madras States after April 1937)
candy	unit of weight, equal to between 500 and 600 pounds
canumkar	tenant of a janmkar
cash	see Travancore rupee
cherikal	shifting cultivation, particularly in forest areas. The concept is also used to denote land outside Brahmanical village control
chitties	Syrian Christian joint funds, later often transformed into Western-style banks
chowrey	variation of *cho(w)key*, a station or post especially for police and/or collection of customs
chuckram	see Travancore rupee
crore	= 100 lakhs = 10,000,000
desam	village
Dewan	first minister
Dewan Peishc(k)ar	government officer in charge of a division
division	largest administrative unit
fanam	silver coin
hacienda	a large estate in present or formerly Spanish-speaking countries
harijan	Gandhi's name for the depressed castes, the untouchables

hundredweight (cwt)	= 112 pounds (lbs) = 1792 ounces (oz) (≈ 50 kilogram)
Huzur (Cutcherry)	office of the Dewan
janmkar	landlord, controlling large tracts of land in eighteenth century Travancore
jenmies	≈ janmkars, concept particularly in vogue with regard to nineteenth and twentieth century southwest India
kangany(ies)	plantation labour recruiter and overseer, especially in south India and Ceylon
karutha ponnu	black gold, i.e. pepper
karuva	a medicinal plant, the bark of which is used in the Ayurveda-Indian system of medicine
khadi	homespun cotton cloth worn by adherents of the movement for autonomy in India instead of the mill-made foreign product
kshatriyas	the varna of warriors
kudikapparan	landless agricultural labourer, who – with permission – resides on a landowner's property
kuris	see chitties
lac/kh	unit of 100,000
lathi	police cane
Maharaja	literally: great king
Maharani	literally: great queen
maistry(ies)	see kangany
mandur	Javanese foreman on estates on Java and Sumatra
M.E.	Malayalam Era (1000 M.E. ≈ AD 1824–25)
par(r)ah	measure of capacity, approximately 80 square inches (≈ 2.5 cm)
pence	see British pound
Peria Durai	'big gentleman', the manager
pice	see British rupee
praja	people
raj	rule, government
Raja	king, ruler
ramadan	ninth month of the Muslim year, when Muslims fast between sunrise and sunset
rickshaw	in Trivandrum a motorized, three-wheeled vehicle

Sessions Judge	judge in charge of a Zillah Court
shaitan	an evil spirit
shilling	see British pound
Sinna Durai	'little gentleman', the assistant manager
sirc(k)ar	government or state
sirdar	plantation labour overseer, especially in north India and in Mauritius
sudras	the lowest category of the varna system, the labouring order
tandil	Chinese foreman on Sumatran estates
Ta(h)sildar	government officer in charge of a taluk
taluk	administrative unit beneath a division
taravad	extended household of the Nayar caste
Travancore rupee (Rs)	= 28 *chuckram* (chs) = 448 *cash* (c)
untouchables	see a-varna
vaishyas	the varna category of traders
varna	one of the four categories of the Hindu social hierarchy, see also brahmins, kshatriyas, vaishyas and sudras
Veda	any or all of the most ancient, sacred writings of Hinduism
Viceroy	highest position within the colonial, hierarchical order in British India
Vir Patom	an 'inferior' kind of tenure
zamindar	landlord, landholder
Zillah Court	district, sessions court
zindabad	Long Live!

One

Introduction

1.1 In Search of Plantations

Trivandrum-Aleppey, 24 December 1991. I had been looking forward to this trip for a long time. No wonder: during the last four months I tried to find my way through the historical data available in Trivandrum, the capital of the Indian state of Kerala. Though enjoying my search for relevant information, often under dusty and chaotic conditions, I thought that travelling through Kerala would refresh my mind. In certain respects, the journey represented a pilot-study. I intended to travel through those areas which were of fundamental importance to my research, making arrangements for future, more lengthy visits. At the same time, the trip was a short holiday. It was almost Christmas and I was happy to be accompanied by my girlfriend and her parents and sister.

Travelling through Kerala is always a fascinating experience. An important reason for this is the large number of people who are eager to exchange views about almost every subject that makes life so interesting, including family life, religious practices and political preferences. That is also what I recall most from my previous trips through south-west India: my discussions with fellow travellers.

And there are a lot of interesting subjects for conversation within reach. Especially at first sight, Kerala's society is complex and packed with paradoxes. For instance, one would expect that the inhabitants of south-west India would stress their 'ethnic' identity as Malayalis. In theory, this might easily be the case. All speak the same language, share certain historical roots and are governed by the state of Kerala. In reality, however, most inhabitants of south-west India emphasize their other identities, in terms of religion, caste and/or class. That Kerala's society is fundamentally divided, in all sorts of ways, is easy to observe. The great number of churches and mosques, as well as the synagogue situated in Cochin, reveals that

Hinduism has no religious monopoly. Christians and Muslims each represent about one-fifth of the total population. In addition, gatherings of the Nayar Service Society (NSS) and the Sree Narayana Dharma Paripalana Yogam (SNDPY) show that caste and communalism continue to be important features of social, cultural and political life. The NSS protects the interests of the Nayars whereas the SNDPY represents the Ezhava community. At the same time, walls painted with representations of Mahatma Gandhi and Nehru and *rickshaws* decorated with the portraits of Marx, Engels, Lenin and Stalin indicate that other movements are ideologically secular in nature. As most people know, these symbols reflect Kerala's political continuum to a very large extent: from the extreme leftist Communist Party of India Marxist-Leninist (CPI-ML) to the right of centre Indian National Congress (Congress Party). This situation engenders most of the apparent paradoxes: Kerala, a unity in certain respects, embraces many unities; Malayalis, while mostly governed by *secular* parties and inhabiting a *secular* state, are deeply influenced by religion and caste in their daily lives, exemplified by Hindus worshipping the gods and goddesses in their places of residence, by Christians attending Church on Sundays, by the strict observance of *ramadan* by the Muslims and the elaborate marriage rituals of all communities.

Making use of the public transport along the palm-fringed coastline, another paradox obtrudes. The buses and trains, often overcrowded and in a parlous state of repair, make their way through beautiful scenery. Indeed, this image provides a suitable metaphor. South-west India, and the Kerala government in particular, is confronted with a high population density, huge unemployment figures and insufficient financial means. For large sections of the population this means poverty, often culminating in a decision to migrate to Madras, Bombay or another Indian city, or to accept work, often under humiliating conditions, in one of the Gulf countries. Yet, Kerala is blessed with highly-fertile regions and a favourable climate, both important conditions for economic development. The rich natural heritage is also reflected in the market-places. Here, the bananas, cardamom, cinnamon, cashews, coffee, ginger, jackfruit, nutmeg, onions, pepper, pineapples, tea and tomatoes on offer are mostly of local origin. Indeed, for people arriving from Tamil Nadu, northern India, or as in our case, Europe, Kerala's natural miracles give the impression that one has walked straight into a dream. Simultaneously, Kerala's social reality is the opposite: a nightmare to many.

Aleppey, our final destination for that day, lacks the prestige which characterized this town in previous centuries. According to my readings, especially in the late eighteenth century the port functioned as the trading centre of the region. Here Rama Varma, the ruler of the rapidly expanding state of Travancore, organized the collection of and trade in pepper. With great success: he had been able to dictate the terms of trade in his negotiations with the European trading companies. Although that age left no visible traces at all worth mentioning, visiting the town allows one to understand why Aleppey could have achieved such importance in earlier times. The town is ideally situated in the middle of Kerala's *backwaters:* the extensive network of rivers, lakes, canals and lagoons that lace the interior coastline. It is therefore by no means astonishing that Aleppey, connecting its hinterland with the Arabian Sea, was once given the function of intermediary by the Travancore monarchy.

Aleppey-Cochin, 25 December. Continuing our way northwards, we are continuously struck by reminders of Kerala's trade history. The backwater area, for example, discloses lots of Chinese fishing nets and Chinese-style junks. Entering Cochin, we pass an old Portuguese church, an abandoned Dutch cemetery, and ramshackle British storage buildings. And during a visit to Cochin's *Jew Town*, the owner of an antique shop tries to sell us some Roman coins. Indeed, these signs indicate that south-west India has always been integrated into long-distance trade networks. Though of different origins, most traders were attracted by the same products: spices, particularly pepper. To a much lesser extent, ivory, sandalwood and – even more occasionally – elephants, tigers and parrots were items of trade.

History books had informed me that the trade in pepper had begun in the port town of Muziris, situated some miles north of Cochin. In those early years traders, especially from China and the Mediterranean regions, including Greeks and Phoenicians, flocked to south-west India. Romans, who considered

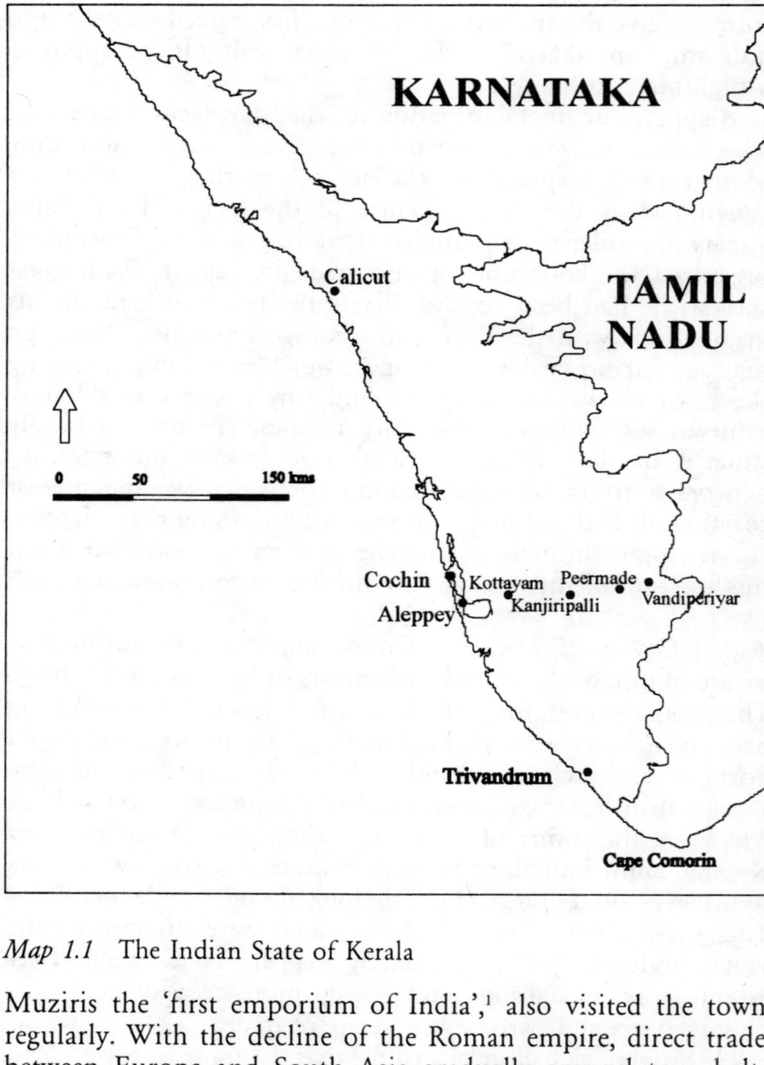

Map 1.1 The Indian State of Kerala

Muziris the 'first emporium of India',[1] also visited the town regularly. With the decline of the Roman empire, direct trade between Europe and South Asia gradually ground to a halt.

[1] The quotation originated from the Roman author Pliny the Elder as far back as the first century of the Christian era (Singh 1986: 8).

Though Arabs and Turks filled the vacuum, the Indian traders did not vary their demands. In return for pepper, they were especially interested in brass, copper and gold coins. No wonder Malayalis nicknamed pepper *karutha ponnu* (black gold) for centuries!

Later, Europe's seemingly insatiable demand for spices and south-west India's fame as an important supplier of these products contributed to the discovery of the sea route connecting Portugal with South Asia. In 1498 Portuguese ships, under the command of Vasco da Gama, reached the Malabar coast, the sea traders' equivalent for south-west India. That event negatively influenced the position of Muziris. In their struggle to control the trade in spices, and bringing their superior military power to bear, the Portuguese brought the prominent role of the Arab traders to an end and made Cochin their stronghold. Only in the mid-seventeenth century did the Dutch replace the Portuguese as the dominant trading community and take over Cochin (1663). During the course of the eighteenth century the British East India Company gradually rose to prominence.

Today, Cochin still serves this function of mediator between external demand and regional production. And quite successfully: Cochin is the most important trading port in Kerala in terms of commerce and trade. One of the places where this is reflected is Willingdon Island, an artificially created stretch of land constructed under the British in the middle of Cochin harbour. A great number of the head offices of plantation companies are situated here, including that of Malayalam Plantations, co-ordinating their contacts with the markets and financiers abroad as well as their communication with the local estate managements. Willingdon Island also accommodates the headquarters of large trading companies specializing in plantation products. Of these, Aspinwall and Company is especially prominent. And lastly, the Cochin Chamber of Commerce and Industries, an organization with the stated aim of facilitating trade relations, is also situated here. It should, therefore, be no surprise that the Association of Planters of Kerala maintains its secretariat nearby as well, though not on Willingdon Island. Therefore, I hurried throughout the day – visiting these organizations, discussing my interests and making appointments

– before I could finally sit down and enjoy our conversation over dinner on Christmas evening.

Cochin-Vandiperiyar, 27 December. After having attended a service of the Church of South India in the early morning, we continued our trip, this time to the east. Our destination for today: the High Ranges, a tea estate situated in the Peermade/Vandiperiyar plantation district.

Making headway with the journey, we realized the enormous changes in the environment, revealed in all sorts of ways. Along the coast, for example, the sequence of huts, houses and other buildings determines the landscape to an overwhelming extent. Most of these settlements are shaded from the sun by a firmament of palm leaves and are girded by paddy fields. Indeed, in these densely-populated regions the cultivation of coconuts and paddy predominates. An intriguing feature, as the two crops represent two opposite forms of production. Whereas the production of paddy is capital-intensive as well as labour-intensive, coconuts are planted mostly by smallgrowers, who only employ outside workers sporadically to harvest the crop.

Another atmosphere surrounds the midlands, the strip of land situated in between the coastal areas and the high regions. Here the rubber industry is omnipresent: collection, drying and transport of latex is visible at every turn. At the same time, the character of the environment gradually changes. To explain it in a nutshell, the further one travels and the higher the altitude, the fewer the people who inhabit these areas and the larger the agricultural enterprises. This is also reflected in the places we pass through on our route. Kottayam, for example, is a fairly large, wealthy city and is of great importance for two reasons: it houses the Rubber Board and is considered the centre of the Syrian Christian community. Here one comes across old Christian churches, the *Malayala Manorama* is printed, the Congress Party has its base and influential entrepreneurs meet in The Planters' Club. Kanjirapalli, some miles east of Kottayam, is already smaller in size. The Central Travancore Agricultural Association, a lobby organization of rubber smallholders, is located here. Our next stop, Mundakayam, is closely connected with large-scale rubber cultivation. Besides the Mundakayam Planters' Association, there are yet other indications that Mundakayam represents estate production *par excellence:* the town is crowded with cheap shops, restaurants and hotels: a place where plantation labourers and migrant workers huddle together. Indeed, Mundakayam resembles nothing so much as a frontier post, marking the boundary between smallholdings employing family members and the agro-industry based on the large-scale employment of wage labour.

Peermade and, some miles further on our way, Vandiperiyar are similar to Mundakayam in many respects. Though situated amongst hills dotted with tea bushes and located at a much higher altitude, the same characteristics obtrude: only a few persons ever stay permanently in Peermade and Vandiperiyar. Most are temporary guests, either estate workers visiting the market place or the desperate unemployed trying to find work. Mundakayam, Peermade and Vandiperiyar have something else in common. They were all founded by British planters.

The estate we visited was also initially opened up by British entrepreneurs. Traces of the European legacy immediately caught our attention: some of the bushes were planted in the late nineteenth century, as a wooden sign informed us, and the bungalow was typically English-style. Yet, many things had changed since the British repatriated. The management is totally Indianized and the estate has become part of a large Christian-owned plantation firm, A.V. Thomas and Company. To our great surprise, however, we observed that despite these alterations a great number of British customs still flourish. The Club, for instance, continues to represent the heart of the planting community. Here, assistant managers meet the drink deep to forget their frustration, induced by hard work and social isolation; here, elderly managers discuss business and have the opportunity to enjoy games and sports; and at the Club the planters' women meet, often exchanging gossip on family affairs. Even more surprising, the colonial ambience is still intact, largely due to the strict division of tasks and authority, which is reflected in living conditions. Managers inhabit large bungalows, with large beautifully-tended gardens, a tennis court sometimes included; staff live in separate quarters, not luxurious but comfortable; while all field workers are crowded into overfull labour lines. In addition, the British experience was prolonged in the *choice of language*. Amongst themselves, the Malayali managers prefer English to their mother tongue, thereby clearly excluding their inferiors from their conversations. Of greater importance, the colonial atmosphere – amalgamated with practices particular to south-west India's caste system – has continued in the *choice of words*. This was most clearly shown by our host, a young assistant manager from a well-to-do background. Thoroughly experienced with the supervision of wage labourers, including that on his family property, he shocked us every time he summoned 'his boy', a lazy, newly-wed Christian Tamil as he informed us. In fact, our host was rather rude in his approach to his servant, shouting for 'Joseph' whenever he deemed it necessary. Having finished our dinner, the assistant manager once again called for 'his boy'. 'Joseph . . . Joseph . . . . . . JOSEPH!'. The servant, however, did not appear this time. Though our host apologized for the particular occurrence, we did not care in the least about the domestic help's disappearance. In fact, we thought Joseph had taken the right decision, finding some better company outside the bungalow, perhaps celebrating Christmas with his wife.

1.2 Plantations Discussed

1.2.1 Historiography on South-west India

Given the fact that the history of south-west India has always been closely related to the production of and trade in export crops, it is not surprising that a substantial body of literature deals with plantation production. There is, however, no agreement about the nature of plantation development in south-west India.

Kershaw, for example, presents a picture in which labourers were eager to work on the plantations, especially in the rubber district of Mundakayam. 'Fortunately for the early planters', Kershaw states, 'labour proved adequate, if not abundant, and labour relations were healthy'.[2] This contrasts with Ravi Raman who emphasizes that the workers neither came to the plantation districts spontaneously nor remained on the estates of their own

[2] Kershaw in: Mayne 1953: 225.

free will. Ravi Raman claims that coercion and brute force were used extensively to mobilize workers and to keep them as resident captive labour.[3]

A second discussion addresses the question of the linkages between plantations and other agrarian sectors, and it is intriguing to note that the United Planters' Association of Southern India (UPASI) is actively involved in this debate. Naturally enough, the planters' association emphasizes that large-scale estates serve as a model for smaller holdings. 'Plantations provide the silver lining to our agriculture, with high levels of productivity, employment and foreign exchange earnings'. In addition, plantations make catalyzing contributions towards rural growth.[4] Uma Devi, however, concludes that the growth of the plantation sector was detrimental to the economic development of Travancore. She argues that plantation development caused permanent damage to the forests, brought about a decline in the livestock population, and changed a paddy-exporting region into one which imported on a large-scale. In addition, Uma Devi argues that the fruits of plantation development were unequally distributed among the various castes and religious groups, which built up social tensions and set another obstacle on the path of economic development.[5]

Another issue under debate is the significance of the government in the emergence and development of plantations. Heather Lovatt, for instance, stresses that the authorities were not prepared to admit responsibility for assisting the industry. Therefore, Lovatt continues, the transformation of the hills is a story of the enterprise, courage and self-reliance of the planting community.[6] Pandian clearly disagrees with Lovatt. In his view, the state was instrumental in accelerating the pace of plantation development. 'Under the dictates of the British Indian Government', Pandian argues 'Travancore provided European planters with extensive areas of land at throw-away prices, constructed roads in the high-ranges (. . .) and taxed the estates very moderately'.[7]

[3] Ravi Raman 1991: 247.
[4] UPASI 1981: 5.
[5] Uma Devi 1989: 181–2.
[6] Lovatt 1972: 1 and 15.
[7] Pandian 1990: 78.

I would like to add here that the above debates are not without political connotations. The planting organization UPASI, the planter Kershaw and planter's wife Lovatt favour estate development, whereas the leftist-oriented Ravi Raman, Uma Devi and Pandian openly criticize plantation growth in general and the powerful planters' lobby in particular.[8]

1.2.2 Debates on Tropical Asia

In fact, the historiography regarding south-west India mirrors the debates with regard to other Asian regions. For example, one group of scholars argues that from the labourers' point of view the decision to work on the plantations seems to have been a deliberate choice.[9] They stress that a high degree of social differentiation in the various Asian societies made low-class people leave their areas whenever an improvement in their position seemed possible. Most plantation workers came from areas where they had only either limited or absolutely no access to the means of production and where many of them were indebted to local landlords and/or moneylenders. For low-caste Indian labourers, the opportunity to work on plantations offered a way out of their depressed conditions in their caste-ridden villages. Labourers who were originally town-based also tried to better improve their position by accepting work on plantations. Other workers left areas plagued by famine. These scholars clinch their argument by saying that the advances and employment offered by the plantations meant *a way out* of difficult circumstances. This is illustrated by the fact that many of the labourers used the advances given by the recruiters to pay off existing debts. Griffiths, for instance, represents this vision very clearly when he argues that in south India the advances made by estates helped to break the system of hereditary serfdom and contributed to ameliorating the indebtedness and depressed conditions of labour.[10] In their discussions of plantation work,

[8] In contrast to earlier planters' publications, UPASI's last officially sponsored book, i.e. Muthiah's *The First 100 Years of the UPASI* (1993), fairly openly discusses the negative, coercive aspects of labour recruitment for the plantations during the colonial period.

[9] See particularly: Shlomowitz 1981; Emmer 1986.

[10] Griffiths 1967: 39.

these writers do not focus on labour discontent and, often implicitly, stress the consensus between the worker, the overseer and the manager.[11]

Others, however, oppose what might be called the 'free choice model'. Their 'coercion vision' diverges at three major points: the nature of labour recruitment; the methods of labour control; and the extent of labour resistance.

They stress, for example, that many workers were misled by recruiters, officials and planters. The labourers were given false information about their destinations and the nature of plantation work. This explains why people with access to land sometimes left for the plantations. But it also means that the choice of the labourers might have been different had they had known the conditions prevailing during transportation to and life on the plantations. Some were even taken captive and shipped off to plantations against their will. In these cases there was absolutely no question of any 'free' choices. In addition, these scholars emphasize that although accepting an advance from a recruiter could mean a way out of bondage, it certainly implicated *a way into* a new one. Breman, speaking about the nature of labour recruitment for plantations in Sumatra in the Netherlands East Indies, stresses that intimidation played a major role. He therefore claims that it is 'nothing other than a colonial fantasy to maintain that a contract was entered into voluntarily'.[12]

These authors also focus on labour control. They stress that the plantations used extra-economic methods in order to determine the lives of the workers. As a consequence, apart from the control exercised during working hours, the social and cultural lives of the labourers were also regulated to an overwhelming extent. This was partly related to the system of recruitment. During the second half of the nineteenth century and the first third of the twentieth century, the system of the recruitment of indentured labour prevailed. Under this, a labourer was bound by a penal contract to serve one particular employer for a specified period. The state provided the penal codes (the Workman's/Criminal Breach of Contract Act in British India and the Coolie

[11] See for example: Graham and Floering 1984.

[12] Breman 1990a: 132.

Ordinance in the Netherlands East Indies) which enabled plantations to track down, sentence, and punish runaway labourers. This, however, does not automatically mean that the labour conditions were much different either before the enactment of these legal measures or after their abolition. Although not bound by any written agreement, even the non-indentured labourers worked and lived in a state of bondage. This sorry state was the outcome of the violent treatment of the plantation labourers by the planters and the overseers (named *kangany* in south India and Ceylon/Sri Lanka, *sirdar* in north India and Mauritius, *tandil* (Chinese foremen) in Sumatra and *mandur* (Javanese foremen) in Java and Sumatra). Also liquor, opium, gambling and prostitution were allowed by the management to keep their work force indebted and bonded. Wages were very often inadequate for survival, the upshot being that workers became indebted to the plantation.[13]

Lastly, these scholars point out that there was no consensus about the social climate on the plantations. To illustrate this argument, they highlight conflicts within the plantations, expatiating on the various forms of labour resistance. This reveals that there was a gradual shift from individual acts (such as desertions and personal revenge) to more organized forms of resistance (like strikes). This change – which was most probably also closely interconnected with the shift from a temporary to a permanent workforce on the estates – was greatly affected by an increased solidarity among workers, which ensued from the rise of (mostly illegal) political parties and labour unions. On many plantations contradictions in class coincided with those of race/ethnicity. In these cases, the class struggle often matched the national liberation struggle. It is striking that resistance was not always directed towards immediate economic advantage. In Fiji, for example, work-gangs of women 'coolies' humiliated their European overseers by capturing them, beating them, immobilizing them and urinating on them.[14] Clearly, many of these acts

[13] Regarding the non-market, coercive aspects of labour control, see for example: Bandarage 1983: 220 (coffee and tea estates Ceylon); Stoler 1985: 45 (tobacco and rubber plantations Sumatra); Das Gupta 1986: PE-3 (tea estates northeast India); Murray 1992: 53 (plantations Indochina).

[14] Kelly 1992: 259.

of resistance were intended to show contempt for the dominant class, race and gender.

As has been briefly alluded to, the second debate deals with planters and peasants: the effects of plantation development on other agrarian sectors, to be more precise. Some scholars have argued that plantations should be considered enclaves. Plantations were characterized by a clearly demarcated territory, an isolated labour force which lived on plantation land, and a market abroad. Manufacturing, transport and marketing of the plantation crop were handled by the plantation companies themselves. Moreover, the industrial products introduced to cultivate land and mechanize production, the technological expertise, the capital investments, and the management decisions came from the Western countries. This means that the recipient country was a 'dual society', a concept which originated with Boeke.[15] Plantations and their surroundings were clearly discrete from each other and had their own economic, social, cultural and political characteristics.

Consequently, underdevelopment in the 'backward' areas did not stem from plantations as these 'modern' agro-industrial enterprises had no direct linkages with the 'traditional' economy. Underdevelopment was spawned by the 'traditional' sector itself as the 'traditional' structures were either not at all or insufficiently superseded by 'modern' ones. In this regard, plantations functioned and continue to function as 'agents of development'.[16] Plantations are beacons of the modern road to enterprise and therefore function as good examples. Graham and Floering argue that in many Third World countries 'the efficiency of plantation estates often contrasted very sharply with the inefficiency of the surrounding agricultural economy'. Indeed, plantations brought to agriculture the corpus of skills typical of modern industry.[17]

Others do not characterize plantations as enclaves nor do they see their host countries as dual societies. Plantations had many linkages with their surroundings. They claim that the emergence

[15] Boeke 1940: 1.

[16] Courtenay 1971: 3–4.

[17] Graham and Floering 1984: 2; see for a similar statement Allen and Donnithorne 1954: 270.

of plantations encouraged the development of capitalism in other agricultural sectors, changed the social hierarchy, and introduced new political ideas.

The linkages were often of a negative nature. The development of the plantation sector was usually at the expense of the other agrarian sectors. Land is a case in point. Good quality land was reserved for plantation agriculture and sometimes peasants were even forcibly removed from their land. The large-scale establishment and development of plantations often meant deforestation and subsequent soil erosion, causing a serious disturbance to the natural equilibrium in these regions. Furthermore, labour was often migratory and even circulatory in nature. Thus, the costs of reproduction of labour were paid for by the peasant communities. Also, capital from local investors was invested in the plantation sector, meaning it was withdrawn from the other sectors. All the more damaging, the state protected the interests of the plantation sector, even when these conflicted with those of the other agrarian sectors. The demands of plantations were given priority with regard to land, labour, capital, and infrastructure.[18]

With regard to the government's role in the emergence and development of plantations in Asia, some would claim that the state, in particular in the colonial period, put its laissez-faire ideology into practice and acted only on the sidelines of the economic arena. These scholars believe that plantations developed primarily under their own steam, aided only marginally by government intervention, and being above all subject to the free working of supply and demand. In fact, plantation development was often hindered by the authorities. Income tax, agricultural income tax, export and excise duties, sales tax and transit taxes placed a heavy burden upon the expenditures of the plantations situated in colonial India.[19]

Others, however, hold the view that the role of the colonial government was of crucial importance. The introduction and/or raising of taxes combined with different kinds of land policies monetized the economy and increased social differentiation, which, in its turn, often intentionally, resulted in the ready availability of cheap wage labour for plantations. Undeniably,

[18] See for example Beckford 1972 as well as Guha 1968 (Assam); Bandarage 1983 (Ceylon/Sri Lanka); and Xaxa 1985 (north Bengal).
[19] Griffiths 1967: 557–76.

the administration gave plantations direct assistance by taking measures which facilitated labour recruitment and labour control. Thus, as far as labour is concerned, by disciplining 'natives' from different parts of Asia indirectly (which facilitated labour recruitment) and directly (to simplify labour control), the state mobilized labourers into conditions which eventually immobilized them.[20] In addition, the government played an important role in facilitating the planters' acquisition of land, access to credit and transportation of products. The colonial authorities thus created the framework with in which the emergence and development of plantations could take place.

1.2.3 A Reappraisal of the Discussions

As the discussion of the historiography concerning south-west India as well as other Asian plantation regions demonstrated, we can distinguish three debates in the field of plantation studies; those regarding (a) plantations as social organizations; (b) plantations and other agrarian sectors; and (c) plantations and the government.

Those scholars who focus on the advantages of the plantations for the labourers recruited (a way out of indebtedness and other rigorous conditions); for the surrounding economy (plantations as agents of development); and for the administration (contributions to the exchequer) can be categorized as the Modernization School, sometimes also designated as the Imperialist or Colonial Group. In contrast to them, those who stress the exploitive and violent character of labour recruitment and control, argue that plantation development took place at the expense of indigenous modes of production, and suggest that the state acted primarily as an executor of the planters' policies are part of the (Neo-) Marxist, Nationalist or Anti-Colonial Tradition.[21]

20 Breman 1990b: Ch. 6.

21 Cf. Rote 1986: Ch. 1; *Journal of Peasant Studies* 1992, vol. 19, nos 3/4, Special Issue on Plantations, Proletarians and Peasants in Colonial Asia; Chatterjee 1993; 27–34. In the field of free and unfree labour, those (Modernist) authors who argue that (plantation) labour contracts were entered into voluntarily are also designated as writers with a Neoclassical Approach, which, by the way, is the dominant viewpoint in the academic discipline of economics at present.

In my view, both schools have more in common than is apparent at first sight. For instance, their studies deal predominantly with the colonial period and with the European-led agro-industries. To put it another way, the focus of most representatives of both traditions is Eurocentric, not only in time but also in space. The unfortunate result is that the European planters and officials are mostly thrust into the foreground, large and clear, whereas the Asian elements (labourers, peasants, state officials and especially planters) are usually relegated to the background, grey and undifferentiated.[22]

Another problem is related to the explanatory model. The historical process is frequently reduced to a one-way traffic, in which one dominant interest group – the European planters – used, misused, destroyed and/or guided the other interest groups – the Asian labourers, state officials, peasants and planters – who had no option but to resist, take counter-measures, follow and/or imitate. I, however, want to emphasize that the other interest groups did not only *react*. They had their own – not reducible only to the Europeans and their influence – backgrounds, motives and goals on which their tactics, strategies and policies were based. Similarly, I want to stress that the European planters did not simply *act*. They responded to local forms of labour recruitment, regional rights to land, and indigenous forms of power, sometimes trying to change these elements, occasionally attempting to fit in with their surroundings, at times successfully but often to no avail.

To sum up the first two arguments, in the debates on plantations insufficient attention has been paid to specific developments which occurred before, during and after the colonial period and which were only partly, or not at all, influenced by the European presence, but which were nonetheless important in shaping the historical reality in which plantation development took place.

My third point of criticism is closely related to the above. Because both the subjects chosen and the explanations provided were often Eurocentric in nature, it seemed that every change

[22] I hereby closely follow Smail (Smail 1961). Unlike Smail, however, I hesitate to use the concept of *autonomous* history, especially because of its implicit suggestion of objectivity (compare also Section 1.4). Smail, by the way, largely based his ideas on the works of J.C. Van Leur (cf. Leur 1983: 145–56).

within the indigenous societies, either 'positive' or 'negative' in nature, was related to the European planters and their political system. In this way, the Eurocentric approach promotes undesirable value judgements. Clearly, it is not the task of the historian to judge the behaviour of the European planters as either 'good' or 'wrong'. Historical events and developments should above all be explained and contextualized.

Fourthly, the often one-sided focus, explanations and value judgements clog the current, stagnating debate. For many decades, the heart of the matter seems to have been whether or not the plantation labourer was 'free' or 'unfree', whether or not the peasants and other agriculturalists profited from or were victimized by plantation development, and whether or not the state was an autonomous actor or a passive instrument controlled by the planters. I am convinced that the complex, multi-dimensional historical reality should not be reduced to its most simple form. Mostly, it is not a question of 'either . . . or . . . ' but of 'and . . . and . . .' As will be shown in my study, plantation workers were both 'autonomous' and 'controlled', peasants were both 'profiteers' and 'victims' and ministers were both 'independent' and 'unfree'. And not only at an abstract level. These outwardly contradictory, dual features also characterized the majority of individual labourers, agriculturalists and government representatives.

Having formulated these critical notes in fairly general terms, I would like to specify my objects in more detail. It is my explicit intention to question the following ideas, misunderstandings which might arise out of the Eurocentric approach: (a) that only the European planters represented a successful planting community; (b) that efforts to involve unfree labour in the production of export crops were restricted to European agro-entrepreneurs and/or the colonial period; and (c) that the close relations between European planters and representatives of the colonial state as observed during the heyday of European imperialism were unique.

Indeed, I would like to offer a third, alternative perspective on the study of plantation development in Asia. Unlike most Modernist and (Neo-)Marxist historians, I hesitate to consider the rise and fall of European plantations as fundamental breaks in history, at least in certain respects. Taking a non-Eurocentric point of view, we might arrive at the conclusion that the European planters were not unique as a particular interest group that, with the help of the agencies of the dominant government (namely the colonial administration), tried to control land and labour for export purposes. Further research may show that the socio-cultural importance of particular communities, the ability to control the cultivation of certain crops, the nature of the labour regimes on the production units and the access to state power were closely interconnected with one another in the pre-colonial, colonial and post-colonial periods.[23]

1.3 Towards a Multi-Dimensional, Long-Term Analysis of Plantation Development

In my study on south-west India, I therefore try to present a complete history of plantation development, paying attention to the strategies of all the interest groups involved. This having been said, I must admit that the nature of plantation development

[23] As some readers may have noticed, the discussion regarding plantation development is closely related to the debate about the nature and implications of colonialism, both in South and South-east Asia. Regarding the former region, I find some of the ideas of C.A. Bayly, D. Washbrook and other scholars of the so-called Cambridge School highly inspiring. Indeed, state formation and capitalist development, like the organization of export production, are continuous processes in the history of South Asia, and in that sense not particularly related to the colonial period.

At the same time, and in this I agree with Ranajit Guha and other members of the Subaltern Studies Group, I realize that for South Asians the *British* Raj and the growth of *Western* enterprises, as well as the dominance of the *European* planters, meant a fundamental break in *their, Indian* history. In other words, without British rule the trajectory of south Asian society, including the development of Indian politics, capitalism and export production, would have been different (cf. Chatterjee 1993: 27–34).

Furthermore, I consider the Subaltern School's focus (the disadvantaged groups, as opposed to the elite) and methodology (reading documents 'against the grain' and using unconventional or neglected sources, including popular memory and oral discourse) refreshing (cf. Guha and Spivak 1988).

On the other hand, I strongly feel that writing history from all sides, showing the activities, strategies and policies of all the parties involved, and based on information obtained from all groups, should be every historian's task. Only then can historical events and developments be 'fully' understood, that is to say to the greatest extent possible.

itself prevents any historian from fully achieving this objective. After addressing the main topics of my research, I will elaborate on this issue in more detail.

The central theme throughout the book is the planters' struggle for land, labour and government assistance. 'Planters' simply refers to those persons who control the plantations, and includes both the category of owners-cum-superintendents as well as the group of non-proprietary estate managers. Moreover, in this book 'plantations' and 'estates' refer to the same phenomenon: large-scale agricultural enterprises, mostly specialized in one particular crop grown for distant markets, employing large numbers of wage labourers working under close supervision in a strict hierarchical organization. In keeping with my aim, I want to emphasize that these particular organizations have been set fairly and squarely in a wider historical context: the production of cash crops for export purposes.[24]

This central theme stimulates an in-depth, multi-dimensional analysis of plantation development and encompasses four different, though closely inter-connected, fields of interest. Of these, the first focuses on the planters. To start with one leading question in this area: did the *British* planters introduce a new kind of export agriculture to south-west India, both in terms of the crops produced and the labour organization involved? And this question raises three other issues: (a) what were the 'indigenous' modes of production, particularly that of pepper in the pre-colonial period; (b) in which ways did the European enterprises 'Westernize' the local forms of agriculture; and (c) how did the South Asian context 'Indianize' the British estates? Other questions focus on the rise of the *Indian* planters. Issues to which this gives rise are: (a) to what extent was the rise of local planters an autonomous process and were British and Indian planters colleagues, competitors, or both; (b) how can the Indian planting community be identified, in terms of ethnic, religious and communal background; and (c) in which way did the Indian planters recruit labour, obtain land and secure government aid?

24 Above all for reasons of clarity, I decided to define 'plantations' fairly specifically, while admitting that a more general definition of the concept 'plantation', meaning all large-scale, export-oriented production, might perhaps be truer to the continuities of history (for a wider definition of plantation agriculture, see for example: Baak 1991: 89–90).

A second cluster of questions relates to the plantation workers. The central question here is: how did the labourers use, or try to use, the estates for their own benefit? In answering this question, again three topics are of central importance: (a) what were their initial circumstances, especially financially and socially; (b) did they at first fully realize the meaning of plantation life, in other words, did they make a rational choice, based on sufficient information, when they accepted plantation work; and (c) how did they try to alter the status quo on the estates, either by destructive acts of resistance (suicide, murders) or by opting for constructive strategies (trade union activism)?

The third group of questions concentrates on the broad category of peasants, landlords and other non-planting cultivators. Here, one basic question is whether this category, or particular groups among them, considered plantation development a stimulant, a threat or both. To provide some sub-questions: (a) were these non-planters initially involved in the opening up of uncultivated areas, and if so, what were the implications of the emergence of plantations in this respect; (b) with regard to labour recruitment and control: did estate development cause a scarcity of labour outside the plantation districts, thereby affecting the bargaining position of employers *vis-à-vis* workers; and (c) did the non-planters consider themselves a separate category whose interests were represented by a particular association?

The last focus of consideration concerns the role of the state. A fascinating area because it is indeed very difficult to speak about *the* state. Taking south-west India as an example, the state of Travancore amalgamated with its northern neighbour into the political entity of Travancore–Cochin in 1949. Its successor, the state of Kerala, was established in 1956. Simultaneously, these entities functioned in wider political contexts: 15 August 1947 marked the transfer of power from the British colonial government to the independent state of India. At the same time, supra-state associations, like the League of Nations (1920–46), influenced regional affairs. There are even more factors which make it difficult to speak about *the* state. Vitally important is that the nature of these political entities changed. For example, the state widened in the sense that more communities were able to enter the political arena.

Initially, the monarch *(Raja/Maharaja/Maharani), Dewan* (first minister) and *British Resident* (called the *Agent to the Governor General, Madras States* from October 1923 and *Resident for the Madras States* from April 1937) largely determined state policies. Later, the Travancore Legislative Council (1888–1932), the Sri Mulam Popular Assembly (1904–32), the Travancore Sri Mulam Assembly and the Travancore Sri Chitra State Council (both from 1933), and the Travancore, Travancore-Cochin and Kerala Legislative Assemblies were established and gradually gained power. Co-incidentally, the state deepened in the sense that government policies focused on spheres which had hitherto remained outside the scope of state intervention. Therefore, the following questions will be answered: (a) what were the attitudes taken by the leading political figures and the policies adopted by the legislative bodies with regard to plantation development; (b) in which ways, including those off the record, did the European and indigenous planters, the other agriculturalists and the workers try to influence the formulation of government policy; and (c) what did the official government policies mean in actual practice, in other words, to what extent were the various interest groups prepared or compelled to follow the instructions given by the government?

1.4 About Academic Ideals and Historical and Present Realities

As has been briefly indicated above, to answer these questions fully would be a hazardous, if not impossible, task. This is primarily due to the nature of plantation development itself: most estate workers were illiterate and therefore left no written materials whatsoever. Nor did most of the 'original' inhabitants of the areas in which the planters opened up estates, the so-called hill tribes, have any knowledge of how to read and write. To sum up the historian's first problem, only the planters and the government left a considerable amount of information regarding plantation development.

That does not mean that no knowledge at all can be gleaned about the predominantly illiterate groups. The sources of planters and state officials *do* provide information about each and every group involved in estate agriculture, including workers and hill tribes. Yet, the published sources especially *do not* provide 'objective', all-embracing knowledge. It should not be forgotten that all publications were written with a particular goal in mind. The planters' official documents, for instance, were meant to convince the reader that planters did their utmost to increase productivity, to improve the quality of their products and to ameliorate the lot of the workers. The same holds true for government documents. These materials were produced to highlight the government's concern for its subjects' prosperity, initially focusing only on the planters' wishes and later also taking into account the interests of the workers. The bulk of the published material is very biased indeed and confronts the historian with a second difficulty.

This problem can be circumvented too, though again not fully satisfactorily. Archival materials present a more multifaceted picture of plantation development than do published, official documents. The planters' memoirs provide a valuable insight into their daily lives: their joys, dilemmas and problems. In addition, the correspondence of the planters' associations sheds light on the planters' lobby, both the official and the more hidden channels. At the same time, the archival files of the Dewan and the British Resident clearly show the factors which influenced government policy. Indeed, all these sources provide information about happenings behind the scenes. Yet, the historian realizes that these materials are also incomplete: here too a process of selection has been applied, firstly by the particular planter, Dewan and British Resident themselves, or by their close associates, and later by the archivists.

For these reasons, interviews with representatives of the various interest groups helped to clear up a lot of matters. Importantly, knowledge was now obtained from plantation workers themselves. They told me about their backgrounds, their daily difficulties and the ways in which they strove to improve their conditions. Some labourers also provided information about particular events, including the strike in the Kannan Devan Hills in 1958. Most helpfully, small rubber peasants clarified their position, particularly in comparison with the large-scale enterprises and *vis-à-vis* the Rubber Board. Here also certain historical events were discussed: the resistance against the International Rubber Regulation Agreements of the 1930s and 1940s in particular. In their turn, two previous chief ministers as well as a whole array of owners and/or managers of estates explained the dilemmas they faced, then and now. E.M.S. Namboodiripad for example clarified the difficulties the leftist movement encountered in its efforts to change Kerala society radically, while endorsing the democratic and administrative principles formulated in the Indian constitution. The planters were confronted with other dilemmas: how to increase the productivity and profitability of the enterprise without disturbing the delicate balance prevailing between management, staff, and field workers.

In themselves, the visits to the estates and plantation districts were as important as the direct reason for these journeys, the interviews. Indeed, the landscape reflects the human setting both as it was and as it is now: the factories and planters' bungalows are situated mostly on top of hills, proudly overlooking the surrounding areas. Simultaneously, the environment betrays symptoms of the pioneer planters' thoughtless and irresponsible methods of deforestation, causing soil erosion and leaving useless tracts of land.

It is for all these reasons that I cannot claim that my book represents 'the complete truth'. Not all the interest groups involved produced historical material, the published sources available often provide one-sided information, a difficulty which could be only partly evaded by using archival data, and the persons spoken to presented their point of view, avoiding particular subjects which could harm their position. Leaving the sources aside, it was the researcher and he alone who selected the historical material, communicated with the informants and visited the plantation regions. It was, moreover, that very same person who interpreted the information thus gathered, and who, like any other historian, consciously or not, carries with him his own interests, influenced by personal experiences, family history, and society at large. Therefore, the author may consider this book successful if the different interest groups could recognize their own as well as their predecessors' tactics and strategies, or, if they are reluctant to admit so publicly, those of their adversaries.

1.5 A Final Note Regarding Periodization

Throughout time, historians have spent much of their energy on discovering turning-points in history, trying to answer what they considered a fundamental question: when, how and why did influential changes occur? Scholars working on south-west India have formed no exception to this general search for periods of historical transformation and, like their colleagues focusing on other parts of the world, they have produced different, often conflicting results.

There are anti-imperialist authors, including E.M.S. Namboodiripad, who pay close attention to 15 August 1947, the date of Indian independence.[25] The constitutional historian M.J. Koshy considered the late nineteenth and early twentieth century a period of innovation. The inauguration of the Travancore Legislative Council (1888) and the Sri Mulam Popular Assembly (1904) paved the way for responsible government.[26] For Robin Jeffrey the second half of the nineteenth century were years of transition. Dealing with the shrinking influence of the dominant caste in his book *The Decline of Nayar Dominance*, he argues that the movement from inherited to achieved status, from the interdependence of castes to the competition of individuals, from traditional authority to modern bureaucracy, started then.[27] K. Saradamoni, interested in the emergence of slave castes, mentions that the end of the eighteenth century was of crucial importance. The establishment of British rule ushered in the gradual disappearance of the old social, economic and political order.[28] Susan Bayly, particularly concerned with the process of state formation, pointed out that major transformations had started even earlier. In her view, it was during the reign of Marthanda Varma (1729–58) and Rama Varma (1758–98) that a wholly new political system was built up in the region.[29]

Taking plantation development as a guideline for the study of historical changes, a most interesting characteristic of south-west Indian historiography comes to the fore. The above-mentioned historians were all both accurate and erroneous. These

[25] Namboodiripad 1968. Basing himself on Marxist models, Namboodiripad in his book *Kerala; Yesterday, Today and Tomorrow* divided south-west Indian history in four main phases: militarist-feudal; colonial-feudal (past); bourgeois (present); and communist society (future).

[26] Koshy 1972: Ch. 1.

[27] Jeffrey 1976: 265.

[28] Saradamoni 1980: 13.

[29] Bayly 1984: 177; cf. Kooiman 1992a: 587; Struben 1992: 1.

scholars were incorrect in the sense that they all paid too much attention to *one* particular turning-point in the history of south-west India, thereby implicitly, and not infrequently explicitly, claiming the 'static' and 'traditional' nature of society before that specific occurrence or development.[30] In my view, history is not characterized by major watersheds: at best, history is characterized by a great number of turning-points, which themselves are part of longer developments and continuous processes. Therefore, these historians were all equally right about 'their' periods of transformation. Indeed, the history of south-west India, and particularly the accounts dealing with the last two and a half centuries,[31] can roughly be divided into five periods: the time of independent Travancore (particularly during the rule of Marthanda and Rama Varma); the establishment of colonial rule (first part of the nineteenth century); the heyday of British authority (second part of the nineteenth century); the 'democratization' of the government (first half of the twentieth century); and Indian independence (since 15 August 1947). As reflected in the composition of the book, these periods paralleled the various stages of plantation development, each stretch of time marked by the entrance of new actors setting the agenda in the political arena: respectively the Varma monarchy (Chapter 2), the British Residents (Chapter 3), the British planters (Chapter 4), the Syrian Christian agro-entrepreneurs (Chapter 5) and the plantation labour unions (Chapter 6). I would like to emphasize, however, that none of these periods should be treated as totally detached from preceding years. History is a tapestry of change and continuity.

[30] Jeffrey, for example, wrote: 'In the last half of the nineteenth century a society which had survived fundamentally unchanged for 700 years came unhinged' (Jeffrey 1976: 265).

[31] This does not logically imply that there were no structural changes in south-west India in the period before 1729. Although I expect a constantly changing, dynamic society there and then, no valid judgement can be given as long as the numerous sources which are available on this subject have not been sufficiently studied. Most of the historical data are kept in West-Fort, a unit of the Kerala State Archives. The sources' language and material – Old Malayalam texts on palmleaves – particularly discourage scholars from studying pre-eighteenth century south-west India.

Two

Society, State and Export in Eighteenth Century Travancore: *c.* 1743–1800

One musquet for every candy of pepper

Proposal made by Crisna Annavy, adviser of the Raja of Travancore, Marthanda Varma, to the British East India Company in 1744[1]

2.1 South-west India on the Eve of the Reigns of Marthanda and Rama Varma

In early eighteenth century south-west India the social hierarchy was to a very large extent based on caste, at first sight nothing unusual in southern Asia. Like in other parts of the subcontinent, the status of the caste into which a person was born greatly influenced that person's place in society. Belonging to a specific caste involved a certain religious status and affected access to land as well as person's share in political power.

To a greater extent than in other parts of the south Asian peninsula, however, in south-west India the caste system was all-embracing. There, Christians and Muslims were also an integral part of the caste structure, a social hierarchy sanctioned by Hindu religious texts.[2] Caste was all-pervading. The ranking of a person's community within the caste system determined access to public roads and temples and even the style of personal appearance adopted. Belonging to a particular caste involved a particular way of dressing and determined whether or not a person was allowed to wear ornaments or use an umbrella. Last but not least, day-to-day life was strongly influenced by the precisely defined rules regarding distant pollution, an issue

[1] Quoted in: Kooiman 1992a: 600.

[2] There are four *varna* categories in the Hindu scriptures: *brahmins* (priests); *kshatriyas* (warriors); *vaishyas* (traders); and *sudras* (menial workers and craftspeople). Those who do not belong to one of these four categories, the *a-varna*, are considered most impure.

which will be discussed in more detail below. Especially because of these reasons – the elaborate and explicit nature of caste ranking in south-west India – it is said that the caste hierarchy was more rigid here than in other parts of southern Asia.[3]

The apex of the hierarchy was occupied by the Brahmins, especially by their most important section, the Malayali Nambudiris. Although the Brahmins formed only a very small proportion of the total population of south-west India, they filled the topmost positions in the religious and economic fields. As Brahmins were believed to be the holiest of men and the representatives of God on earth, only they could act as priests for the Hindu community. More mundanely, they functioned as landlords.

Since Kshatriyas were relatively few in number, the Nayars, who were nominally Sudras, came next.[4] They dominated the military, economic and political life. The Nayars provided the majority of the military leaders and warriors and they were the most important landowning caste. When they were not landlords themselves they held 'superior' tenancies from Brahmins – who often led a secluded life – and/or from other wealthier Nayars. Although possessing a lower religious status than the Brahmins, the Nayars actually ruled south-west India.

The Syrian Christians also enjoyed a relatively high status, though perhaps slightly below that of the Nayars. As a specific Hindu vaishya caste was absent in the area, the Syrian Christians provided the traders. Furthermore, particular families possessed large tracts of land. Though some were wealthy landlords, most Syrian Christian landholders were tenants of Brahmins or Nayars, usually holding 'inferior' forms of tenancy. They also were specialized as skilled warriors. An interesting aspect of the Syrian Christians is their claim to be the descendants of Nambudiri Brahmins who were converted by Thomas, 'the doubting apostle'. This is most intriguing for two reasons. Firstly, the Syrian Christians' maintained Nambudiri genealogy, no matter whether true or false, shows the importance they attached to birth as the organizing principle of society, an intrinsic feature of the caste ideology. The Christians' 'Hindu' identity also

[3] Alexander 1989: 368; Kooiman 1989: 15.

[4] In other texts, Nayars are sometimes referred to as Nairs.

explains why they observed the same rules of ritual purity as the upper Nayar subdivisions.[5] And secondly, Thomas landed in Muziris,[6] the 'first (pepper) emporium of India'. Thus, the (claimed) origin of the Syrian Christian community was closely connected with the trade in export products, particularly pepper.

Like the Syrian Christians, the Muslims were engaged in commercial activities. Apart from being traders, Muslims also functioned as landlords, tenants and landless labourers, reflecting a kind of caste hierarchy within the Muslim community. In general, however, Muslims – like Syrian Christians – were accorded a respectable status.

In the caste hierarchy a fairly low religious and economic position was assigned to the Ezhavas[7] and the Shanars.[8] Their 'traditional' occupation was toddy-tapping, the extraction of an alcoholic – and thus impure! – beverage from palm juice. They were also involved in other trades connected with the coconut palm, including rope-making as well as weaving. They also functioned as agriculturists, often holding lands under 'inferior' tenancies from Nayars and Syrian Christians, and as agricultural labourers.

The Pulayas and Parayas formed the lowest strata of society. The other, higher-ranking castes regarded them the most impure of all human beings, if indeed human at all, for they were regularly referred to as slaves, their children as calves and their grain as chaff.[9] Indeed, Pulayas and Parayas were considered serfs and were bought, sold and mortgaged in the same manner as the land on which they dwelt. They were also offered by their masters as presents to friends or as gifts to temples. Extra humiliation was heaped on the Pulayas and Parayas, in that the members of these so-called slave castes were required to observe gestures of subordination which accorded with their ritual status. Maintaining a minimum physical distance from higher castes was one of these requirements. If a man or woman did not do so, and approached a member of a higher caste transgressing

[5] Bayly 1984: 184.

[6] Doshi and Liebl 1988: 43.

[7] Ezhavas are sometimes described as Iravas.

[8] In 1921, the caste-title of Shanar was officially changed to Nadar (Sarojini 1994: 23).

[9] Alexander 1989: 369.

the 'pollution distance', the former was accorded severe punishment by the latter, sometimes resulting in the serf's death.[10] Thus the lowest castes were not only considered *'untouchables'*, whose presence and touch caused ritual impurity to members of higher castes, but were virtually regarded as *'unseeables'*, induced by the idea of transmission of pollution over long distances.[11]

If the social order was characterized by 'relative rigidity', the political reality was very different, perhaps even the complete opposite. Though studies dealing with this subject are scarce, the information available discloses a complex, fluid and loose political system. Broadly speaking, the region consisted of numerous, mostly highly militarized societies. Indeed, the coastal areas and the more interior parts of south-west India were dotted with innumerable petty principalities, their rulers acting in a constantly changing, often openly violent political arena.

Perhaps conflicting with initial expectations, these polities were essentially different from the states which emerged in south-west India during the course of the eighteenth, nineteenth and twentieth centuries. In the early eighteenth century the political entities were vaguely defined in terms of geography and government. Though all possessing different rights over their territories, the various powerholders shared the fact that they did not create any 'hard' territorial boundaries. Nor were these kings, queens and others primarily concerned with the introduction of any 'clear-cut' policy fundamentally affecting agriculture or trade.

[10] This was customary for a very long period, as shown by subsequent statements. In the beginning of the sixteenth century, Duarte Barbosa, who accompanied Magellan on his trip around the world (Wolf 1982: 45), wrote: 'When they [the Nayars] walk along a street or road, they shout to the low caste folk to get out of their way; this they do, and if they will not, the Nayre may kill him without punishment' (cited in: Jeffrey 1976: 3). And in 1860, the wife of a missionary wrote: 'a Nair may approach but not touch a Nambudiri Brahmin; a Chogan Irava must remain 36 paces off, and a Poolayen 96 steps distant. A Chogan must remain 12 steps away from a Nair, and a Poolayen 66 steps off, and a Pariar some distance farther still' (quoted in: Jeffrey 1976: 9).

[11] With regard to the social hierarchy in south-west India see: Jeffrey 1976: 9–36; Fuller 1976: 53–68; Saradamoni 1980: 13–41; Tharamangalam 1984: 53–5; Kooiman 1989: 14–20; Alexander 1989: 367–70.

Both attitudes were also observed in the rulers' daily pursuits, dominated by their constant feuding with each other. Though often leading to violent clashes, these struggles were usually not directed towards outright conquest. Powerholders generally sought to assert a measure of authority by building temples or endowing existing shrines within a neighbouring domain, but these extensions of ritual authority were not necessarily a prelude to any attempt to control commercial or agricultural resources.[12]

Shifting our attention to the long-distance trade networks embracing early eighteenth century south-west India, the situation was much more simple, at least at first sight. In sharp contrast to the great number of well-matched competitors in the 'internal' arena, the Dutch East India Company[13] dominated the 'external' sphere, the domain of exports and imports.

Dutch 'rule' in Malabar lasted *circa* one and a half centuries (from 1663), roughly the same period of time as their predecessors had 'controlled' the area. The Portuguese traders had dominated the region since 1498, first breaking the powerful position of the Muslim sea traders and later enforcing treaties including stipulations regarding obligatory and exclusive delivery of pepper upon indigenous powerholders. Dutch hegemony started in early 1663 with the capture of the Portuguese stronghold of Cochin. Stationing its headquarters in that town, the Dutch East India Company increased its power by setting up a chain of forts, military outposts, factories and lodges along the Malabar coast, from Cape Comorin in the south to Cannanore in the north. At the same time, the Dutch introduced a pepper monopoly, entering into exclusive alliances with local men of influence, thereby aiming to keep out other European as well as non-European competitors. Dutch hegemony was challenged drastically only during the second quarter of the eighteenth century.

It should be clearly understood, however, that the supremacy of the Dutch East India Company, even at its peak, was extremely attenuated and fragile. Inheriting a problem the Portuguese had

[12] Panikkar in: Bayly 1984: 181.

[13] In other readings, the Dutch East India Company is called the Dutch United East India Company or the VOC. All concepts refer to the *'Vereenigde Oost-Indische Compagnie'*, literally meaning the United East India Company.

been confronted with in earlier times, the Dutch were unable to prevent what they called 'smuggling': fruitful attempts by Malabar merchants to sell their goods to the highest bidder, 'irrespective of any agreements their princes may have concluded above their heads'.[14] Equally importantly, the other European powers did not mean to let the Dutch monopoly of trade go unchallenged. They too settled on the spice coast of India; the British – the main adversaries of the Dutch – in Anjengo and Tellicherry, the French in Calicut and Mahe, and the Danes in Colachel and Edava.

Considering the fragmented political map of south-west India, the loose authority structures among local rulers and indigenous traders, and the great interest shown by other European traders in Malabar pepper, it is far from surprising that the Dutch East India Company was unable to control south-west India fully and enforce its pepper monopoly. At the same time, the Dutch were not prepared to simplify the political landscape and to establish direct control over the production of pepper. Indeed, the Dutch East India Company was unwilling (and possibly also unable) to conquer these small principalities. Instead of waging expensive wars, the Dutch preferred to intervene indirectly in the internal affairs of these chiefdoms. They supported one powerholder against another, and, when trading prospects required it, vice versa. Clearly, they tried to manipulate the political situation of constantly feuding principalities as much as possible to their own commercial advantage.

The reverse was also true. The indigenous powerholders used the Dutch presence as much as possible to their own advantage. This meant that most of the principalities only permitted Dutch influence to the extent it was deemed desirable, admissible or unavoidable. In these cases, the rulers presented only a proportion of the pepper under their control to the Dutch in exchange for certain military, financial or political favours; the remainder was sold to the highest bidder. This could be a Malayali trader who was oriented to local markets, an Indian trader who focused on other markets of the south Asian subcontinent, or the trading companies of the British, French or Danes making profits in Europe and Asia.

[14] Das Gupta 1976: 14.

2.2 Control over the Pepper Trade and Travancore State Formation

South-west India's political system, international position and social order changed fundamentally during the course of the eighteenth century. In this process, Marthanda and Rama Varma played an enormous role. Together they ruled Travancore for a total of 70 years, 1729–58 and 1758–98 respectively. What was their significance?

First and foremost, both Rajas were able to ensure a steady income. Though this may seem nothing spectacular in our eyes, that was an extraordinary accomplishment for eighteenth century south-west India. In this period, most rulers of Malabar could not claim any land revenue. They lived off their own lands and derived an unpredictable income from customs. In the first step towards gaining financial stability, Marthanda Varma succeeded in establishing considerable control over the levying of the various tolls and duties, at least to a greater extent than the other rulers of south-west India had done. Later, he tightened his grip on local affairs. In 1743 the pepper trade was declared a state monopoly.[15] For this purpose, a Commercial Department was set up, regulating the buying, warehousing and selling of pepper. Moreover, it was decreed that the pepper cultivators were strictly forbidden to sell their produce to unauthorized persons. Consequently, the previously independent merchants of Travancore had two, but more probably three, options: either to become an authorized trader, to trade illegally or possibly to combine both kinds of businesses. Though some of the traders successfully chose the second or third option (and the state never established full control over the trade in pepper as a result), for the Varma monarchy the pepper monopoly was extremely successful, and not only from a financial point of view.

As explained above, the political situation in south-west India was typified by constantly feuding principalities. Perhaps more shrewdly than their rivals, Marthanda and Rama Varma realized that a well-equipped army was an absolute prerequisite for their political survival. At least to a greater extent than the other

[15] Though of minor importance, the monopoly also included other commodities such as cardamom, salt, tobacco and timber during subsequent years.

rulers of south-west India, both rulers of Travancore utilized the existing political situation to their advantage: they exchanged pepper, the most important product of the Malabar coast for the European traders, directly for weaponry, the European product on which the Malabar rulers most wanted to lay their hands. To give an idea about the conditions and volume of this trade, on 22 August 1744 Marthanda Varma proposed that the British East India Company exchange 'one musquet for every candy of pepper'. As the bearer of the royal messages informed these Europeans, the Travancore ruler was also interested in the supply of other military equipment, including ammunition. Some days later, the king sent his exact request: the annual supply of 15,000 flints, 30 candies of lead, 50 candies of saltpetre, 50 candies of brimstone, 200 shells for mortars, 200 hand grenades and 500 cannon balls.[16] On the foundation of their control over the trade in pepper, the Travancore rulers were able to build up a strong army, well-equipped and well-organized.[17]

The Travancore rulers' military power simplified the political situation which earlier had been complex, fluid and loose. Abandoning ideas of 'traditional warfare' in south-west India, their armies annexed many of the neighbouring petty principalities by outright conquest. The defeated powerholders were their pensioned off or imprisoned, their armies disbanded and their lands expropriated. These conquered lands were converted into state-owned or *sircar* lands and became an important source of revenue for the Travancore state. Other more ruthless methods of subordination and humiliation were used as well. In one fell swoop Marthanda Varma captured the recalcitrant chiefs of eight leading Nayar families, had them executed and sold their women and dependents to low-caste fishermen.[18] In order to administer their new territories the Varma monarchs

[16] Kooiman 1992a: 600.

[17] One interesting sideline of history is that the modernization of the Travancore army was entrusted to Eustache Benoît de Lannoy in the period 1741–77. De Lannoy deserted from the Dutch East India Company around the time of the Dutch defeat by the Travancore army at Colachel (1741), reorganized the Travancore forces and became known as the *'Valiya Kappittan'* (Great Captain). To a greater extent than was customary in south-west India, he recruited non-Nayars for his army (Lannoy 1986).

[18] Jeffrey 1976: 3.

surrounded themselves by loyal and obedient Tamil Brahmins. To sum up the military-political achievements of Marthanda and Rama Varma: during their reigns the once tiny and weak chiefdom of Venad, centred on the village of Tiruvidamkod, expanded to Cochin in the north and Cape Comorin in the south, was transformed into a strong power and earned widespread respect as Travancore.

Indubitably, the rise to power was an cumulative process. Because the Rajas of Travancore firmly controlled the trade in pepper, they secured a stable income, and at least as importantly, could acquire weaponry, both being crucial prerequisites in their struggles for political authority and military expansion, which again enlarged their control over their main source of power: pepper. This contributed to two related developments during the eighteenth century. In the case of supply: the shift from a whole range of pepper traders (individuals, merchant groups and different rulers) to one dominant pepper trader, namely the Varma monarchy. And in the case of demand: the change from one dominant unit, the Dutch East India Company, to various units, including the British, French and Danes. That was a marvellous chance: as the Dutch found it increasingly difficult to play off the various indigenous interest groups against each other (on account of the waxing strength of the Travancore monarchy), they became less interested in Malabar, thereby providing new opportunities for other European traders. There is no shadow of doubt that the international pepper trade and Travancore state formation were closely related with one another.

These events had enormous repercussions on the social hierarchy in south-west India. It seems that the Tamil Brahmins replaced the Malayali Brahmins at the apex of society. As mentioned above, Tamil Brahmins in particular were appointed to influential posts within the Travancore state. At the same time, the influence of Brahmins in general was diminishing. This eclipse was related to a very important religious ceremony, a political masterstroke. At the end of his reign, Marthanda Varma dedicated his whole kingdom to the Sri Padmanabha temple in Trivandrum. He simultaneously appointed himself the sole manager of the temple and all its landed properties and made clear that any protest or disturbance against his rule would be tantamount to a rebellion against the deity. The Nayars also saw their influence recede. The administration was headed by non-Nayars, namely Tamil Brahmins, and new military equipment (European weaponry) and methods of warfare (including

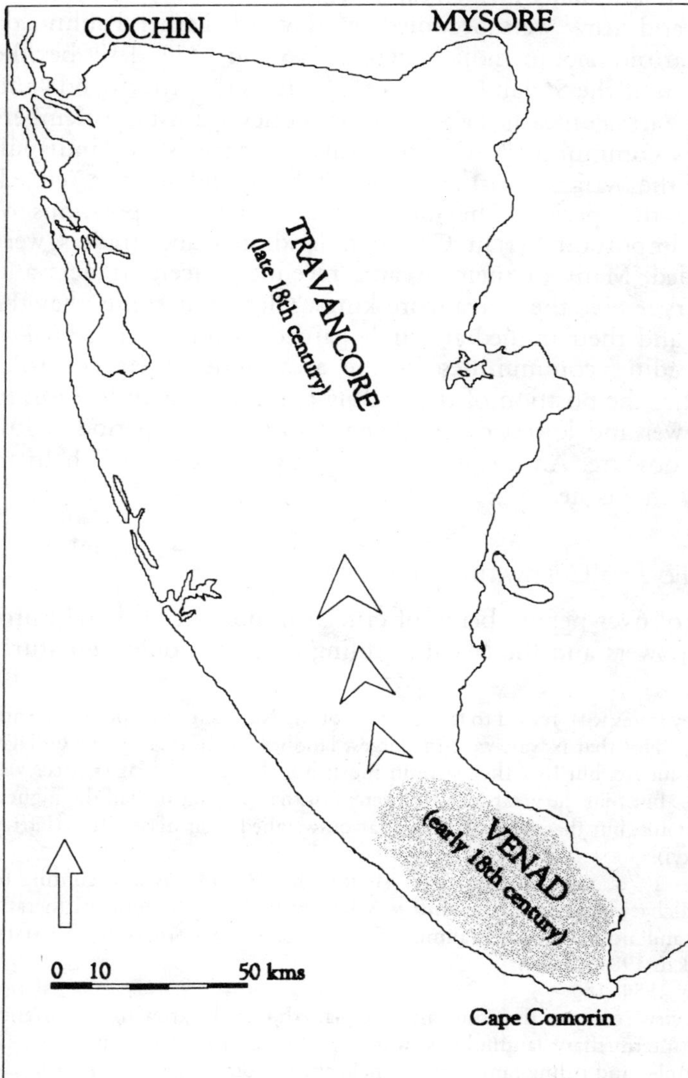

Map 2.1 The Expansion of Travancore in the Eighteenth Century

the recruitment of large numbers of non-Nayars) undermined their importance in the armed forces. However, they were still the dominant community in society; they controlled most of the land and many influential positions within the state bureaucracy and army were occupied by Nayars.[19] Father Paulino da San Bartolomeo, in India between 1776 and 1789, describes the position of the Syrian Christians. He states that Marthanda and Rama Varma unleashed a persecutory policy against the members of this community.[20] Although this evidence is not indisputable,[21] the Syrian Christians most probably did suffer a setback during this period. The independent, privileged positions of some important Syrian Christian landlords and traders were curtailed. Many of them became revenue officers, traders and soldiers *serving* the Travancore kings. Clearly, it seems that the Rajas and their immediate circle gained in importance *vis-à-vis* the leading communities in the area under their control.[22] Whether the position of the leading communities in relation to the lower and lowest castes changed in the same period is still fairly obscure. An analysis of pepper production will help to clarify this issue.

2.3 Forced Cultivation of Pepper

Control over pepper being of crucial importance for the European powers and the Malabar ruling elites, it would be natural

[19] Jeffrey states with regard to the position of the Nayars in the mid-nineteenth century: 'Not that Nayars were the largest landholders in every *desam* (village) in Travancore, but that they were in most; not that every village officer was a Nayar, but that the great majority were; not that the Nayars had the highest ritual status, but that their privileges far outweighed their disabilities' (Jeffrey 1976: xvi).

[20] Fuller 1976: 56; Oommen 1977: 41; Joseph 1988: 135. Also according to P.K. Michael Tharakan, 'the shift was largely in favour of non-aristocratic Nairs, and non-Malayali Brahmins from outside the borders of the state' (Tharakan 1984b: 1961).

[21] Bayly 1984: 180.

[22] This view confirms the observation of Ganesh that the growing importance of the intermediary landholders, who actually controlled the land, *vis-à-vis* the temples and ruling families, who held the customary ownership of lands, in the sixteenth and seventeenth centuries was reversed in eighteenth century Travancore (Ganesh 1991: 315).

to expect a lot of information about the production of the crop. Unfortunately, this is not the case. Among other factors this lack of historical knowledge is related to the nature of the archival sources of the European trading companies. These materials mention pepper cultivation only marginally as European merchants were not seriously interested in the production of the crop; they concentrated their efforts on the trade in pepper. Equally importantly, many Old Malayalam records, most probably containing important information about pepper cultivation, still await exploration. Although there is not much information available regarding pepper production in general and during the rule of Marthanda and Rama Varma in particular, I will try undauntedly to 'reconstruct' the historical reality.

It is certain that the pepper was not produced by the important Brahmin and Nayar landlords, the *janmkars*. They were not directly involved in the cultivation of the land. Nevertheless, it is conceivable that the pepper was produced by that group of people which held 'superior' tenancies from the janmkars, the *canumkars*. This group, which included predominantly Nayar but also some Syrian Christian tenants, employed landless labourers to do the actual cultivation on their lands or on parts of their possessions. However, nor does it seem that the canumkars produced the bulk of the pepper. Most of the pepper was produced by smallholders. This view, which is the consensus of most historians,[23] is based primarily on information provided by the archives of the Dutch East India Company. In these sources the pepper producers are often referred to as *'arme landsman'* (a poor countryman) or as *'kleine landsman'* (little countryman). Although these historical sources give some indication of the area of the land on which the pepper was produced, the data do not provide any illumination about the ownership of these holdings. By taking into account the social hierarchy of the region, it becomes evident that these holdings were cultivated mainly by relatively rich sections of the lower castes (like Ezhavas and Shanars) as well as by some relatively poor members of the leading castes (Nayars and Syrian Christians), and were held on 'inferior' forms of tenancies from

[23] Das Gupta 1976: 23; Jeffrey 1976: 96; Prakash 1987: 4; Ganesh 1991: 305.

canumkars or their sub-tenants. Without any doubt, some of these pepper producers called in the help of the lowest castes, the Pulayas and Parayas, to plant, look after and harvest the pepper.

Fortunately, more information is available about the relations between traders and pepper producers. At the beginning of the sixteenth century, Duarte Barbosa mentions that the Vyapari Nayars traded extensively in the interior parts of south-west India. He describes them as frequently giving advances to Nayars and other landholders and buying up all the available ginger and pepper.[24] The relation between pepper traders and pepper producers was more extensively described by Francis Buchanan in the early nineteenth century. In north Malabar he observes:

> All the gardens are small, and all the cultivators have other property. In June, July or August, the traders go round to the cultivators, and advance them money, on condition, that in January or February the cultivators shall deliver their pepper at a given place. The money advanced is in proportion to the wants of the cultivator. If he be pressed for money, not above two thirds of the value will be given. In other cases, where the cultivator is not so necessitous, the money advanced is nearer the true value of the pepper. The condition of the bargain is also such, that, if the cultivator does not deliver the stipulated quantity of pepper, he must pay for the deficiency at the Calicut price, which is considerably greater than the common rate of the interior parts of the province. The advance is frequently made in cloth or other goods; but most commonly in old *Fanams*, worth 1/4 rupee. There are, however, several men, chiefly of the *Moplay* caste, who are prudent enough to be able to wait for their money until the produce of their gardens is ready for delivery. These, in general, let their gardens on *Vir Patom*, where the cultivator is in fact nothing more than a superintendent for the proprietor, who furnishes every expense, and allows a small percentage on the produce for the support of the tenant. The traders who make the advances to the farmers are mere factors for the merchants residing in the great towns, who furnish them with goods and cash to make the advances, and who have them in as great subjection as they have the inconsiderate cultivators.[25]

These two quotes are stacked with important information. It is

[24] Kieniewicz 1969: 73; Das Gupta 1976: 21. According to Kieniewics, Vyapari Nayars or Biabares 'were living at the foot of the mountains'.

[25] Cited in: Das Gupta 1976: 21–2. The concept Moplay originated from *Maha Pillay*, the honorific title of an influential, high-caste landlord.

most striking that in the relationship between the individual pepper merchant and his supplier, the advance given by the former to the latter played a crucial role. Pertinently, most of the advances were given to the actual producers, although sometimes the advance or payment was given to their landlords, respectively the Nayars as mentioned by Duarte Barbosa or the Moplays as reported by Francis Buchanan. In the latter cases, the tenants most probably paid part of their rent in pepper. Most importantly, the relations between traders and suppliers were of an unequal nature, especially in those cases where there was a direct connection between merchant and actual producer. Buchanan describes this phenomenon in great detail. 'If he [the trader] be pressed for money [by the cultivators], not above two thirds of the value will be given. In other cases, where the cultivator is not so necessitous, the money advanced is nearer the true value of the pepper'. Clearly, most of the actual cultivators were in such desperate need of money, either in order to buy food and/or cloth or to refund already existing debts, that they, in contrast to wealthier landholders, were not 'able to wait for their money until the produce of their gardens is ready for delivery'. Though aware that accepting the money, cloth or other goods meant only a temporary relief from misery, the poor cultivators were not in a position to resist the advance offered and thus obliged themselves to deliver a certain amount of pepper for an unfair price. Buchanan was absolutely convinced that the merchants were fleecing the pepper producers. He estimated the profit of the traders at between 60 and 80 per cent.[26] This also explains why the well-to-do class of canumkars did not produce much pepper. Less indigent, these richer and more influential landlords could easily turn their back on the advances offered, thereby avoiding obligatory and exclusive delivery of pepper. Undoubtedly, as the crop was produced and delivered by the poor peasants below 'the true value', that is below the market price, cultivating pepper was unattractive to the canumkars.

Probably there was yet another reason which dissuaded the canumkars from producing large quantities of pepper. As shown in Barbara Watson Andaya's study concerning seventeenth and

[26] Das Gupta 1976: 22.

eighteenth century Jambi (Sumatra), pepper growers ran great risks.[27] Especially during the initial stages of planting, pepper cultivation was highly labour intensive. The young plants being quite vulnerable to variations in weather patterns, pepper required close supervision. Cultivators had to spend much of their time on watering the plants during the dry season and preventing floods during the monsoon. The risk involved in pepper production was heightened by the fact that pepper plants did not mature for four to seven years (and ceased to bear after thirty years). It is, therefore, no wonder that only those willing to accept the traders' advances for reasons of sheer survival, the poor smallholders, ventured into pepper production, most probably ending up in situation of debt slavery in later years.

Although no historical sources deal in detail with the relations between pepper merchants and suppliers during the reigns of Marthanda and Rama Varma, there are no reasons to assume that the advance system was not in force in Travancore during these years. Having said this, the nature of the relation between trader and pepper supplier was most probably different in the Travancore of Marthanda and Rama Varma. It is my conclusion that the position of the pepper cultivator with respect to the trader grew even more vulnerable. This stemmed from various factors. To begin with, from 1743 the producers were allowed to sell their pepper only to authorized traders who offered a price, read: an advance, of which the amount was determined by the Commercial Department. Those who violated the pepper monopoly were punished severely. Additionally, the merchants functioned as state agents after the introduction of the monopoly. The traders were appointed by the ruling aristocracy of Travancore and were backed by the judicial and military agencies of the state. And the army – there were said to be 150,000 fighting men serving the state by 1780 – was not only used for territorial expansion but was also useful for settling internal affairs. Paulino da San Bartolomeo noted that 'the troops are always marching up and down through the country to change their cantonments, to enforce the collection of taxes, and to preserve peace and tranquility'.[28] Thus in sharp contrast to the

[27] Watson Andaya 1993: 99.
[28] Quoted in: Bayly 1984: 187.

supply side, the demand side was monopolized, well-organized and backed by powerful state agencies. This also explains why the pepper monopoly was so successful. Smuggling could be prevented to a very large extent – the European trading companies regularly complained about the efficiency of the Travancore trade monopoly – and the price/advance offered could be kept low,[29] perhaps be pushed even lower.[30]

Concluding this chapter, I will now turn to the principal matter of the study: plantations as situated in the wider historical context of export-oriented cash crop production. The first pertinent observation deals with plantations and discontinuity. As perhaps known to scholars interested in pre-colonial south-west India, A. Das Gupta once referred to the pepper units as *plantations* and Ajit Mani mentions pepper as *'plantation' produce*.[31] Starting from the concept of plantation as defined in the introduction of the book, it can hardly be said that there were plantations in eighteenth century Travancore. The pepper plots were not large-scale agricultural enterprises and therefore differed fundamentally from nineteenth and twentieth century estates.

This first remark must be followed up by a second comment, concerning export production and continuity. Though strictly speaking pepper cultivation was not plantation production, a great number of actual cultivators were working in an authoritarian hierarchical organization, producing pepper for distant markets. Indeed, in that sense history was an uninterrupted process, putting eighteenth century pepper cultivators on the same footing as the plantation workers of later centuries. In keeping with this interpretation, the eighteenth century export elite, the Varma monarchy, was a forerunner of the nineteenth and twentieth centuries planting elites. The growing importance of both Marthanda and Rama Varma was closely related to their ability to control the trade in and cultivation of one particular

[29] Kooiman 1992a: 599.

[30] K.M. Panikkar writes about the trade monopolies, of which the pepper monopoly was the most important one: 'The trade and economic policy of Travancore was a crude kind of state enterprise. A system of monopolies was introduced which from the point of view of the present day was a gross exploitation of the people' (cited in: Ramachandran Nair 1979: 73). See also: Tharakan 1984a: 1917.

[31] Das Gupta 1976: 21; Mani 1984: 791.

agricultural crop: pepper. Conversely, their influence within the administrative, judicial and military state agencies greatly affected the production of the crop. Seen above, the position of the pepper growers was extremely weak during the reign of both these kings. The Commercial Department, through its authorized merchants, controlled the trade in and the price of pepper. Showing some striking similarities with prominent planting communities in later times, the Varmas' ability to control the trade in and production of pepper and their political authority were closely interconnected with one another. For these reasons, pepper production in eighteenth century Travancore can be considered a proto-plantation system.[32]

[32] I fully realize that the use of the concept 'proto' is not without difficulties. Most importantly, the term 'proto' might suggest that one particular historical phenomenon (like pepper production in eighteenth century Travancore) would definitely lead to its ultimate outcome (in this case plantation production in nineteenth and twentieth century south-west India). However, I do not subscribe to such a teleological, evolutionist point of view.

On the other hand, I do use the concept of 'proto' for the following reasons: (a) *a lack of alternatives*. Labels such as 'traditional' or 'pre-modern' indicate wrongly that south-west India was originally static, and that Western colonizers finally introduced change and growth. Also, I find it insufficiently rewarding to characterize pepper production in eighteenth century south-west India simply as 'commercial agriculture'. Though the production of pepper was oriented towards a commercial market, the actual pepper producers were not able to offer their produce to the highest bidder.

And (b) *the possibility of paying attention to both continuity and change*. 'Proto', and as used in the concept of 'proto-plantation system' in particular, provides an opportunity to emphasize certain long-term developments (like labour functioning in an authoritarian, hierarchical framework and an export elite using its political influence to organize production), thereby preventing the historiography from becoming totally compartmentalized. At the same time, the use of 'proto' also hints at fundamental change, like the shift from peasant cultivation to estate production (cf. Perlin 1983).

Three

Imperial Initiatives in Export Crop Production: *c.* 1800–Late 1850s

As I commenced the cultivation by the recommendation of Colonel Newall who, with the Dewan in 1825 personally and in the presence of the District authorities, pointed out 1 or 2 situations he considered best adapted for the growth of coffee etc., and in consequence a public order was issued by the Dewan to the Tahsildar *to allow me to take such ground as might be necessary and for which I was promised a grant, as soon as the extent required and situation could be determined, which could only be ascertained after the experience of two or three seasons had enabled me to judge if the soil and situation were sufficiently favourable*

Letter of Huxham to the British Resident on 17 September 1836[1]

3.1 The Decline of Varma Dominance and the Establishment of British Rule

If the Varma monarchs were the initiators of a proto-plantation system in the eighteenth century, it was the British Residents who took over the leading role in (re)organizing export-oriented agriculture in the first half of the nineteenth century. They were the instigators of the *destruction* of the old economic order, a system in which certain state agencies (the Commercial Department and the army) firmly controlled the production of and trade in particular crops, especially those of crucial importance to the independent functioning of the policy (pepper in particular). Furthermore, the British Residents were largely responsible for the *creation* of a new, imperialist order, reshaping state and society in such a way that individual agriculturalists, if possible British, could begin to cultivate those crops which were highly profitable on markets controlled by British trading firms. Coffee, of course, was such a product *par excellence*. All the above

[1] PTSMC 1926.

developments had their origin in the political field: the transfer of power from the Varma monarchy to the British East India Company.

At the very end of the eighteenth century, the financial situation of the ruling elite of Travancore deteriorated considerably. This economic collapse was closely related to the growing influence of the British East India Company on the Malabar coast. Stepping up their activities after Dutch Cochin was taken in 1795, the British went from strength to strength in pursuing their policy of establishing a pepper monopoly. Having accomplished their aim of fencing off the Malabar coast from European competitors, the British East India Company adopted an increasingly aggressive policy towards the Travancore state, even 'terrorizing' its ports.[2] For these reasons, Travancore's Commercial Department lost its initiative in the determination of the conditions of trade, selling its pepper to the highest bidder.

In addition, Rama Varma 'volunteered' to cede some of his military autonomy to these very same British. The areas under his control needed to be safeguarded from the expansionistic armies of Tippu Sultan, ruler of Mysore. To achieve this, the Raja entered into treaties with the East India Company in 1788 and again in 1795, agreeing that two and three British battalions respectively would be stationed in his territories, the expenses thereof being met by the Travancore state.

The military alliance meant more than a loss of political sovereignty. It also aggravated the difficult financial straits in which the Travancore royal family found itself. The price to be

[2] The way in which the British East India Company tried to break the influential position of the ruler of Travancore is described fairly exactly by the Portuguese Governor-General at Goa. On the eve of the nineteenth century he wrote: 'The English make a monopoly (of pepper) from Mount Deli to Cape Comerin, wishing to claim for themselves in the first place all the produce from which the king (of Travancore) derives any profit (. . .). The said English fearing that they will not get from the said King as large quantity as they desire, have provided themselves with certain light war vessels which go round the ports of this kingdom every day in order to terrorise over the said King so that they alone and no other nation can obtain pepper' (quoted in: Department of Public Relations 1976: 76).

paid for military aid – predominantly in cash but also in pepper – proved to be too high for the Raja of Travancore, forcing him to borrow large sums of money.[3] Though the king raised loans from merchants, from bankers and even from the Bombay government, he could not defray all the expenses incurred in the battles against Mysore and for continued military protection as had been agreed earlier. Plagued by lack of money, subsidy and loan repayments fell into arrears and the Travancore state was dragged into the vortex of a financial crisis.

When Rama Varma died in 1798, his successor Bala Rama Varma inherited a state which was almost bankrupt. Unfortunately for the new Raja, the British East India Company did not show the slightest sympathy for his financial difficulties. Even after Tippu Sultan was finally defeated in 1799, the British insistently demanded the clearing of the arrears of tribute and the delivery of the stipulated quantity of pepper.

In 1800, the British changed their policy of exerting pressure from the outside. They sent a British Resident to Travancore in order to see that the state's obligations were met. For Travancore's political elite the implications were immediately tangible. The officer appointed as the first official representative of the British government, Major Colin Macaulay, regularly intervened in Travancore politics. It was he who played a major role in appointing Velu Thampi, an influential aristocratic Nayar, Dewan of Travancore. Although the British Resident's influence in the internal affairs of Travancore was considerable and although Velu Thampi adeptly improved the state finances, Macaulay, as indeed his British superiors, believed that the treaty with the Raja should be revised: the influence of the British Resident in the internal affairs of the Travancore state should be enlarged and formalized.

Soon after an insurrection of about 10,000 Nayar soldiers –

[3] Madhava Rao, Dewan of Travancore half a century later, was aware of the serious nature of Rama Varma's financial problems during the last years of his reign. He noted: 'Even extraordinary taxation [like imposing more state monopolies and transit duties] failed to yield adequate means to answer the incessant call for funds. The Maharaja felt that he could not, without exciting dangerous discontent among his subjects, increase their burdens, and saw no alternative but that of raising loans' (quoted in: Department of Public Relations 1976: 76).

which was directed towards the liquidation of Macaulay and Velu Thampi and the restoration of the old order under which the Nayar troops were well paid and were indispensable to the defence of Travancore against external enemies – was suppressed (1804), a new treaty was imposed on the Travancore government. The treaty, which was signed on 12 January 1805, confirmed the treaty of 1795, stating that the defence of Travancore against foreign enemies would be the responsibility of the British East India Company (Article 1). Moreover, the new treaty obliged Travancore to pay to the Company a substantially raised tribute, amounting to a total of 800,000 rupees annually (Article 3). The revised treaty served to undermine the monetary foundation of the Travancore state even further. Most damning of all, the treaty of 1805 diminished Travancore's autonomy considerably. It forced the Raja 'to pay at all times the utmost attention to such advise as the English Government shall occasionally judge it necessary to offer to him' (Article 9).[4] In short, the new treaty dismantled Travancore militarily, financially and politically. It reduced that state, though formally independent, to the position of a subordinate power.[5]

Understandably, influential members of the Nayar community in particular did not accept British paramountcy lying down. The Nayars had already surrendered their military and political supremacy to the Europeans and were afraid that their position would be weakened even further. Therefore, it is no surprise to discover that all anti-British rebellions in Travancore in the period immediately following the new treaty were led by Nayars and were aimed at restoring the old order. In 1809 an insurrection led by Dewan Velu Thampi, who had gradually became fed up with the manipulation of Travancore politics by the British, was reactionary in nature as well. Velu Thampi warned the Travancoreans against the British who 'put their own guards in

[4] Subsequently, Article 9 mentioned the issues concerned: 'with a view to the economy of his finances, the better collection of his revenues, the administration of justice, the extension of commerce, the encouragement of trade, agriculture and industry or any other subjects connected with the advancement of His Highness's interests, the happiness of his people and the mutual welfare of both states'.

[5] The treaties of 1795 and 1805 are reproduced in respectively: Sankarankutty Nair 1977: 86–91 and 93–8.

the palaces, Sircar buildings, and the fort gates, destroy the royal seal (. . .), put up crosses and Christian flags in pagodas, (and) compel intermarriages with Brahmin women without reference to caste or creed'.[6] Although these revolts were crushed fairly easily, the British grew more circumspect about the community background of the aspirant Dewans. The upshot being that, after 1817, mostly non-Malayali Brahmins held the dewanship and only one Travancore Nayar was ever to be Dewan again.[7]

The subordinating alliance ensured that the political system in Travancore in the period 1805–58 was dominated by the British Resident. Although in theory the Raja was there to rule, the Dewan to execute and the Resident to advise, in actual practice the tripartite division of authority was often reversed. In matters of crucial importance to the British, the Resident ruled, the Dewan executed and the Raja sanctioned. In this way, the imbalance in strength between Travancore and the British East India Company was reflected in the importance of the leading political figures of south-west India.[8]

In brief, the events in the late eighteenth and the early nineteenth century altered the position of the royal family of Travancore completely. The Varmas, who headed a very strong and well-organized army and were sovereign in internal affairs during the second half of the eighteenth century, lost ground in almost every possible sense: financially; militarily; politically. Filling the power vacuum, a new political elite, the British bureaucrats, firmly established themselves in the political arena during the first half of the nineteenth century, thereby fundamentally affecting the nature of export production.

3.2 The Disintegration of State-Controlled Pepper Production

All these events had major repercussions for the trade in and production of pepper. The Travancore government lost both its influential position in the international pepper trade and its

[6] Quoted in: Jeffrey 1976: 5.

[7] Jeffrey 1976: 6. N. Nanu Pillai was the Dewan in question. He held the dewanship in the period 1877–80. M. Krishnan Nair, Dewan in the period 1914–20, originated from directly-ruled Malabar.

[8] Cf. Jeffrey 1975: 263–8.

grip on the indigenous pepper suppliers. Turning to the former first, in 1805 the political elite of Travancore had to cede its independence in external trade altogether. In order to control the foreign trade of the state, the British stationed a Commercial Agent, in charge of Travancore's Commercial Department, in Aleppey. Although the Commercial Department retained its position as the sole agency authorized to collect pepper, the British did not allow the department subsequently to sell its products to the highest bidder: the British East India Company became the sole purchaser of pepper from the Travancore state.

Even after the enactment of the British Charter Act of 1813, which abrogated the monopoly of the British East India Company in Indian trade and enabled private merchants and trading companies to enter into direct relations with the Commercial Department, the Travancore government was not able to profit greatly from its pepper monopoly. Much of this stemmed from the changed attitude of the Travancore administration itself. Apart from purely financial reasons, the Travancore ruling elite was no longer particularly interested in its pepper monopoly. In this period, in which the interests of Travancore were subordinated to those of the British, it was impossible to exchange pepper for arms and the trade was therefore no longer synonymous with military power and political independence.

The Commercial Department's failing control of pepper was not merely the consequence of the changed attitude of the indigenous political elite. Two other factors were significant. Most vitally, the state traders, in contrast to those functioning during the rule of Marthanda and Rama Varma, were not backed and controlled by a well-organized and well-equipped army. To a much greater extent than before, the military kept itself aloof from the local arena, including the spheres of production and trade. Smuggling was therefore much easier for the actual pepper producers and for the authorized and non-authorized traders during the first part of the nineteenth century. Of almost equal weight was the fact that the Commercial Department was not able to advance/pay high amounts of money because of the depleted state treasury. It was simply not in a position to compete with the traders who offered the pepper suppliers better terms. The gradual rise in the price of pepper played an important role in this regard. Eventually, less and less pepper was delivered to the Commercial Department – in the period 1845-9 it received no pepper at all – and ever-mounting quantities of pepper were smuggled out of the country.

In the early 1850s, the Madras government, pressured by British trading companies and the government of India, frequently informed the Resident that it considered the pepper monopoly a 'practice which violates principles of the Free Trade of the contemporary period'. Consequently, in August 1855 Resident W. Cullen wrote to Dewan V. Krishna Rao about the 'expediency of abolishing the pepper monopoly altogether and immediately'. The Dewan, who was not in favour of abolishing the pepper monopoly, proposed improving the production of pepper by raising payments to the cultivators. The government of India, after being informed about the Dewan's opinion, resolved to approve his proposal only as a temporary measure. In early 1860, the newly-appointed Resident, F.N. Maltby, was asked to submit a report on the question of abolishing the pepper monopoly. He advised the government of Madras to force the Raja of Travancore to abrogate the monopoly and to introduce an export duty of twenty rupees per *candy*. The government of India, again having been informed of the matter, recommended that the government of Madras fix the duty at fifteen rupees per candy. The unremitting pressure exerted by the British and the non-conservative attitude of the new Dewan, T. Madhava Rao, altered the political situation. In November 1860, the pepper monopoly, once the chief source of revenue of the kingdom, was abolished and replaced by an export duty of fifteen rupees per candy.[9]

3.3 British Residents, Huxham and the Introduction of Plantation Production

The role of the British officials was by no means restricted to the downfall of the 'original' political and economic structures. They also stimulated the creation of a new politico-economic climate which would attract European planters and stimulate the production of 'Western-oriented' crops by indigenous cultivators.

[9] Quotations in: Ramachandran Nair 1979: 70–84; cf. Nagam Aiya 1906: 566; Mahadevan Nadar 1980: 217–18.

The first attempts to grow coffee in south-west India had already been made by the trading companies of the European powers in the early eighteenth century. The British East India Company tried to cultivate coffee in the vicinity of Anjengo. Unfortunately for the British, they were confronted by great difficulties. On 19 October 1727 Alexander Orme, Chaplain Whitehill, J. Salisbury and N. Goodwin requested the British settlement in Tellicherry to 'provide us a few coffee plants to be sent down by the cruizers that will speedily be with you, for these two years past we have used all possibilities endeavours to raise some, but have been unsuccessful'. They therefore requested to 'send us directions about raising them and an account of the nature of the soil which best agrees with them'. The British claimed that their Dutch competitors were more effective in cultivating coffee. They informed their fellow countrymen in Tellicherry about 'the large advances the Dutch have made in their plantation at or about Papinate; they have this year likewise put up a plackat at Tutocorine'. It is interesting to note that the Dutch tried to restrict the cultivation of coffee by indigenous peasants, 'prohibiting the Paravas and others to plant coffee trees'.[10]

The first individual effort to cultivate coffee in Travancore was made by W. Huxham, who started planting in the hills east of Quilon in 1824.[11] His first plantation was named Camapalle.

It should be emphasized that Huxham was an exception in southern South Asia. Most of his European colleagues started coffee cultivation in other parts of the region, particularly in Coorg, Mysore, Nilgiris, Wynad (all four areas forming parts of the Madras Presidency) and Ceylon (also under direct British rule). The political factor was of major significance in the planters' decision not to settle in Travancore. In contrast to all areas mentioned above, Travancore was formally independent. Colonial government aid was more likely to be forthcoming in the directly-ruled areas than in Travancore.[12]

Huxham, presumably like most of his fellow planters striving for personal wealth, therefore had to be convinced by the British

[10] Records of Fort St George, Letters to Tellicherry, vol. 1, 1726–8.

[11] Letter of Huxham to the British Resident on 17 September 1836 in: PTSMC 1926; cf. Jeffrey 1976: 98; Tharian George and Tharakan 1986: 201.

[12] Mysore came under direct rule in 1831. In 1881, however, Mysore again became a formally independent 'Native State' (Hettne 1978: prologue).

Resident that Travancore in general and the political climate in particular were well suited to the opening up of plantations. As already noted in the quotation at the beginning of this chapter,[13] the British Resident, Colonel Newall, played a crucial role in this regard. Newall encouraged Huxham to start an agro-enterprise specializing, amongst other crops, in coffee, advised him about the selection of suitable land, and told the Dewan and the Tahsildar in question to provide the necessary legal assistance for Huxham's undertaking.

Colonel Newall, who 'became so ill as to be almost incapable of attending to public affairs and finally embarked for Europe',[14] was right in at least one important respect. Soil and climate were extremely suitable for the cultivation of coffee. After twelve years of planting experience Huxham made an observation about 'the valley itself being peculiarly well adapted for spice and coffee cultivation'.[15] Antithetically, the political context did not prove to be so reliable. This became apparent particularly in the fields of both fixity of land tenure and labour recruitment.

The correspondence between Huxham, the British Resident, and the Dewan illustrates the pioneer's difficulties regarding the legal acquisition of land. In early May 1836, Huxham informed the then British Resident, Colonel J.S. Fraser, that 'the Dewan has repeatedly promised to execute the grant for the land I commenced cultivating in 1832 but he has hitherto omitted to do so'. Huxham disclaimed any desire to cause difficulties by

[13] In another letter, Huxham mentions: 'In 1825 the late Dewan (Venkata Rao) issued an order to the authorities of the Pathanaverum District to allow me to commence clearing and planting the land in such situations as might be found fitting for the growth of coffee and spice. This order was issued at the instance of the then Resident Col. Newall who first advised me to commence the spice cultivation in the jungles and who with the Dewan personally selected a spot of ground which he considered applicable. In consequence of which five or six situations in the District were tried to a limited extent (. . .)' (letter of Huxham to the British Resident on 25 July 1836 in: PTSMC 1926).

[14] Letter of Huxham to the British Resident on 17 September 1836 in: PTSMC 1926.

[15] Letter of Huxham to the British Resident on 25 July 1836 in: PTSMC 1926.

submitting yet more applications and asked Fraser to put in a good word at a suitable opportunity. 'Either it has escaped his attention hitherto, or I have some false friend at the *Huzur*', Huxham concluded his letter.[16] Perhaps unfamiliar to some readers, Huzur refers to the *Huzur Cutcherry*, the office of the Dewan.

The British Resident sprang immediately into the breach to help the planter. A few days later Fraser reported to Huxham: 'I have spoken to the Dewan on the subject of the grant and he tells me that the necessary papers are in preparation'.[17]

Only two months later, the Dewan, named Soobrow, explained his objections to the British Resident. 'His having cultivated ground at that place (. . .) seems to have been without any regular sanction, consequently not legal'. However, the Dewan, aware of his subordinate position, was not able to enforce his point of view by taking counter-measures. Soobrow therefore continued the letter thus: 'in consideration however of the labour and expense that he might have incurred on account of it, I think that such portion of the jungles which he has already cleared at Shalakerry may be granted to him'. Soobrow had in mind a separate lease, which 'I intend to propose Mr Huxham after I have been favoured with your sentiments on this score'.[18]

This gave the British Resident the opportunity to influence the terms of the agreement and he wrote to Huxham three days later, on 22 July 1836, 'to request you will be so good as to favour me with any remarks you may have to offer on the subject (. . .) before it receives my reply'.[19]

Though Huxham and Fraser were seemingly in control, the Dewan remained unwilling to sanction the grant for two more months. On 14 September 1836 the British Resident informed Huxham once more about the Dewan's reluctance to act. As the Resident rephrased the Dewan's message, 'the delay to which

[16] Letter of Huxham to the British Resident on 6 May 1836 in: PTSMC 1926.

[17] Letter of the British Resident to Huxham on 10 May 1836 in: PTSMC 1926.

[18] Letter of the Dewan to the British Resident on 19 July 1836 in: PTSMC 1926.

[19] Letter of the British Resident to Huxham on 22 July 1836 in: PTSMC 1926.

you alluded has arisen solely from your including in the proposed grant a considerable extent of ground which was never authorized by the Sirkar and was brought by you under cultivation merely on some private understanding between yourself and a former Dewan or Resident'.[20]

Subsequently, Huxham, tired of waiting, lost his temper and wrote the Resident an ill-natured letter. 'As I presume it is still the wish of this Government to encourage the introduction and establishment of agricultural concerns, I trust he [the Dewan] is disposed to afford all due and necessary support and assistance to the continued progress of my undertaking'. Accusing the Dewan, Huxham informed the British Resident that the former seemed to be under a misapprehension that the land at Pathanapuram was only being cultivated because of an old agreement between the late Dewan and a former Resident, 'whereas the contrary is the case'.[21]

Further historical evidence suggests that the Dewan, most probably strongly pressured by the Resident to take immediate measures, finally complied with the British wishes, making all the necessary arrangements to meet Huxham's request.[22] This occurred only in mid-November, i.e. six months after Huxham's

20 Letter of the British Resident to Huxham on 14 September 1836 in: PTSMC 1926.

21 Letter of Huxham to the British Resident on 17 September 1836 in: PTSMC 1926.

22 As the Dewan explained to Huxham: 'I beg to state that the Sirkar was not aware that you had cultivated lands at Sholakerry (until the receipt of your letter of the 30th June last in reply to mine dated the 27th of that months) nor have I been able to find anything upon the records giving you authority to that effect. In consideration however of the expense you must have incurred in clearing a portion of the jungles at that place, the Sircar condescends to you a separate lease for this land but as to the rent to be fixed for it and the period to which the lease is to run I intend communicating with the Resident on the subject; and after I shall have been in possession of the opinion of the Resident, I shall lose no time in preparing the necessary agreement to be concluded between you and the Sirkar relating to the land in question. In the meantime, I shall direct the Tahsildar of Pathanapuram and Kottarakaray to ascertain the extent that you have cleared at Sholakerry by measurement and define it boundaries and I request you will be so good as to depute some person to be present on your part on this occasion' (letter of the Dewan to Huxham on 16 November 1836 in: PTSMC 1926).

initial approach to Resident Fraser, and presumably following years of disagreement with Dewan Soobrow.

This case is highly illustrative of the political context of the European plantations in the first half of the nineteenth century in Travancore. Most, if not all, British Residents favoured the development of a European plantation sector. The Dewans for their part were hesitant to take any measures which would attract British planters. As the above correspondence illustrates, Dewan Soobrow opposed the earlier agreement made between Huxham, Colonel Newall and Venkata Rao, but, because of his subordinate position, was not able to take any counter-measures. His political manoeuvrability was extremely limited. As head of the bureaucracy, he could do no more than delay the process.

Huxham's difficulties with the Travancore bureaucracy, and with Dewan Soobrow in particular, have been described in a fair amount of detail above. However, the fixation of tenures of the areas in his possession were certainly not his only problems. Most probably, labour recruitment and control were his prime concerns.

Unfortunately, there is very little information available which deals directly with Huxham's plantations. We do know, however, that in 1851 the European planter supervised nine estates, varying in area from 20 to 611 acres. In the same year, the total area of land brought under cultivation amounted to 2333 acres.[23] There is unequivocal evidence which suggests that the chain of command on the plantations was quite rigid. Huxham, in his letters, spoke about 'the Chief Manager' and about 'a garden under my own Superintendents'.[24] Unfortunately, no direct information is given about the geographical and social background of these overseers or of the labourers under their command.

Therefore, it is useful to make a comparison in time and space. In 1797, the British East India Company opened up an experimental garden in Randatara in the Malabar district of the Bombay Presidency, trying out such crops as pepper, cotton and coffee. Murdoch Brown, who was appointed overseer and manager, experienced extreme difficulties in getting any grip on the

[23] CSF. No. 11247.
[24] CSF. No. 7253.

local inhabitants and obtaining the 200 to 300 workers he required daily. Brown complained about a lack of authority and requested the Bombay government to appoint him the local magistrate and revenue collector. The British officials complied with his request and gave Brown the omnipotence he found absolutely necessary to run the plantation. When Malabar passed from the Bombay Presidency to the Madras government, Brown acquired the ownership of the enterprise. It was then that he used his judicial and administrative authority to the full in favour of the estate. The size of his plantation increased from 200 to 3,000 acres, largely by the usurpation of the property of local people. Not content with the land, he obtained a large degree of control over the labourers working on his possessions. Apart from workers from the so-called slave castes, some of whom Brown had stolen from Travancore,[25] the planter used labour from the Nayar and Mapilla[26] communities. It is no great surprise that the enterprise was seen by the local inhabitants as extremely oppressive and that during the so-called Pazhassi Revolts all plants and trees were systematically destroyed.[27]

Most probably Huxham faced the same problems in the recruitment and disciplining of labour in Travancore as Brown did in Malabar district. Both planters functioned in a labour-market in which most of the agricultural labourers were bonded to a particular landlord, an issue which will be dealt with in detail in the following chapter. There was, however, one important difference between Huxham and Brown. The planter in Travancore was not, in contrast to his colleague in Malabar district, simultaneously both magistrate and revenue collector. To be sure Huxham was aided by the British Resident in Travancore whereas Brown himself was the local representative of the British government. It was therefore much more difficult for Huxham to settle local affairs to his own advantage than this was for Brown. Small wonder then that Huxham's enterprise was less successful than that of Brown. In 1852, three years after all Huxham's separate plots were incorporated into one single agreement (the so-called Ten Square Miles Concession), Huxham

[25] Kooiman 1991: 63.

[26] See also page 31, note 25 regarding Moplay and Maha Pillay.

[27] Kurup 1984: 188-9. The political backgrounds of the Pazhassi Revolts are discussed in Menon 1991: 268–71.

transferred his leasehold rights to Binny and Company. The unhealthy financial position of his enterprise had forced him to take this step.[28]

3.4 Dewans and Indigenous Coffee Cultivators

Though reluctant to help Huxham, the Dewans did not hesitate to take measures which encouraged the development of coffee cultivation by local producers. As the preamble of a proclamation, dated 27 Mithunom 999 *M.E.* (July 1824), stated: 'we wish that our citizens should cultivate more and more Coffee and *Karuva*'. For that purpose, the regulation stipulated that 'those who are interested in developing plantations (. . .) will be provided with such land and that the land (. . .) will be exempted from tax for the first 4 years; but from the 5th year onwards a marginal amount of tax will be levied'.[29]

Apart from the offering of attractive land and tax concessions to potential growers, a European agriculturalist-cum-administrator, named E. Winson, was asked to draw up a plan to encourage coffee cultivation by indigenous cultivators. Among other measures, free coffee seeds were distributed among peasants.[30] In addition, the Travancore sircar enlisted the help of another European, Mr Winson, in implementing the scheme. A certain Mr Lowe was appointed as 'the supervisor who is to accompany and help Lord E. Winson when the latter goes to inspect the areas of cultivation, and who is to submit the monthly account of coffee to the collectorate', being paid twenty-five rupees a month. Taking into consideration the wage level (rupees eight per month) and the essential knowledge of local languages (Malayalam and Tamil), it was a local person who was appointed as interpreter-cum-accountant in the project.[31]

[28] PTSMC 1926; CSF. No. 11247.

[29] Translated from Malayalam into English, in: TLRM, vol. 5, 1916: 362–3.

[30] This is evident from Huzur Order No. 784 dated 10th Edavom 1013 *M.E.* (May 1838). It stated: 'It is seen that there is immense scope for coffee cultivation in this country and the tenants and traders are likely to make a profit out of it. As such Lord E. Winson has made a rule regarding the implementation of such a programme. The printed copies of that rule together with the coffee seeds are being sent free of cost to the tenants and others' (translated from Malayalam into English, in: TLRM, vol. 5, 1916: 598).

[31] TLRM, vol. 5, 1916: 598.

An intriguing light is thrown on the situation because the coffee seeds which were to be distributed among the indigenous cultivators of Travancore were bought from and through Huxham, the European planter. He acted as a middleman between the Travancore sircar and a European trading company, most probably Binny and Company of Madras. On 3 May 1837, he wrote to the then Dewan R. Runga Rao: 'I have the honour to acknowledge receipt of your letter of the 25th . . . and will as you request . . . (send) 100 *parahs* of coffee seed to be delivered to the Tasildar of Quilon when it is ripe which will be in October and November next'.[32] The same document reveals that these were no idle promises. Jane (?) Top (?), his supplier of coffee seeds, stated: 'I have the honour to inform you that I have delivered to the Tasildar 100 parahs of coffee seed, according to the Sircar parrah which the Tasildar sent for the purpose of measuring it'. Nor did Top omit to send some instructions along. 'I have directed the bearer of it to guard it from the sun and wet, and as soon as it arrives, it should be unbaled, spread in a room and stirred every day'.

Huxham obviously supported the sircar's policy of encouraging Travancorean cultivators to plant coffee, doing his utmost to profit from the programme. In the same letter of 3 May 1837, Huxham also informed Runga Rao that 'should it be required, I shall be happy to let you have a supply of coffee plants, to the extent of one *lac* (. . .) by which measure the cultivation of coffee may be introduced throughout Travancore at once'.[33] Most probably the Dewan did not show much interest in Huxham's proposal. When Runga Rao was succeeded by R.R.R. Venkata Rao, the European planter tried to convince the new Dewan to buy three year old coffee plants – five lakhs this time – instead of younger plants or coffee seeds. He wrote to the Dewan expatiating on their advantages. He pointed out, for instance, that 'many of the plants should become the property of careless and idle people' and that 'the inhabitants finding that the trees give them a return in the following year after planting are more likely to take care of them'.[34] Trying to

[32] CSF. No. 7253.

[33] CSF. No. 7253.

[34] According to Huxham, the other advantages of three-year old coffee plants

convince the Dewan to buy large numbers of coffee plants he continues that 'a speedy benefit to the country can only be expected by dispensing the plants liberally and it is only by being able to export extensively that the Travancore coffee will become known to the commercial world'.[35] For that purpose, he advised sending '100 bags of Travancore coffee to New York in America (. . .) for the Americans are the greatest consumers of coffee in the world, and as their import duties upon coffee are very low they would give more for it than any other nation'.[36]

Nor did the British Residents oppose the Dewans' policy. On 13 October, Huxham wrote to Venkata Rao about 'a few plants that were distributed by Colonel Fraser and myself in June and July 1837 to the people living about Tycaud'.[37] This would seem to offer sound evidence that the British Resident saw no problem in distributing coffee plants to Travancore cultivators. This is all the more evident since these coffee plants, at least partly, originated from the Residency Garden at Tycaud.[38]

The ultimate goal of the Dewans was to establish a monopoly in the trade in coffee.[39] Most probably, they planned to copy the pepper monopoly which had once been so successful. If so, their ultimate goal was a system in which powerful, authorized traders would acquire the coffee from petty, quasi-independent cultivators. The Commercial Department would sell the coffee obtained from their agents to the highest bidder among the foreign merchants.

This coffee monopoly remained an illusion. There were two main reasons for this. In the first place, during the middle of the nineteenth century the British government was trying to abolish the existing trade monopolies in Travancore, like those on pepper (see Section 3.2) and tobacco (abolished in 1863).

---

were: (a) 'The plants are sufficiently matured to resist the shock of transplanting and the effects of dry weather'; (b) 'The plants being strong and healthy will produce good seed'; and (c) 'The duties that will be derived from the exportation of coffee, will in a few years repay the whole outlay with interest at eight per cent' (CSF. No. 7253).

35 CSF. No. 7253.

36 CSF. No. 7253.

37 CSF. No. 7253.

38 CSF. No. 7253.

39 Cf. Jeffrey 1976: 98; Kooiman 1992b: 63.

The British Residents would therefore certainly not approve of creating a new one, all the more so because in the British empire the production of coffee and the trade in that crop proved to be extremely profitable for both individual British planters and British companies. The British officers understood clearly that the establishment of a coffee monopoly in Travancore would be detrimental to the British commercial interests.

Turning to the second reason, the endeavours of the Dewans to encourage Travancoreans were largely a failure. As can be seen in Table 3.1, the quantity of coffee exported did not increase significantly in the period 1843–59/60, a clear indication that the production of coffee did not grow substantially in these years.

TABLE 3.1
Quantity of Coffee Exported from Travancore in the Years 1843 and 1852/53–1859/60 (in cwts)[40]

Year	*Quantity*
1843	910
1852/53	1,033
1853/54	453
1854/55	1,244
1855/56	830
1856/57	1,332
1857/58	929
1858/59	1,174
1859/60	587

Huxham provided an explanation as to why the indigenous cultivators did not take up coffee cultivation despite the government's encouraging policy. In his view, 'carelessness'[41] on the part of the Travancorean agriculturalists played a major role. Huxham remembered his experiences when he and Fraser, the British Resident, distributed a few plants to some of the peasants

[40] TAR 1042: 41; Mateer 1883: 225. For the figures concerning the export of coffee in later years, see Table 4.1.
[41] CSF. No. 7253.

living around Tycaud. 'The inhabitants would not pay that attention requisite for the healthy growth of the plants and those that did not die (. . .) will grow into strictly woody trees producing little or no fruit'. He regretted deeply 'having distributed the growing coffee plants that I did but probably the few that will live may grow up healthy for they were very carefully distributed'.[42] Clearly, the explanation attributed by Huxham – carelessness – is inadequate. I believe that the Travancorean cultivators were not very much interested in the cultivation of coffee primarily because of the low return on the crop, especially in the period 1844–9.[43]

Considering the early nineteenth century as a whole, various related processes, both coinciding and conflicting, can be observed. At first sight, it looks as though the historical processes of the waning Varma prominence and the introduction of British rule merely matched 'harmoniously'. Contemporaneously, it appears that the larger economic framework of state monopolies gradually developed into one of Free Trade. Taking a less abstract point of view, however, we arrive at a different conclusion. Both elements – those trying to hold on to 'old' ideas and practices (the Dewans in particular) and those whose intention it was to introduce 'new' structures (the British Residents and Huxham) – conflicted with one another, resulting in a situation in which the old system did not fully disappear and neither were new ideas easily adopted. This led to a disruption in the production of cash crops for export purposes. The pepper monopoly, which had lost all its political and economic significance since the introduction of British rule around the beginning of the eighteenth century, was only abolished in 1860. At the same time, the British Residents' policy of encouraging European planters as well as indigenous cultivators largely failed.

[42] CSF. No. 7253.
[43] Bandarage 1983: 75.

Four

The British Planters' Raj: Late 1850s–Early 1900s

I cannot too strongly urge that the Sircar should at once declare the policy which it proposes to adopt with regard to these jungles and make the gentlemen who have applied aware whether they can cultivate these jungles. Everyday is of importance, and I receive repeated requests from gentlemen to know when they may expect answers to their applications

Letter of F.N. Maltby, British Resident, to T. Madhava Rao, Dewan, on 30 January 1862[1]

4.1 European Planting Hegemony

4.1.1 The Emergence of a Plantation Sector

Compared to the earlier period of imperial initiatives, the nature of plantation development changed radically in the second half of the nineteenth century. The greatest change was that a European (predominantly British) plantation sector emerged, its owners and managers setting the political agenda to ensure growth and profitability. They demanded regulations which facilitated the acquisition of fertile soil, the recruitment of a sufficient number of workers and the export of estate products. This was the period in which the role of the (still very influential) British Resident altered from being the initiator in the development of export-oriented agriculture to that of the European planters' assistant. The Dewans, reluctant to help Huxham in the first half of the nineteenth century, now also encouraged the development of a European plantation sector. All these are good reasons for designating this stage in the history of plantations the period of the British Planters' *Raj*.

The development of an estate *sector* was largely a phenomenon of the late 1850s and early 1860s. The first to begin were

[1] CSF. No. 3964.

P.D. Devasahayam[2] and some European planters-cum-(ex-)missionaries, like John Cox, who started plantation production in the Ashamboo Hills, an area situated in the south of Travancore. John Grant, a Scottish entrepreneur who had won his planting spurs in Ceylon, launched large-scale cultivation in the same region in 1864.[3] John Daniel Munro was the pioneer in the Peermade range of hills. He opened the Hope Estate in 1862, and was swiftly followed by General Stevenson on Woodlands, Robert Baker on Stagbrook and F.G. Richardson on Twyford.[4] In 1877 in the Kannan Devan Hills, in the north of the state, planting was begun by the same J.D. Munro, together with Henry Gribble Turner of the Madras Civil Service and the latter's half-brother A.W. Turner.[5]

Before entering into a discussion of the planters' lobby, some major transitions within the plantation sector need to be dilated upon. At first, coffee cultivation prospered. Table 4.1 illustrates the quick rise in the production of coffee after 1859–60, the export of coffee increasing from 587 cwts in 1859–60 to 49,503 cwts in 1876–7. However, in the late 1860s coffee cultivation was threatened by stem borers,[6] and in the 1870s, by a leaf disease.[7] Unfortunately for the planters, the latter menace devastated most of the crops on their enterprises during the late 1870s and early 1880s. To make matters worse, during the decade 1879–88 there was a depression in the price of coffee caused by the introduction of relatively cheap Brazilian coffee on to the world market.[8] Mateer, a missionary with a direct interest in planting,[9] wrote his well-known book *Native Life in Travancore* at the height of these difficult years. He witnessed the overall disappointment in the European planting community at close quarters. In many individual cases, disillusionment and bankruptcy led to the abandonment of estates.[10] Table 4.1 also shows

[2] Mateer 1883: 226; Jeffrey 1976: 132.
[3] Mateer 1883: 228.
[4] Nagam Aiya 1906: 71.
[5] Griffiths 1967: 158.
[6] Griffiths 1967: 158.
[7] Mateer 1883: 231.
[8] Tharian George and Tharakan 1986: 202.
[9] Jeffrey 1976: 132.
[10] 'Times were sadly changed. Where formerly two or three hundredweight

TABLE 4.1
Coffee Exported from Travancore, 1843 and 1852–1900: Quantity (cwts) and Value (Rs)[11]

Year	*Quantity*	*Value*	*Year*	*Quantity*	*Value*
1859/60	587	8,397	1879/80	45,700	883,100
1860/61	1,513	22,612	1880/81	29,611	599,400
1861/62	1,941	37,370	1881/82	40,481	835,200
1862/63	2,567	51,336	1882/83	19,561	389,135
1863/64	2,979	59,644	1883/84	13,708	225,704
1864/65	3,965	80,221	1884/85	22,209	446,085
1865/66	4,777	120,787	1885/86	11,099	221,225
1866/67	9,653	211,542	1886/87	5,407	100,714
1867/68	14,140	243,000	1887/88	7,950	159,007
1868/69	12,865	251,300	1888/89	6,620	130,908
1869/70	16,991	338,800	1889/90	5,161	103,286
1870/71	27,093	544,800	1890/91	2,087	42,061
1871/72	29,597	596,800	1891/92	5,434	109,508
1872/73	41,597	839,000	1892/93	3,562	71,549
1873/74	29,407	584,000	1893/94	3,501	69,191
1874/75	26,031	524,000	1894/95	5,946	112,909
1875/76	39,335	786,696	1895/96	2,644	52,477
1876/77	49,503	990,058	1896/97	2,659	52,954
1877/78	39,737	803,700	1897/98	2,314	141,467
1878/79	18,781	374,600	1898/99	6,693	111,325
			1899/1900	3,978	71,239

of coffee per acre was expected even off two years old plants, or, at least, repayment of all expenses, there are now so many estates producing almost nothing that the whole average produce will not come to two cwts per acre. A certain estate which cost Rs 10,000 for up-keep in 1879, produced only eighty cwts of coffee, value less than Rs 4,000. Other estates only repay the annual expenditures. Some are quite abandoned after long and patient waiting and struggling in hope of better times. The total acreage thus abandoned we have not been able to discover; the subject is evidently too painful for free publication' (Mateer 1883: 232).

[11] Mateer 1883: 225; TAR 1035–75. For the period 1870–80, the ten-year average of coffee in relation to the total value of commodities exported from Travancore was approximately 7 per cent (Panicker et al. 1978).

the end of the coffee boom. In 1886-7 the quantity of coffee exported had already been reduced to 5,407 cwts. In subsequent years there was no substantial revival in the cultivation of coffee. To take one example, in the year 1899-1900 the amount of coffee exported was only 3,978 cwts.

Understandably, during the late 1870s the planters began to consider alternatives. Various crops were tried to find the best substitute for coffee. Initially, the cultivation of cinchona was the most profitable.[12] According to Lovatt, a planter's wife who described the early years of the Peermade/Vandiperiyar district on the basis of interviews and personal experiences, cinchona was widely planted between 1870 and 1880 when quinine was selling at 10 to 12 shilling an ounce.[13] Cinchona was also the most popular crop in the Kannan Devan Hills during the 1880s. Table 4.2 illustrates that all plantations in north Travancore cultivated cinchona in 1885.

Table 4.2

The Kannan Devan Hills in 1885 and 1901[14]

Year	*Total Number of Estates*	*Total Acreage*	*Estates Growing Coffee*	*Estates Growing Cinchona*	*Estates Growing Tea*
1885	17	2,247	2	17	6
1901	43	17,136	17	14	27

It was in the early 1890s that tea became the most important plantation crop. Tea was the most suitable crop for the Peermade/Vandiperiyar elevation. The decision to change over to planting tea was most probably given an extra fillip by a glut in the world market of cinchona at that time.[15] Tea also took over as the most prominent crop in the Kannan Devan Hills, which is shown in Table 4.2. In 1901, twenty-seven plantations were producing tea, while only fourteen estates were still cultivating cinchona. Of greatest significance, the increase in acreage in the Kannan Devan Hills, from 2,247 acres in 1885 to 17,136 acres in 1901, was

[12] It is from the cinchona tree that quinine is obtained.

[13] Lovatt 1972: 15.

[14] Papers regarding the Kannan Devan Hills in: CDS. RF. 546.

[16] Lovatt 1972: 16.

brought about almost solely by the development of tea. Tables 4.3 and 4.4 illustrate the rapid growth of tea cultivation in Travancore and the sharp increase in the value of tea exported from the state.

TABLE 4.3
Tea Cultivation in Travancore, 1885-1904[16]

Period	*Average Area in Acres*	*Average Production in Pounds*	*Average Production Per Acre in Pounds*
1885–89	3,352*	342,718*	102
1890–94	7,186*	1,735,803*	242
1895–99	15,682*	3,505,938*	224
1900–04	24,931	6,107,429	245

* includes figures for the Cochin state.

TABLE 4.4
Tea Exported from Travancore, 1891-1900[17]

Year	*Value (Rs)*
1891/92	660,556
1892/93	653,865
1893/94	812,813
1894/95	1,074,268
1895/96	1,287,488
1896/97	1,536,059
1897/98	1,491,459
1898/99	1,477,021
1899/1900	1,241,047

The shift from the cultivation of coffee to tea had many far reaching consequences for plantation development. Geographically, it meant that the relative importance of the various plantation districts was modified. During the formative stage, south Travancore formed the backbone of the European plantation

[16] *The Indian Trade Journal* 1921: supplement 10.
[17] TAR 1067–75. To give an indication of the importance of tea in Travancore's export packet, the value of tea exported amounted to 10.5 per cent of the total value of goods exported in 1896–7.

sector. Most plantations in Travancore state were situated on the Ashamboo Hills, in the districts of Ashamboo, Shencottah and Poonmudi. It was here that the planters of south Travancore established the Travancore Planters' Association, the first, and for several years the only, planters' organization in the state.[18] During these early years, there were not enough planters in the Peermade/Vandiperiyar area to form their own association. Instead, these planters sent their representatives to the annual general meeting of the Travancore Planters' Association, which was held in Trivandrum, the capital of Travancore, during the so-called Planters' Week, a social event which included races and a dinner with the Travancore monarch. The Peermade Planters' Association, representing the planters of Peermade and Vandiperiyar, was not constituted until 1874. Both south and central Travancore planting districts were seriously affected by the appearance of the coffee leaf blight. However, because of its more suitable altitude the planters of the Peermade/Vandiperiyar area were able to turn from coffee to tea more successfully. This meant that the balance shifted from south Travancore to Peermade/Vandiperiyar (see Tables 4.5 and 4.6). To underscore the importance of their district, the Peermade Planters' Association was renamed the Central Travancore Planters' association in 1895. Yet it was the Kannan Devan Hills which developed into the most important tea producing area of Travancore (see Table 4.6). Again this was also reflected in the growing importance of its association. The Kannan Devan Planters' Association, established in 1888, became the most influential planters' organization in the state.[19]

The change from coffee to tea also heralded fundamental alterations in the ownership pattern of the plantations. During the pioneering phase, plantations were often opened up, managed and owned by the same individual. In other words, in this period many of the planters were both the owner and the manager of their estate, although sometimes ownership and management were in different hands. Many of these planters were either (ex-)missionaries (like John Cox), their sons and/or brothers (Robert and George Baker were sons of Rev. Henry

[18] The exact date of foundation of the Travancore Planters' Association is unknown.
[19] CTPA 1928: 4; *UPASI Planting Directory* 1974: 520.

TABLE 4.5
The Plantation Sector in Travancore in 1893[20]

District	*Number of Estates*	*Acreage Cultivated with Cinchona*	*Acreage Cultivated with Coffee*	*Acreage Cultivated with Tea*	*Total Area Cultivated*
Ashamboo	13 (14%)	-	1,176 (27%)	1,516 (12%)	2,692 (14%)
Shencottah	9 (10%)	-	842 (19%)	2,250 (18%)	3,092 (17%)
Poonmudi	13 (14%)	-	74 (2%)	1,803 (14%)	1,877 (10%)
Central Trav.	23 (24%)	81 (6%)	887 (20%)	4,085 (32%)	5,053 (27%)
Kannan Devan	36 (38%)	1,364 (94%)	1,357 (31%)	3,135 (25%)	5,856 (32%)
Total	*94 (100%)*	*1,445 (100%)*	*4,336 (100%)*	*12,789 (100%)*	*18,570 (100%)*

TABLE 4.6
Tea Cultivation in Travancore in 1904[21]

District	*Number of Estates*		*Acreage Cultivated with Tea*	
South Travancore*	18	(27%)	3,830	(15%)
Central Travancore	24	(36%)	8,415	(34%)
Kannan Devan Hills	24	(36%)	12,466	(50%)
Total	*66*	*(100%)*	*24,711*	*(100%)*

* Ashamboo, Shencottah and Poonmudi.

Baker Sr. and brothers of Rev. Henry Baker Jr., both influential missionaries in nineteenth century Travancore), (ex-)government servants (like General Stevenson and Henry Gribble Turner), or

[20] UPASI Proceedings 1893; see also: TAD 1895: 380-3. Throughout the book, the various percentages were calculated by the author. In all cases, these figures were either rounded off upwards (.5) or downwards (.499).
[21] Nagam Aiya 1906: 79.

their sons and/or grandsons (like J.D. Munro who was a son of Urban Verres Munro, the first Conservator of Forests in Travancore, and a grandson of Colonel John Munro, British Resident in the period 1810–19). The predominance of individual planters was made possible because, as Lovatt noted, 'coffee planting needed little capital, land was cheap, cultivation and manufacture simple and labour seasonal'.[22]

It is actually quite a difficult undertaking to characterize these early European planters. In sharp contrast to the relatively full account of Devasahayam's career (which will follow in Section 4.1.4), historical sources provide either rather general or else too specific, usually irrelevant, information about the lives of the European pioneers. Without any doubt, however, all these men were very adventurous individuals. Those who opened up an agro-enterprise in the midst of forest areas dared to take great financial risks and took social isolation from family and fellow countrymen for granted. They lacked even the most basic facilities including shelter and paid no heed to their personal health (malaria!). In striking contrast to those of their successors in the late nineteenth and twentieth centuries, the lives of early planters were neither secure nor luxurious. Working and living side by side with their labourers while clearing the jungle, the social distance between the planter and the labourers was relatively small in these early days.

We know that the majority of these pioneer planters were not rich, particularly not in the initial phase of their enterprise. The clearing of forests entailed great expense and immature plants did not yet yield any returns. This forced them to contract loans, sometimes from family members in Europe. Nevertheless, most of these European adventurers were well-off in a socio-political sense. Their social networks, particularly their family relations, were their greatest assets. Because many of the European planters were related either by birth or marriage to influential officials working in Travancore and the Madras Presidency, the planting community was well-informed about and could influence political affairs. Figure 4.1 shows that missionary activities, planting interests and political influence 'converged' in the Baker-Munro family, greatly affecting the path of plantation development during the mid-nineteenth century.

[22] Lovatt 1972: 17.

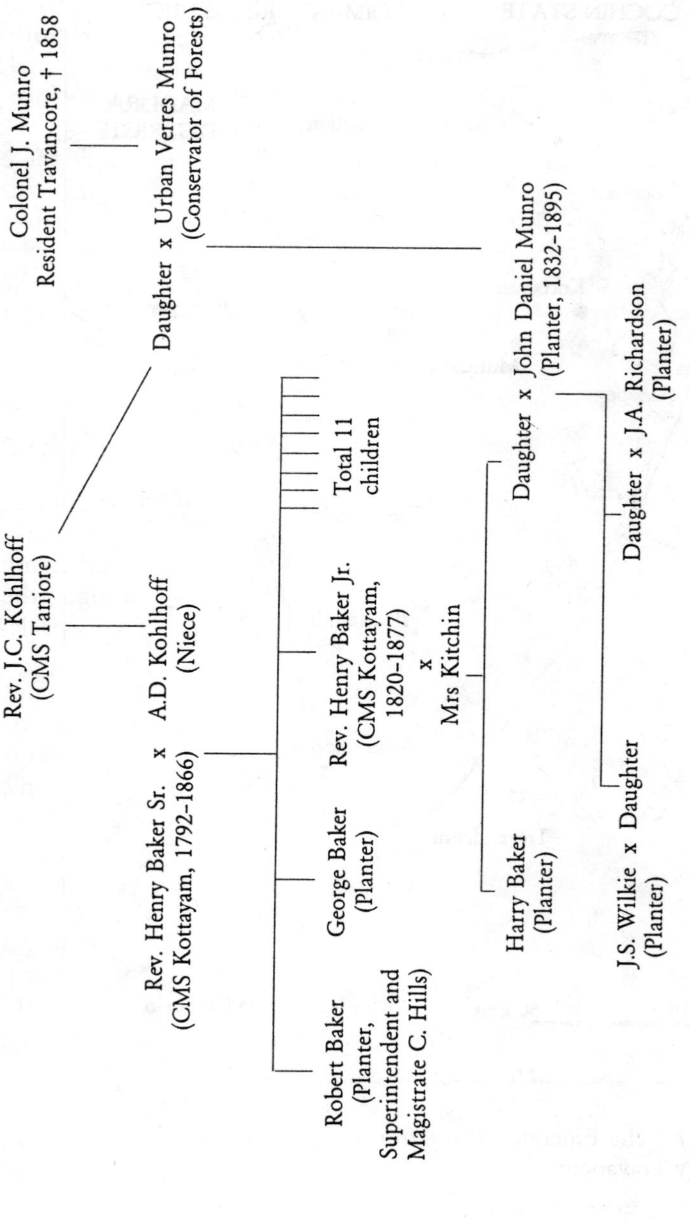

Figure 4.1 The Baker-Munro Family

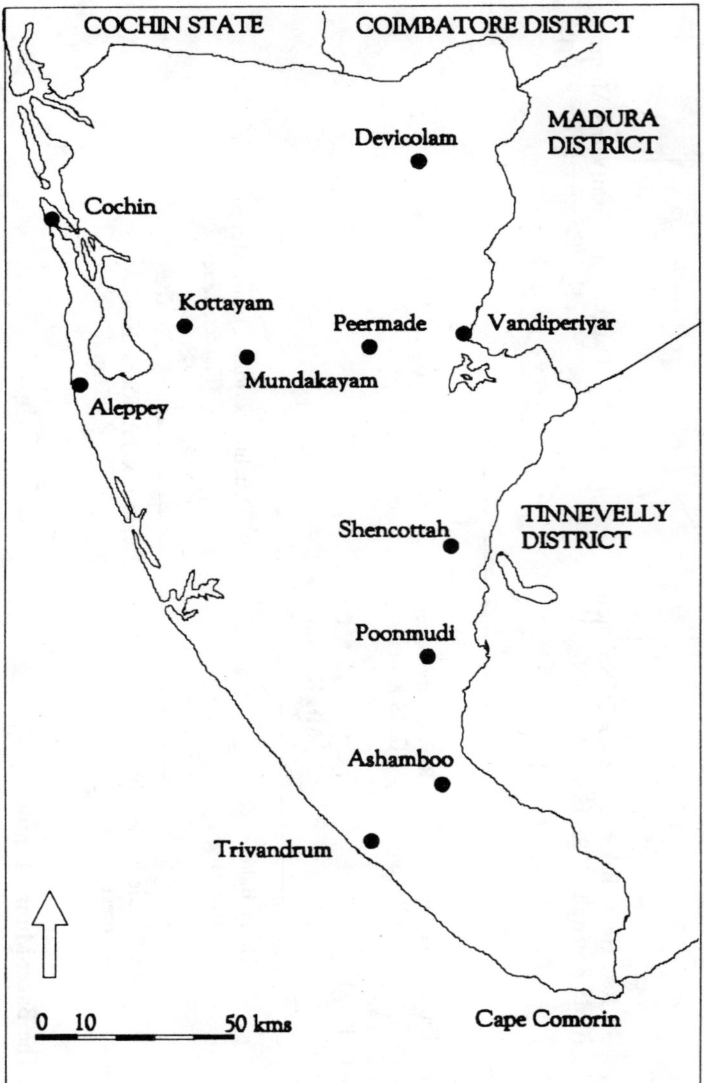

Map 4.1 The Emerging Plantation Districts in Mid-Nineteenth Century Travancore

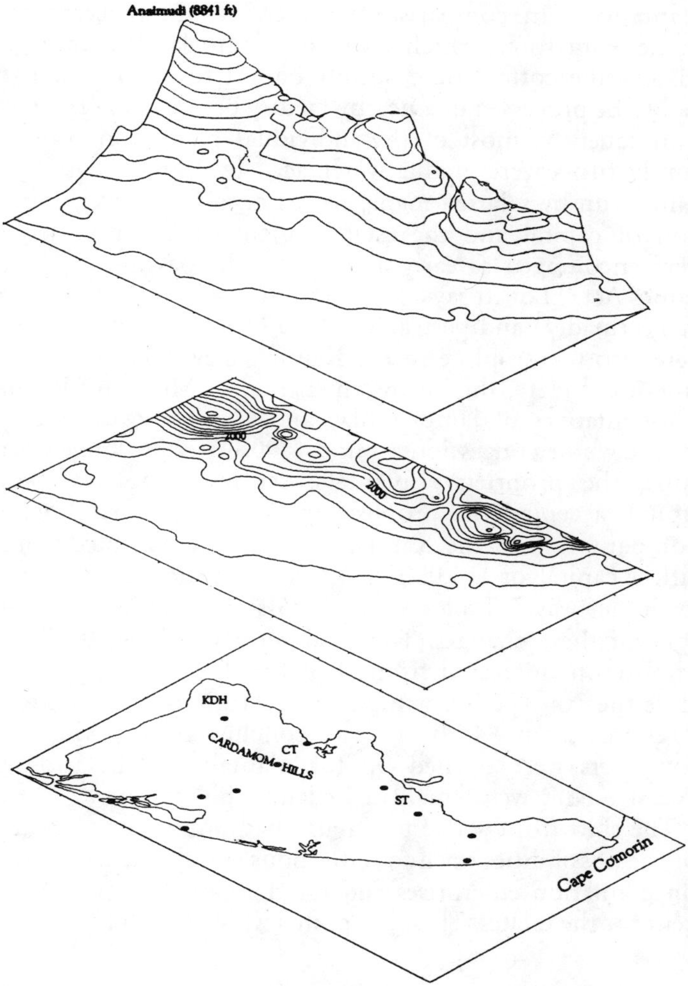

KDH = Kannan Devan Hills
CT = Central Travancore
ST = South Travancore

Map 4.2 Two and Three-Dimensional Projections

Tea was quite a different undertaking. Large investments were necessary for the opening up and exploitation of the plantations. In comparison with coffee, tea required a more stable work force, which I will look into later in more detail. Also, while coffee could simply be left to dry in the sun, tea had to be processed in a nearby factory before being transported. Consequently, most of the individual plantation owners and family firms were gradually replaced by limited liability companies run by salaried managers. At an official level, concentration of capital and the establishment of larger undertakings were encouraged as early as 1862 by the passing of the Companies Act.[23] Lovatt says the first joint stock companies appeared in Peermade/Vandiperiyar in the 1890s but the same changes were also taking place in the Kannan Devan Hills at the same juncture.[24] Here, the businessmen Finlay, Muir and Buchanan, representatives of James Finlay and Company, made an offer for the whole concession around 1890. After prolonged negotiations, the proprietary planters, seriously short of working capital, accepted the proposition. In 1897, James Finlay and Company floated the Kannan Devan Hills Produce Company with a capital of £ 1,000,000. After that, development rapidly got under way.[25] Tables 4.5 and 4.6 show some of the implications of these changes. The total acreage cultivated with plantation crops increased from 18,570 in 1893 to 24,711 in 1904, while the same period witnessed a decrease in the total number of estates from 94 to 66. The conclusion is obvious: many new tracts were opened up for plantation cultivation while several estates were amalgamated into bigger ones.

The shift from coffee to tea also had important consequences for the possibilities open to indigenous entrepreneurs to own and run plantation enterprises and for the nature of labour recruitment for the estates. These are points which I shall return to later.

4.1.2 *A Favourable Land Policy*

The British Residents continued to play a crucial role in the development of plantations during the second half of the

[23] Tharian George and Tharakan 1986: 213.

[24] Lovatt 1972: 17.

[25] Mackay 1984: 7.

nineteenth century. Their initial role was to provide the planters with knowledge and planting materials. This is clearly illustrated by W. Cullen, British Resident in the period 1840–60. He owned many experimental gardens – at his death he left at least five gardens in Travancore – which were cultivated with such crops as cotton, coconuts, nutmeg, cocoa and cloves.[26] Cullen also reintroduced coffee and tea to the region;[27] ultimately they were to become the most important plantation crops in Travancore during the second half of the nineteenth century. This British Resident is also credited with the discovery that the Ashamboo and Cardamom Hills were regions which were extremely suitable for the cultivation of coffee and tea respectively. His experimental garden at Ashamboo was the forerunner in the development of large coffee estates in the area. Equally interested in tea, Cullen experimented on a plot situated near the Tinnevelly frontier at an elevation of 3,200 feet.[28] Cullen was not the only British Residents to show a great personal interest in planting. Most of his colleagues were like-minded. They cultivated various plantation crops in their Residency Gardens and regularly exchanged their experiences with European planters.

Even more cogent to the development of plantations were the unremitting efforts of successive British Residents to secure a favourable land policy. As had been their practice in the first

[26] R/2/897/351; Mateer 1883: 226–7.

[27] Both crops were cultivated by Huxham before.

[28] Cullen reported in October 1859: 'The tree thrives well in Travancore territory, both at the level of the sea and altitudes of 1,800 and 3,200 feet. I first met with it in the coffee plantation of Mr Huxham in the year 1841, on the road from Quilon to Courtalam, at a factory called Caldoorty, about forty miles inland and six or seven hundred feet above the sea (. . .) I procured plants from Mr Huxham and put them down in an experiment spice garden which I established some twelve years ago at 1,800 feet on a hill in the South of Travancore near Oodagherry. They are now trees of 20 to 30 feet high, growing vigorously; and I have four hundred plants procured from their seeds growing on another hill near the Tinnevelly frontier, at an elevation of 3,200 feet. There can be no doubt therefore of the facility of its introduction, although from the moderate altitude and great atmospheric moisture of the localities hitherto selected, they may possibly be considered to grow more luxuriantly than is desirable; but which if a defect at all, can probably be easily remedied by selecting ground more to the eastward, at great altitudes, and with less humid climate' (quoted in: *UPASI Planting Directory* 1974: 7–8).

half of the nineteenth century, in the second half too the British Residents did all in their power to sponsor the growth of the European plantation sector. In the latter period, however, their task was less arduous.

Their paths were smoothed in two ways. By this time, many European entrepreneurs had themselves become interested in opening up estates in Travancore, thereby rapidly taking over the reins of plantation development from the British Resident, thereby setting the political agenda with regard to land, labour and taxes. Hugh Crawford, the Commercial Agent in Aleppey who applied for a particular tract of land for the cultivation of coffee in 1856, was the first European entrepreneur to act on his own initiative. Crawford, unlike Huxham, operated without the decisive mediation of the British Resident but both cases show an important similarity. Crawford's efforts to acquire proprietary rights in 1856, like Huxham's endeavours in the 1830s, were counteracted by the Travancore government. Crawford wrote referring to this that 'after a number of years and a deal of correspondence it ended the ground having been taken up for Sirkar purposes'.[29]

There was yet another significant shift. The opposition previously shown by the Travancore administration to the development of a European plantation sector disappeared with the appointment of T. Madhava Rao as Dewan in 1858. Of course, in this period in which many European planters were applying for land, the British officials ascertained for themselves whether the bureaucrats to be appointed as Dewans were likely to cast a favourable eye on the development of European plantations. But this was not the only factor which explains the changed attitude of the Dewans and Travancore bureaucrats. In the political field, the Travancore government was anxious to improve its relationship with the British rulers, who in the years immediately preceding had threatened to annex Travancore. Although this threat disappeared after the Mutiny (1857), it was felt that the political situation would change for the better if British economic and political interests were met. Financially, the sircar was keen to improve the impoverished

[29] Letter from H. Crawford, Commercial Agent, to G.A. Ballard, British Resident, on 26 October 1872, in: CSF. No. 1769.

state of the treasury. The state revenues which could be obtained from European planters through the sales of land and the levying of land taxes and export duties, were seen as attractive prospects. The Travancore state also hoped that the opening up of uncultivated, backward regions would stimulate the economic development of these areas.[30] Creation of employment for Travancoreans was also stressed.[31]

Taken together, all these groups and their discrete motives – namely: the European planters in their efforts to create a legal framework which encouraged plantation development; the British Residents helping the planters in their dealings with the Travancore government whenever necessary; and the Dewans, for the reasons mentioned above – meant that all parties concerned were in accord with the main lines of the policy concerning land suitable for the cultivation of plantation crops. Nor is there any lack of evidence to support this statement.

On 30 January 1862, British Resident F.N. Maltby explained to Madhava Rao the expediency of adopting a policy favouring plantation development. 'I cannot too strongly urge', Maltby informed the Dewan, 'that the Sircar should at once declare the policy which it proposes to adopt with regard to these jungles and make the gentlemen who have applied aware whether they can cultivate these jungles'. 'Everyday is of importance', he stressed.[32]

The Travancore bureaucracy reacted swiftly to the Resident's letter. The applications were processed immediately. Of greater significance is the fact that the government policy towards land suitable for the cultivation of coffee was enacted not two months later in the Memorandum of 8 March 1862: the Rules Regarding

[30] As a *Travancore Administration Report* stated: 'The Sircar feels satisfied that the country would largely benefit by the introduction of the capital, skill and enterprise of European gentlemen in utilizing tracts of valuable land, which, for the most part, would otherwise be untouched for generations. It has accordingly been the anxious desire of the Sircar to afford facilities to planters' (TAR 1038: 13).

[31] 'This enterprise promises to be the means of giving employment to many subjects of Travancore. (. . .) All the planters are gentlemen sincerely solicitous to deal fairly with their labourers, and to rely upon good treatment and good wages alone for attracting labour' (TAR 1040: 22–3).

[32] Letter of F.N. Maltby, British Resident, to T. Madhava Rao, Dewan, on 30 January 1862, in: CSF. No. 3964.

Grant of Land for Cultivation of Coffee. Clauses 6 and 8 of these rules were the most stimulating for plantation development. Clause 6 stated: 'should the applicant so desire, land for planting Coffee will be granted free of assessment for five years, on condition that quarter of the land shall be cleared and planted within the first three years'. Clause 8 was perhaps even more significant. It stated: 'The lands applied for shall be granted in perpetuity as heritable or transferable property'. Thus, in contrast to Huxham, who *leased* his land directly from the Travancore state, the new generation planters were legally entitled to *own* their estates. This was a much more attractive prospect for the European entrepreneurs. Two of the advantages included were that the transfer of an estate to a third party was less complicated and the Travancore state could not interfere in the negotiations. Conversely the interests of Travancore state were offered protection by Articles 1 (regarding land tax), 2 (which safeguarded against the large-scale felling of valuable timber, like teak, ebony and sandalwood, which were important sources of government revenue) and 10 (with regard to export duty).[33]

Shortly afterwards, on 15 December 1864, Madhava Rao heralded new legislation regarding land suitable for the cultivation of plantation crops, stating 'that in consequence of several parties applying for the same tracts of waste lands, (. . .) waste lands suited for the cultivation of Coffee shall be sold by public Auction to the highest bidder, under rules which will be shortly published, at an upset price of one British rupee per acre'.[34]

The law thus previewed became known as Regulation II of 1040, the so-called Rules for the Sale of Waste Land on the Travancore Hills, dated 24 April 1865.[35] It regulated the way in which 'waste lands, in which no rights of private proprietorship or exclusive occupancy exists, and which may not be reserved (. . .), may (. . .) be sold' (Article 1). Clause 2 stipulated the different procedures to be followed by local and European candidates, proclaiming that 'applications for lands shall be addressed to the Dewan of Travancore, directly in the case of

[33] TLRM, vol. 2, 1915: 1–3.

[34] TLRM, vol. 2, 1915: 3.

[35] TLRM, vol. 2, 1915: 3–11. The concept '*waste* land' is extremely misleading. As will be shown later, often these so-called waste lands were used for shifting cultivation.

natives, and through the British Resident, in the case of Europeans'. This would be followed by a survey carried out by the sircar authorities (Section 5). And, 'if (. . .) the Dewan has reason to believe that the lot applied for is saleable under these Rules' (Clause 6), 'he shall then advertise the lot for sale on a given day' (Article 8). Moreover, 'the advertisement shall be in English and in Malayalam and shall specify the locality, extent and boundaries of the lot, the annual assessment, and the place, time and conditions of sale'. It also required that the announcement 'shall be posted for three months at least on the land itself, as well as in the neighbouring villages, in the offices of the Dewan, the *Dewan Peishkar* and the Tahsildar of the *taluk*, and in the nearest police office'. 'A notice of the intended sale shall also be inserted in the Travancore Gazette' (Section 9). Finally, 'the lot shall, as advertised, be put up to auction and sold to the highest bidder above the upset price' (Clause 14). These rules were plainly intended to ensure that the applicant would not acquire the land applied for as a matter of course and at no real cost, as had been the case before the enactment of these regulations. Or as Madhava Rao informed the British Resident on 30 November 1864, 'you are aware that many circumstances indicate strongly the desirableness of selling land suited for coffee cultivation instead of granting it as hitherto for the asking'.[36] Table 4.7 shows that the auctions for the sale of coffee lands most probably took place annually up to 1874. It also indicates that the amount of land offered for sale at these auctions gradually increased. After 1874, due to the failure of coffee, there was no auction for twenty-one years. Auctions for the sale of government waste lands were again resumed in 1895, and then were held more frequently than they had been in the period 1865–74.

Other pieces of legislation determined the finer points of the rights and obligations of the planters and the Travancore bureaucracy. The Rules for the Grant of Grass Land to Proprietors of Coffee Estates of 31 July 1877 ratified that the so-called 'grass-lands', i.e. estate property used either for 'dwelling-house, compound, cooly lines, out-houses, etc.' ('homesteads') or for 'cattle-pen and other buildings' ('farmsteads'), acquired a special legal

[36] Mahadeva Iyer 1914: 2.

status. Dewan A. Seshiah Sastri (1872–7) agreed that a lower land tax be levied on such lands (Clause xx). Should the land be brought under cultivation the planter was legally bound to inform the Travancore government about such a change, stating the nature and extent of the cultivation (Clauses xxi-xxv).[37] The Rules for the Sale of Waste Land on the Travancore Hills for Coffee and Tea Cultivation, dated 7 July 1898, updated and integrated the earlier legislation on this subject, with one important modification. Article 2 stated that 'Applications for lands shall be addressed to the Dewan'. K. Krishnaswami Rao, Dewan in the period 1898–1904, thus terminated the British Resident's role in this regard, at least in a strict legal sense.[38]

TABLE 4.7

Auctions for the Sale of Lands for Plantations until 1897[39]

Notice Number	*Date Public Auction*	*Lots Offered for Sale*	*Estimated Acreage Offered for Sale*
6	not available	at least 9	at least 185
7	not available	at least 6	at least 25
9	not available	at least 1	at least 200
10	not available	at least 34	at least 1,391
11	not available	at least 56	at least 4,786
12	1874, 14 October	55	2,328
13	1895, 30 October	11	1,550
14	1896, 25 February	7	2,025
15	1896, 8 April	8	1,425
16	1896, 8 April	13	330
19	1897, 7 July	16	4,450

4.1.3 Manipulating the Travancore Bureaucracy

Without any doubt, these legal measures, agreed upon by the European planters, the British Residents and the Dewans, stimulated the emergence and development of plantations. However,

[37] TLRM, vol. 2, 1915: 13–20.

[38] TLRM, vol. 2, 1915: 109–12.

[39] CSF. No. 1925; CSF. No. 11684; CSF. No. 11798.

the illegal activities of the planters, working against the interests of the Travancore government, contributed to an even larger extent to the growth of the plantation industry.

Their stratagems were numerous. At the public auctions the planters often joined forces in order to reduce expenses. Dewan Seshiah Sastri was aware of this subterfuge and complained about it. He informed the British Resident G.A. Ballard that 'from information which I am in possession of and from my knowledge of what actually took place behind the scenes on the last sale of coffee lands, I have no hesitation in saying that there is every probability of, and facility and inducement for a combination among the bidders to keep down the price'.[40]

As another expedient, many planters bought large tracts of land – which were sold by the Travancore government for the explicitly stated purpose of plantation cultivation – without any intention of cultivating their newly acquired properties in the near future, often without any intention at all of using the lands for productive purposes. This is illustrated by the fact that in 1893 only 18,570 acres out of the total of 73,042 acres owned by the estates were cultivated. In other words, in this year 75 per cent lay fallow (see Table 4.8). The planters had sundry reasons for choosing to carry on their business in this particular way. Foremost among them being that they secured land for future cultivation and restricted competition considerably. Speculation also loomed large and many tracts of land were bought expressly for this purpose, which often did indeed prove extremely profitable. Dewan Seshiah Sastri was well aware of this. He wrote to the Resident that 'it is a fact that forest lands bought at an upset price of Re 1 per acre have been sold between planter and planter for as much as Rs 15 last year, and that offers of Rs 25 or 30 per acre have been refused in some cases'.[41] In fact, landed property sometimes changed hands for much higher prices, especially when it was thickly cultivated with good quality plants which were at the productive stage. De Quadros, an experienced

[40] Letter of Seshiah Sastri, Dewan, to Ballard, British Resident, April 1874, in: CSF. No. 1772.

[41] Letter of Seshiah Sastri, Dewan, to Ballard, British Resident, April 1874, in: CSF. No. 1772.

planter, informing Dewan V. Ramiengar (1880-7) about the value of his land, said that: 'It is well known that similar Coffee Estates have changed hands in Travancore at Rs 500 per acre'. Even in 1878, in the depths of the depression, 750 rupees were paid for 4 to 6 acres of land cultivated with coffee.[42]

TABLE 4.8
The Travancore Plantation Sector in 1893 (in acres)[43]

District	*Area Cultivated*		*Area Uncultivated*		*Total Area*	
Ashamboo	2,692	(47%)	3,005	(53%)	5,697	(100%)
Shencottah	3,092	(37%)	5,326	(63%)	8,418	(100%)
Poonmudi	1,877	(20%)	7,576	(80%)	9,453	(100%)
Central Travancore	5,053	(46%)	5,843	(54%)	10,896	(100%)
Kannan Devan Hills	5,857	(15%)	32,722	(85%)	38,579	(100%)
Total	*18,570*	*(25%)*	*54,472*	*(75%)*	*73,042*	*(100%)*

Playing on the strength of their position many planters paid their taxes either too late, only in part, or not at all. To give an example of the last, many planters, who abandoned or sold their estates, left Travancore without having paid the assessment for years. The new owners, perhaps in good faith at the time of the transaction, refused to pay the outstanding tax arrears left by their predecessors even when they were legally bound to do so. Another form of tax evasion was practised in depressed years when planters, pleading poor mouth, were not prepared to pay any land tax.[44] The partial, delayed or nonpayment of taxes were also related to two other illegal tactics employed by the planters: encroachment on government lands and the improper use of the Rules of 31 July 1877.

An examination of the first of these allegations shows that

[42] Letter of De Quadros, planter, to the Dewan on 4 April 1881, in: CSF. No. 1777.
[43] UPASI Proceedings 1893, see also: TAD 1895: 380-3.
[44] CSF. No. 1826; CSF. No. 1827; CSF. No. 1831; CSF. No. 1055; CSF. No. 10661; CSF. No. 11225; CSF. No. 11672.

the planters community gained thousands of acres of land through the fraudulent appropriation of government lands. Usually fully aware of the illegal nature of their conduct, the planters annexed parts of the government lands surrounding their estate without notifying the Travancore Survey Department, thereby acquiring large tracts of land without much expense: they did not pay for the land, nor were they obliged to pay assessment. The planters were well aware that the number of officers employed to protect state properties was insufficient. The planters also realised that the Travancore government was not in a position to punish them severely if the agents of the Survey Department did discover their trespasses. Meagre excuses like 'I was unaware that the portion was not the property of the Estate'[45] and 'I beg to say that the work on the clearing was nearly finished before I found out the mistake'[46] settled the question, at least the penal part of it. Prepared to compensate the Travancore state, the writer of the latter letter consequently proposed: 'we are able and willing to return to government the same acreage of equally good forest as that we have accidentally taken'. Fine words but even if the same quantity and quality of land was returned to the state, the Travancore bureaucracy lost in many respects. Apart from its loss of credibility, the officers of the Survey Department, who had to measure the acreage and borders of the estate again, did not work for charity. At the end of the nineteenth century, the costs for reinspecting the plantations in the Peermade district, which became absolutely necessary because of the encroachments of the Europeans, was estimated at twelve annas an acre. It might be interesting to mention that in 1874 the Dewan, against the explicit wish of the British Resident, ordered a temporary end to the sale of waste lands. The Survey Department needed time to catch up with its inspecting tasks, which had shot up because of the constant trespassing of the planters.[47]

Looking at the second allegation it is obvious that the planters

[45] CSF. No. 11730.

[46] CSF. No. 1830.

[47] CSF. No. 1727; CSF. No. 1830; CSF. No. 1873; CSF. No. 10101; CSF. No. 10106; CSF. No. 10107; CSF. No. 10115; CSF. No. 10129; CSF. No. 10512; CSF. No. 10657; CSF. No. 11226; CSF. No. 11671; CSF. No. 11675; CSF. No. 11690; CSF. No. 11716; CSF. No. 11730.

made improper use of the Rules for the Grant of Grass Land to Proprietors of Coffee Estates, dated 31 July 1877. After the Survey Department had acknowledged particular plots of land as 'grasslands', the 'homesteads' and 'farmsteads' were often planted with coffee or tea, without any mention being made of this change to the Travancore bureaucracy, the planters thereby paying less land tax than they were legally obliged to.[48]

Unfortunately for the Travancore government, the extra-legal activities of the planters were not restricted to those mentioned above. Most planters did not act in accordance with those regulations which were intended to preserve the natural balance within the state either. They thoughtlessly denuded many hill ridges, banks of streams and large areas of land.[49] It will certainly amaze those scholars who believe that care for the natural environment only became an issue in the twentieth century that the protection of nature was already an important task for the Travancore government in the latter half of the nineteenth century. There is ample evidence for this. In the Memorandum of 8 March 1862 it is stated that 'no exclusive rights of water beyond what is necessary for the use of the plantation shall be considered to be conveyed by the grant' (Clause 4). The contents of the original article were broadened in the Regulation of 24 April 1865. Section 23 stated that 'the existing and customary rights of the Sirkar, of other proprietors, and of the public, in existing roads and paths and in streams running through or bounding lands sold under these Rules, are reserved, and in no way affected by the sale of such lands'. Also, it was laid down that 'reserves of grazing and forest land, of land for the growth of firewood, (. . .) for the growth of timber, etc., or required for other special purposes, are not to be sold under these Rules without the express sanction of His Highness the Maha Rajah' (Article 18). In December 1869 and again in July 1871 Madhava Rao, most probably requested to do so by the Madras government, decreed that no plantations could be opened in those areas of which the drainage flowed into the neighbouring British

[48] CSF. No. 10112; CSF. No. 11716; CSF. No. 13700.

[49] See for this subject for example the letter of J.C. Hannyngton, British Resident, to the Madras government dated 21-10-1878, in: R/2/879/14.

[50] Notices of the Huzur Cutcherry dated December 1869 and 14 July 1871, in: TAD 1879: 68.

districts of Tinnevelly and Madura.[50] The surviving evidence suggests that Seshiah Sastri, the successor of Madhava Rao, was even more concerned about the protection of nature. He corresponded with the British Resident on this topic and wrote with great foresight in 1874: 'though a few hold to the theory that, independent of forests, abundant and scanty rainfall in India alternate in cycles of years, it is I believe the generally accepted opinion that they exercise a potent influence on rain clouds, and not only conduce to the fall of rain but to its distribution, the thick vegetation growing up between the trees impeding the mad rush of torrents to the plains below'. He argued that it would be bad policy to allow the forests to be stripped, claiming that any restoration would be the work of a century and more. 'I would therefore reserve all our forests above an elevation of 500 feet, granting only a small patch here and there', the Dewan concluded.[51] Articles 18 and 23 of 1865 were later also included in the Rules of 31 July 1877 (Clause XXVIII) and of 7 July 1898 (Clauses 18 and 23). As just mentioned, most planters ignored the rules intended to protect the natural equilibrium, and the consequent changes in the environment also caused severe problems for the planters themselves. This did not fail to attract the attention of the European community and the Travancore bureaucrats. They agreed that the failure of coffee cultivation was to a large extent caused by the irresponsible behaviour of the planters themselves. In this regard, Mateer noted that 'lands were incautiously selected or carelessly planted'[52] and the author of the Travancore Census Report acknowledged that 'the jungle cutting was too much and thoughtlessly made, helping the rains to wash the hill slopes of all the manure'.[53] More recently, W.K.M. Langley, for many years the director of the influential British trading company Peirce Leslie and Co., argued likewise. He observes that the estates were rendered subject to the natural enemies of the coffee tree – like the stem borer and leaf disease – because of unskilled cultivation. He claims that much of the coffee was planted in the open, and this, followed by soil erosion, caused large tracts to fall out of cultivation.[54]

[51] CSF. No. 11800.

[52] Mateer 1883: 231.

[53] *Census* 1891: 592.

[54] Langley 1962: 28.

Lastly in this catalogue of planter delinquencies, to acquire the property rights of the Kannan Devan Hills, the area situated in the north of Travancore, J.D. Munro, H.G. Turner and A.W. Turner did not bother to negotiate with the Travancore bureaucrats who claimed that the state was the legal owner of these lands. Instead, in 1877 they negotiated with the actual power-holder, a local *zamindar* who was referred to as the *Pooniat Chief* or *Pooniat Rajah*.[55] This clash of interests was ignited when J.D. Munro, who became acquainted with the wider district of Peermade in his functions as coffee planter and Superintendent and Magistrate of the Cardamom Hills, represented the Travancore State in a boundary dispute with regard to the Kannan Devan Hills between his government and the Madras authorities. After visiting the area, the bright potential for plantation development in the district which was officially described as 'practically unexplored regions covered by thick fever-haunted forests, the abode of elephants, (. . .) having no means of communication' dawned on him.[56] On 11 July 1877, Munro, assisted by the Turner (half-)brothers, acquired this vast area of 215 square miles for only 5,000 rupees and an annual rent of 3,000 rupees from the Pooniat Chief (embodied in the so-called first Pooniat Concession). At the outset, the Travancore government disputed the right of the local zamindar to make this grant. Undauntedly, A.W. Turner managed to get access to the government records in Trivandrum and found a document which stated that the land in question *did* belong to the Pooniat Chief. Consequently and 'after much discussion and personal consultation with the Resident',[57] the Travancore government had to recognize the transactions. The Pooniat Concession was then given added legal weight in the Kannan Devan Concession, Deed of Ratification, dated 28 November 1878. For those interested in this particular planting district, it is worth noting that on 17 November 1879, namely only four months after a second concession with the Pooniat Rajah was signed, Munro, Turner and Turner formed the North Travancore Land Planting

[55] The first European planters in Peermade most probably also obtained land through negotiations with the Pooniat Chief (cf. Mahadeva Iyer 1914: 2).
[56] Quoted in: Griffiths 1967: 158.
[57] TLRM, vol. 2, 1915: 29.

and Agricultural Society, which leased out land to individual planters, and that in 1893, because of the severe financial difficulties in which the planters found themselves, the concession rights were sold to the influential Scottish tea producing and trading enterprise James Finlay and Company. As stated earlier, this British firm established the Kannan Devan Hills Produce Company in 1897.[58]

4.1.4 *Indigenous Planters Excluded*

Regulation II of 1040 stipulated that advertisements had to be bilingual in Malayalam and English, which shows that the Travancore government wanted and expected indigenous planters to open estates as well. This sets the stage for the course of the career of P.D. Devasahayam, who became the most influential local planter of south Travancore, Vishakham Tirunal, the First Prince of Travancore, discussed Devasahayam's background at length in a lecture delivered to the Trivandrum Debating Society on 26 September 1874.

Born of very poor parents at Nagercoil, and losing his father early, he was placed in the Mission Boarding School by the late Rev. C. Mault, and was there brought up and educated by him. When about 18 years old, he was attached to the Mission as a Catechist, with a monthly pay of 5 Rs. As such he continued for some years; but in 1844, at the instigation of a friend of his who had just returned from a coffee planter in Ceylon with some little savings, he played the truant and went to that island in hope of large gain. At first, he was employed as a conductor on a salary of 20 Rs under a coffee planter. He continued in Ceylon for about 20 years, during which he served several planters, and by his diligence and sterling honesty, gave satisfaction to all. During the last few years he was employed in a large plantation and drew 100 Rs. Being allowed a few months' leave every year, he was able to spend it in his native land, where, with his savings, he invested 2,000 Rs in a good house and another 2,000 Rs in paddy lands. In 1858 he took it into his head to try coffee planting in the Asambu range of Hills, not far from Nagercoil. His hopes were

[58] Papers regarding the Kannan Devan Hills in: CDS. RF. 546; TLRM, vol. 2, 1915: 21–5 and 29–38; APT, vol. 1, 1935: 658–9; Velu Pillai, vol. 4, 1940: 475–90; Griffiths 1967: 158–9; Ganapathy 1978: 8; Jeffery 1982: 242–3; Mackay 1984: 5–6.

strengthened by the sight of the coffee trees groaning under the weight of scarlet berries in the experimental garden of that keen and unwearied student of nature, the late General Cullen. In spite of very strong dissuasion from friends, he applied for land, obtained, and at once seriously threw himself into the venture. Suffice it to say that the results exceeded the fondest hopes. I have myself seen the Victoria Estate belonging to Mr Devasahayam; and I may say that it is one of the best chosen, best managed, best looking, and best paying coffee estates I have ever seen.[59]

By 1877, Devasahayam was the owner-cum-manager of Joaja (forty acres cultivated), Lower Victoria (ten acres cultivated) and Victoria (fifty-two acres cultivated). In the same year, he also managed the Dennis Estate (eight acres cultivated), owned by P.P. Joseph, and the Assamboo Estate (two hundred acres cultivated), owned jointly by the First Prince of Travancore and Madhava Rao.[60]

Shining though it may be, the success story of Devasahayam was an exception. Although in a strictly legal sense the indigenous entrepreneurs had the same opportunities to open up and run an estate as the Europeans, in actual practice it was mainly Europeans who started plantation enterprises. Turning to figures, in south Travancore only thirty-three of the one hundred and one planting undertakings were owned by local inhabitants in 1877. According to the same source, no indigenous planter owned an estate in the Peermade district (Table 4.9). Given that this may be true, other sources indicate that local cultivators were seriously interested in opening up estates in Peermade in the mid-1870s, though their attempts probably came to naught.[61]

[59] Varma 1874: 12–13.

[60] Ferguson 1879: 159–62.

[61] CSF. No. 10129; CSF. No. 11626 and Haridasan 1986: 3–4. C.J. Eappen, C.J. Matthai, C.P. Potheu, C.E. Matthoo and C.M. Thommy jointly 'owned' the Felicia and Eerie Estates in Peermade. 'Owned' indeed as the plantation was most probably never started at all. Haridasan mentions that a certain Thomas Matthai, with some of his friends, organized a partnership to start a coffee plantation in the High Ranges of Travancore around 1875. Given the same surname (Matthai) and the same particulars concerning the undertaking (joint ownership), crop (coffee), date (1875) and result (failure), Haridasan is probably describing the same effort. For further particulars regarding Eappen and his associates, see also the next two paragraphs.

TABLE 4.9

Indigenous Planters' Share in the Plantation Sector of Travancore in 1877[62]

District	*Total Number of Estates*		*Number of Estates Owned by Local Planters*	
South Travancore	101	(100%)	33	(32.7%)
Peermade	30	(100%)	0	(0%)

The disproportion between indigenous and foreign entrepreneurs in the plantation industry was even much larger than the figures above suggest. Table 4.10 illustrates for example that the 33 undertakings owned by indigenous entrepreneurs in south Travancore in 1877, though representing 32.7 per cent of the total number of estates in the district, cultivated only 1,834 acres. In other words, in that year these holdings represented only 11.5 per cent of the total area cultivated with plantation crops in the region. It stands to reason then, that the tracts of land owned by the indigenous planters were much smaller than those of their European counterparts. As shown by the same table, 55.6 acres on average were cultivated with plantation crops on the local holdings. This figure is in sharp contrast with the situation on the British estates, where hundreds of acres of land were usually planted. Some plots of land possessed by local inhabitants were so small that one may rightly question defining these holdings as plantations and their owners/managers as planters. It is likely that these lands were not cultivated by wage labourers but by family members. For this reason, it is justifiable to consider these cultivators as coffee peasants.[63]

[62] Ferguson 1879: 156-7 and 159-62.

[63] It would be possible to draw a line of demarcation between estates (wage labour) and smallholdings (family labour) at, for example, ten acres. At the same time, this distinction remains highly arbitrary: some owners of land measuring less than ten acres would employ wage workers whereas other owners owning relatively large plots of land would employ only family members, depending upon the size and employment opportunities of the families concerned. In this context it should be remembered that coffee estates were employing 0.88 workers per acre on an average in India in the early 1930s (RCL 1931 Evidence Vol. 11: 383). For a quantitative definition of tea and rubber smallholdings and estates, see also Section 5.2.1).

Table 4.10
Total and Average Acreages Cultivated with Plantation Crops by European and Indigenous Entrepreneurs[64]

District	*Total Acreage Cultivated with Estate Crops*	*Total Acreage Cultivated Owned by Foreign Planters*	*Total Acreage Cultivated Owned by Local Planters*	*Average Acreage Cultivated Foreign Estates*	*Average Acreage Cultivated Local Estates*
South T.	15,890 (100%)	14,056 (88.5%)	1,834 (11.5%)	206.7	55.6
Peermade	4,046 (100%)	4,046 (100%)	0 (0%)	134.9	0

There are two major factors which explain the comparatively small number and limited size of the indigenous undertakings. The first element lies in the attitude of the European planting community towards the local planters. To secure sufficient land of good quality, no less than to control labour and to monopolize government support, the European planting community created a political and legal framework which discriminated against indigenous planters. Local owners of plantations were, for example, not admitted as members of the various district planting associations. Nor were they warmly welcome in the United Planters' Association of South India.[65] The Europeans took very good care to see that only they were able to put forward their applications for land through the British Resident, whose influence, as has noted many times before, was enormous. The European planters also availed themselves of other, less indirect, methods of land control. They encroached on lands belonging to indigenous planters. A local planter called Eappen, probably at his wits' end because of the unvarnished hostility displayed by his European neighbours towards him, his associates and their enterprise, turned to the Dewan in search of help. He wrote in 1876 that 'as regards our (. . .) Estate Eerie I beg to state that its boundaries are for the most part lands

[64] Ferguson 1879: 156–7 and 159–62.
[65] Cf. Mayne 1953: 10.

belonging to Europeans and that we have reason to fear that our commencing work there before the boundaries are worked out by the Sirkar will involve us in serious disputes, to avoid which we have hitherto been obliged to refrain ourselves from doing any thing there'. Eappen therefore requested 'that you will be pleased to order the Revenue Surveyor to have both our Estates surveyed on his next visit to the Peermade Hills, and, kindly, considering our weakness when compared with our European neighbours, give him special instructions as to the careful and impartial making of the boundaries of the latter (Eerie)'.[66] It is highly unlikely that Eappen obtained the assistance he sought from the Travancore bureaucracy. The estate was never a financial success and was finally abandoned.[67] Apart from the above-mentioned conflict, in 1902 H.M. Knight claimed land which Abraham Pillai said was legally his property.[68]

The second factor affecting indigenous planters was related to the nature of the indigenous plantation segment itself. Like the cultivation of pepper in the eighteenth century, coffee was produced mainly on lands possessed by relatively well-to-do sections of the lower castes. The names of the local inhabitants owning or leasing a plantation obtained from materials consulted in the Kerala State Archives (see Appendix 4) or given by Ferguson (see Appendix 5) demonstrate that most of the indigenous planters, like Devasahayam, were Shanars who had converted to Christianity. The Christian names of Benjamin, Daniel, Emanuel, Isaac, Jacob, Joseph, Luke, Moses, Nathaniel, Noah, Paul, Peter and Samuel likewise point to the prominence of low-caste people among the indigenous planters. Many low-caste Travancoreans changed their original Hindu names for new Christian ones after their conversion by the missionaries.[69] Other, higher castes were represented in the plantation sector as well, although to a much lesser extent than the Shanars. Among the indigenous planters there were Brahmins (Madhava

[66] Letter of C.J. Eappen, Secretary Felicia and Eerie Coffee Company, to A. Seshiah Sastri, Dewan, on 15 December 1876, in: CSF. No. 11626.
[67] CSF. No. 10129.
[68] CSF. No. 11226.
[69] See also Kooiman 1989: 127–35. It should, moreover, be mentioned that during that period of time there were hardly any or no Syrian Christians inhabiting the south of Travancore.

Rao), Syrian Christians (C.J. Eappen, C.J. Matthai, C.P. Potheu, C.E. Matthoo and C.M. Thommy jointly 'owned' the Felicia and Eerie Estates in Peermade),[70] Muslims (H.H. Sait) and other high-caste individuals (M. Pillai, T. Pillay, and the First Prince of Travancore). The predominance of comparatively wealthy sections of the lower castes in the indigenous segment of the plantation sector explains the relatively small number and limited size of the indigenous undertakings. Although those local low-caste inhabitants who opened an estate were relatively well-to-do compared to the other members of their communities – many of them were educated by the missionaries and knew English, a language which had a great commercial value as was most clearly observed in Devasahayam's career – they were relatively poor in comparison with their European counterparts. The capital investments which were essential for large-scale cultivation of coffee were much more of a problem for aspirant local entrepreneurs than for European planters, which provides a ready explanation for why the share of the indigenous plantation segment decreased even further with the shift from coffee to tea. Quite apart from the hostility of the European planters, the high investments necessary for the cultivation of tea greatly circumvented the opportunities for indigenous entrepreneurs to own such a plantation.

Realizing the declining significance of the indigenous planting segment, the Travancore government tried to prevent the tea sector from becoming the sole domain of the Europeans. In Notification No. 8178 dated 8th Mithunam 1063 (July 1888), it was stated that 'tea cultivation is flourishing in certain parts of the Peermade High Ranges and that our citizens cannot go there and stay there for cultivating tea, due to the difficulty in reaching there'. This cut them off from the benefits of tea cultivation. Though incorrect in its analysis of the cultivators' incapacity to start tea undertakings, the government sincerely sought for a solution, trying to help the cultivators fiscally and by other means. Noting that there was a plentitude of uncultivated fertile lands in the valleys and low-land areas suitable for cultivation, the Notification states: 'Such lands, if cleared

[70] Letter of C.J. Eappen to A. Seshiah Sastri, Dewan, dated 10 July 1875, in: CSF. No. 11626.

by the tenants for agricultural purpose, will be given to them on free ownership for 5 years, i.e. they are exempted from tax for 5 years'. It also added that the area would be limited to ten acres per person.[71] Unfortunately for the Travancore government, the encouragement given to local planters failed. It foundered on two major oversights, namely that the cultivation of tea should be on the high ranges, in contrast to 'the valleys and low-lands', and this had to be on a large scale, not confined to a mere '10 acres'. Crucial was the fact that the expertise and capital essential for processing tea were not available to 'tenants'. According to an enumeration of plantations by the United Planters' Association of South India in 1893, there were no indigenous planters left in Travancore in that year.[72] Although these figures might not be entirely congruent with the reality, there is no shadow of doubt that in the last two decades of the nineteenth century the plantation sector was completely dominated by the European planters.

4.2 The Labour Question

4.2.1 The Creation of 'Free' Labour

If the first precondition for success for the planters was to acquire vast areas of land for low prices, their efforts to recruit and control sufficient numbers of workers were to be at least as important to their prosperity. As this aim was often rather difficult to achieve, planters referred to this problem as 'The Labour Question'.

The Labour Question recalls the historical debate about 'free' and 'unfree' labour presented in the introductory chapter. In this research, 'free' refers to 'the freedom whether or not to choose one's employer and therefore one's labour conditions, or to choose one's means of production'.[73] 'Unfree', therefore, indicates all the factors which restrict one's choice of employer, such as legal obstacles (like slavery or indentured labour), financial hindrances (indebtedness), a segmented labour-market

[71] Translated from Malayalam into English, TLRM, vol. 5, 1916: 903–4; and cf. TAR 1063: 75.

[72] UPASI Proceedings 1893.

[73] Lucassen 1993: 3.

(based on, for example, educational and/or ethnic/communal characteristics) and geographical immobility.[74]

As the long-term historical developments in this study will show, free and unfree labour alternated each other. In a strictly legal sense, *unfree labour* (slavery) was followed consecutively by *free labour* (abolition of slavery in 1855), *unfree labour* (introduction indentured labour by the Criminal Breach of Contract Act in 1865) and *free labour* (abolition of that same act in 1935). The study will also demonstrate that elements of freedom and unfreedom were both present in any labour relation in any period of time. The labourers were both free and unfree, resulting from the often conflicting strategies of workers, planters and politicians. Indeed, the abolition of slavery in 1855 did not usher in a period of *free* plantation labour, the introduction of the Criminal Breach of Contract Act in 1865 did not establish a completely *unfree* work force, and the abolition of the Criminal Breach of Contract Act in 1935 did not fully *free* the estate workers.

Taking first things first, the phase of slavery and its abolition will be examined. Those protests against slave labour which had a vast impact on Travancore politics were not voiced by the first planter in the area – Huxham – but were expressed by those *who would become* the most influential planters during the end of the 1850s and the beginning of the 1860s: the European missionaries.

In 1847 the missionary societies presented a joint memorial to the Raja in which the abolition of slavery was requested. The document refers to the formal abolition of slavery throughout the Indian territories governed by the British East India

[74] I realize that the free – unfree opposition is strongly rooted in Enlightenment and post-Enlightenment (including the Neo-Classical and Marxist) discourses, implicitly and often explicitly suggesting (a) that unfree labour is negative and free labour positive, and (b) that history is characterized by the gradual elimination of unfree labour relations (cf. Prakash).

I would like to emphasize, however, that in this study it will be shown that free wage labourers were not better off *per se* than unfree slaves (see for example Section 4.2.2). Neither was there a clear-cut trajectory from unfree to free labour. For instance, both the elements of freedom and unfreedom are present in the labour relations on the estates nowadays (see Section 7.2.2).

Company. The most efforts to counteract the influential missionary lobby were made by the high-caste landlords, who were afraid that the enforcement of the abolition of slavery would upset the existing social balance and would create a shortage of labour, especially for paddy agriculture, which was very labour intensive, requiring labour inputs the year round.[75]

Their protests were heeded: in 1853 the missionaries' request was only partially complied with. A proclamation was enacted which abolished slavery formally but in which was stated that 'the slaves as well as all others should clearly understand that there shall be no change in the customs and untouchability in force according to the respective caste etiquettes'. Nor did the regulation include provisions for the emancipation of all slaves in Travancore.

In contrast to the landlords, the missionaries had little reason to be satisfied with this proclamation. On 26 July 1855, the evangelists informed the Governor in Council of the Madras Presidency that they, 'the petitioners, (. . .) pray for the appointment of a commission of well qualified Europeans' for 'this province of Travancore (. . .) is now bordering on utter disorganisation, and in this state of things, the sufferers are the great mass of people'.[76]

This petition considerably strengthened the missionaries' position *vis-à-vis* the Travancore landlords. In that same year, additional political pressure brought to bear by the British Resident and the Madras government resulted in the enactment of a second proclamation by the Travancore government. Slavery was abolished totally from that time, at least in a strict formal sense.[77] As we will observe later, in actual practice the new rule did not change labour conditions in most areas for many years to come.

In the historiography dealing with Travancore, the question of whether the lobbying by the missionaries and the British officials for the abolition of slavery was motivated by humanitarian/Christian ideas or by capitalist considerations is much

[75] Tharian George and Tharakan 1986: 207.

[76] Signed by among others John Cox, quoted in: Yesudas 1977: 87–8.

[77] Kurup 1984: 193–5; Kooiman 1991: 62.

debated. Mahadevan Nadar emphasized that slavery was abolished 'by the benevolent attitude of the Christian missionaries',[78] whereas Kurup observes that the particular regulation 'was closely related to the requirements of the emerging capitalists in Indian agriculture'.[79] Kooiman, who based his research on the archival materials of the London Missionary Society (LMS), combines both points of view. Not willing to overemphasize the distinction between the humanitarian and the capitalist motives, Kooiman states that 'for the LMS missionaries there was nothing more humanitarian than to enable the people of Travancore to share the blessings of a capitalist development'.[80]

I would like to add my own contribution to the above discussion by arguing that the linkages between the abolition of slavery, the religious and economic activities of the missionaries and the development of plantations were even more direct than most of the literature suggests. The missionaries were not only lobbying for the abolition of slave labour *before* they themselves became personally involved in the plantation industry, they were also fighting bonded labour *during* the period in which they combined their original profession with that of planter. There is the case of 18 July 1859 when some missionaries, including John Cox, who was in this period one of the most important planters of the area and presided for several years over the Travancore Planters' Association,[81] requested the Governor in Council of the Madras Presidency for 'the entire removal of slavery,[82] [and] all kinds of forced labour'.[83] Clearly, the missionaries' religious and humanitarian aversion to slave labour initially only coincided with *what would become* their economic interests (which they perhaps did not know), but gradually suited their *contemporary* capitalist interests (of which they were well aware).

[78] Mahadevan Nadar 1980: 217.
[79] Kurup 1984: 196–7, and cf. 193–4.
[80] Kooiman 1991: 63.
[81] CSF. No. 1760.
[82] As will be discussed in the next section, the 1855 proclamation did not change the situation of the slave labourers immediately.
[83] Cited in: Yesudas 1977: 100–1.

4.2.2 Unwilling Wage Workers

The planters needed large numbers of labourers to work on their enterprises. Lovatt wrote that in the Peermade district in the coffee period the number of labourers employed was about 1,400, which swelled to 4,000 during the harvesting of the crop.[84] With the rapid expansion of the plantation industry in the tea period many more workers were required, a fact reflected in the official figures. In the Report of the South India Planters' Enquiry Committee, for example, it is stated that the number of workers on the estates of Travancore totalled 14,497 in 1896.[85] According to Nagam Aiya, the author of the Travancore State Manual, the number increased to 18,410 in 1904. Of this figure 3,432 were employed in the south, 5,590 in central Travancore, and 9,388 in the north of the state.[86]

Although these figures seem to suggest the opposite, the planters experienced severe difficulties in recruiting 'free' labour. The local hillmen who lived in the plantation areas were too few in number to satisfy the planters' needs. They totalled between 8,000 and 10,000 at the beginning of the twentieth century,[87] while at least 18,000 labourers would have been needed during the same period.

Quite apart from their lack of sufficient numbers, the hillmen could not be transformed into a compliant work force.[88] In south

[84] Lovatt 1972: 12.

[85] RSIPEC 1896: 8.

[86] Nagam Aiya 1906: 79. Unfortunately, the number of labourers working on the Travancore plantations as provided by the *Census of Travancore* of 1901 is totally unrealistic. According to these figures, 2,574 men and 1,515 women were employed by the estates (*Census* 1901: 266). In fact, the number of workers was much higher. This becomes evident when Tables 4.3 and 4.6 are consulted again. According to these figures, tea was cultivated on an average on 24,931 acres in the period 1900–4 and on 24,711 acres in 1904. Bearing in mind that nowadays every experienced plucker can cover about 1 acre, in 1901 4,089 workers could never have plucked about 25,000 acres. To pluck more than 6 acres a worker is a virtual impossibility. The figures provided by *The Report of the South India Planters' Enquiry Committee* and by *The Travancore State Manual* should therefore be considered as much more reliable.

[87] Nagam Aiya 1906: 407.

[88] In some of the other plantation districts of south India planters sometimes employed tribals to a considerable extent, although often with great difficulties (Kurup 1984: 191, Ravi Raman 1991: 247). Also in Jalpaiguri district, and in

Travancore most of them would not work for the planters.[89] In the Peermade district, the majority of the hill tribes were not prepared to serve as plantation labour either. They refused to work for hire or to carry headloads.[90] The story was the same kind in the Kannan Devan Hills, although in this region at least some of the hillmen used their knowledge of the jungle to guide the planters in their search for fertile land, in the journey to the coastal areas of Travancore, and on the hunt for game.[91]

There are some obvious reasons which explain the refusal of most hillmen to accept the work offered by the planters. Quite naturally, many of them preferred their way of life to that of the plantation labourers. In the south of Travancore, the Kanikarans practised shifting cultivation.[92] This form of agriculture, also named *cherikal*, involved clearing a particular patch of the jungle – trees were felled and brushwood was burned off – to use the land thus opened for the cultivation of food crops, especially paddy. Often the plot was planted twice with the same crop while the third cultivation consisted of sugar-cane or tapioca. When the soil was exhausted the hillmen moved on to a more fertile spot. Because of the migratory, often circular nature of this form of production, many of these cherikal lands were cleared once in every twelve years.[93] In the Peermade Hills, the Mannans and Uralis also practised cherikal cultivation. Lovatt says that these hillmen exchanged a portion of their paddy for cloth and salt, and that they relied on whatever food the jungle provided them, like game, fish, roots and berries. In her view, another hill tribe, the Pandarams, were 'even more primitive' than the Mannans and Uralis. They were seldom seen by strangers and lived in caves and in the hollow trunks of trees,

---

the area known as the Dooars in particular, tea planters employed relatively large numbers of people who were tribal in origin. As in south India, the tribesmen – of whom most originated from Bihar – did not volunteer for plantation work. The introduction of land ownership based on written deeds did the most to uproot the adivasis' traditional agrarian system forcing them to accept work in the tea gardens of north Bengal (Bhowmik 1981: Ch. 2).

[89] Kooiman 1989: 139.

[90] Lovatt 1972: 12.

[91] Martin 1931: 10; Mackay 1967: 8.

[92] Pandian 1990: 79–80.

[93] Uma Devi 1989: 55–6

using bark for clothing and eating snakes and lizards if nothing else was available.[94] There was an even greater variety of hill tribe cultures in north Travancore. Apart from the Mannans, who have already been mentioned, this area was the home of the Ooralers, Palians, Pooliahs and Muduvans. This area also offered a greater diversity of life-styles. Hunters-cum-gatherers, shifting cultivators and permanent cultivators were all represented in the district.[95]

The tribesmen considered themselves the lords of the soil and regarded the forests as their home.[96] The planters were therefore thought to be unwanted intruders. For many centuries these forested regions had been left almost untouched, sparsely inhabited by cultivators and/or hunters-cum-gatherers, marginally administered by the Pooniat Raja, and visited sporadically by petty traders. The natural, social, economic and political balance was fundamentally disturbed by the emergence of plantations. Large tracts were denuded, increasing numbers of planters and labourers arrived, influential trading companies penetrated the area, and officials of the Travancore state sought recognition of their administration.

The combined interests of the European planters and the Travancore officials frequently contrasted with those of the hill tribes. For instance, in 1868 the government determined that shifting cultivation needed to be restricted so that 'more land would be available for coffee cultivation than would be in a few years if the burning of forests be permitted to go unchecked'.[97]

In addition, land being used for shifting cultivation was often sold to planters, without the cultivators being informed beforehand about the sale. It is no surprise that these often illiterate peasants were not in position to defend their interests against those of the planters and the government. This was also the case in the Kannan Devan Hills in 1896. During his tour through the district J.D. Ree, who held the British residentship three times during the 1890s, noted that the acquisition of 400 acres of land by Sir John Muir, one of the leading directors of James

[94] Lovatt 1972: 7.
[95] Martin 1931: 50.
[96] Nagam Aiya 1906: 407.
[97] Quoted in: Pandian 1990: 79.

Finlay and Company, was 'being complicated by the claims of hillmen, who went to Trivandrum lately to plead their cause with results of which I am ignorant'.[98] It is interesting to observe that these hillmen, like the local planter Eappen, called in the help of the Travancore bureaucracy in their struggle against the European planters, most probably with similarly little success. Thus often without being acknowledged and, in the sporadic cases of which they were informed, frequently without their consent, the hillmen lost the land which they cultivated and which had been cultivated by their ancestors for centuries.

Apart from the land which was confiscated for plantation cultivation, some of the hillmen's paddy lands were transformed into cart roads. In 1892, when A. Martin tried to find out the best trace to link the Kannan Devan Hills with the Madras Presidency, the European planter met a lot of resistance. After discovering the purpose of Martin's tour, the inhabitants refused to assist the European planter and his gang of labourers in any way. They were given neither water nor any food, even after Martin offered them money.[99]

This impasse meant that labour had to be recruitment from the plains. This proved no easy task. At first sight, it might have been expected that the demand for labour by European planters, who were, according to the government, 'gentlemen sincerely solicitous to deal fairly with their labourers, and to rely upon good treatment and good wages alone for attracting labour'[100] and for whom, according to Mateer, 'caste was nothing',[101] would induce huge numbers of labourers, especially those with a low-caste background (including the ex-slaves), to seize the opportunity to escape from their poverty-stricken situations in the caste-ridden villages. However, the contrary was true. Only in infrequent cases, when for example a famine occurred in the Madras Presidency, did wage workers volunteer *en masse* for plantation labour, for which the years 1866 and 1877 provide evidence.[102]

The reluctance of the agricultural labourers to accept

[98] R/2/879/41.
[99] Martin 1931: 15.
[100] TAR 1040: 22–3.
[101] Mateer 1883: 236.
[102] Travancore Government 1883: 22.

plantation work is all the more remarkable since reliable historical material suggests that the wages offered to them were indeed considerably higher than those paid by the landlords in the lower parts of south India. Mateer, for example, noted that in 1859 in south Travancore agricultural labour could be hired for an anna per day and that the Public Works Department, which was carrying out canal works in the region, paid a little over two annas. 'The planters, however, were obliged (. . .) to give at least four annas to induce ordinary coolies to brave the dangers of the hills'.[103] Table 4.11 shows plantation wages in later years.

TABLE 4.11

Daily Plantation Wages in Central Travancore and the Kannan Devan Hills in 1893 and in Travancore in 1896[104]

District	*Year*	*Men*	*Women*	*Children*
Central Travancore	1893	4 annas	not known	not known
Kannan Devan	1893	4.5 annas	2 annas	not known
Travancore	1896	4 to 5 annas	3 to 3 annas, 6 pies	2 to 3 annas

In their efforts to account for the labourers' unwillingness, some scholars have incorrectly stressed the 'traditional', almost 'passive' and 'irrational' attitude of the agricultural workers. Saradamoni, referring to the freed ex-slaves in the context of plantation work, states: 'By habit and tradition they were unused to leaving their huts and hamlets'.[105]

However, there were many factors which have to be taken into consideration to explain the refusal of agricultural labourers to contract for plantation work. Although slavery was completely abolished in a strictly legal sense in Travancore in 1855 and in the directly-ruled territories of the Madras Presidency in 1833,[106] various forms of labour bondage lingered on in south India during the latter half of the nineteenth century. Though legally

[103] Mateer 1883: 235.
[104] UPASI Proceedings 1893; RSIPEC 1896: 12.
[105] Saradamoni 1980: 114–15; and cf. Jeffrey 1976: 100.
[106] Kumar and Raychaudhuri 1991: 323 and 485.

liberated, the relations between these ex-slaves and their former owners often did not alter structurally. The agricultural labourers usually remained bound to a particular plot of land and its owner, particularly through indebtedness.[107] In contrast to some other authors who have dealt with this subject,[108] I would like to argue that these relations between employer and employee, despite their inherent inequalities, were not beneficial only to the patron. It should also be acknowledged that these 'unfree' labourers, despite often being cruelly treated, were, at least in some respects, better off than the so-called 'free' labourers. The former, in particular those attached to paddy lands, were ensured of employment and food during the greater part of the year, sometimes being allowed a loan in times of a family crisis.[109] The latter category was indeed 'free' to offer its labour to the highest bidder but could claim neither work nor sustenance when they were unemployed or in times of scarcity of food. They were 'free to starve'. For this cogent reason, many attached labourers saw no reason to break the bonds with their landlords, lacking any better alternative.

On the employment front, many agricultural labourers were able to improve their conditions on the plains. The abolition of slavery and the demand for labour from the planters and the Public Works Department (PWD) – its establishment in 1860 being closely related with the development of European coffee plantations[110] – gradually strengthened their bargaining positions *vis-à-vis* their landlords. This again led to better working conditions, including higher wages.[111] Indeed, the rise in wages on the plains (see also Table 4.12), at least partially

[107] Griffiths 1967: 397; Tharian George and Tharakan 1986: 207; George 1987; Ravi Raman 1991: 246.

[108] E.g. Saradamoni 1980.

[109] George 1987: A 144.

[110] Lovatt 1972: 51; Menon 1978: 514–15; Pandian 1990: 84.

[111] The Travancore Administration Report of 1870–1 stated for example that: 'In fact till June 1855, predial slavery existed in Travancore fully recognized by law. (. . .) The condition of those who thus regained their freedom has since been gradually improving. (. . .) The law extends to them the same protection as to other members of the community. (. . .) Compulsion is utterly impossible in these days. (. . .) The various public works in progress in different parts of the country and the operations of Coffee planters have been instrumental in ameliorating the conditions of this class of people,

brought about by the rise of plantations in the hills, was a major factor in the reluctance of many agricultural labourers to opt for estate work.

TABLE 4.12
Daily Wages of Agricultural Labourers in Some Travancore Taluks in the Period 1880–1895[112]

Taluk	*1880-85*	*1885-90*	*1890-95*
Vaikom	2 annas, 8 pice	3 annas, 5 pice	4 annas
Ettumannoor	2 annas, 8 pice	3 annas, 6 pice	4 annas, 6 pice
Kottayam	2 annas	2 annas, 10 pice	3 annas, 6 pice
Changanacherry	2 annas	2 annas, 10 pice	3 annas
Meenachil	2 annas, 3 pice	3 annas	4 annas

Fear also played a role. For most of the agricultural labourers the journey to the estates, which took about two to three days on an average,[113] was considered extremely abhorrent. As Heather Lovatt, a planters' wife, recalls: 'the hills with their dark, dripping forests seemed a fearful place, where only a lucky man could escape the dangers from wild beasts, snakes and *shaitans'.* She went on to mention that 'only extreme poverty and a scarcity of work in their own villages, such as experienced in the two terrible famine years in Madras in the mid-70s, could drive them [i.e. the labourers] to this alternative'.[114]

inasmuch as the demand for labour thus created enables them to procure better terms from their old employers' (TAR 1045: 39). In 1883 Mateer observed the same development. 'The labour of the previously enslaved castes', he stated, 'which had hitherto been almost valueless, being remunerated only by a few measures of rice daily, became of as great money value as that of others; caste was nothing in the eyes of the European planter. (. . .) The Sudra masters complained of the planters taking away their labourers. But this competition and demand for labour largely ameliorated the condition of the poor. (. . .) Numbers of the poorest classes of the population were thus [by accepting work on the European estates] removed from the degradation under which they had suffered, and relieved from oppression; while everywhere the landed proprietors were reminded of the necessity of fair and kind treatment to retain the services of their dependents' (Mateer 1883: 236-7).

[112] Velu Pillai, quoted in: Tharian George and Tharakan 1986: 211.

[113] Arnold 1986: 153; Uma Devi 1989: 81.

[114] Lovatt 1972: 12.

Enlightened by those plantation labourers who returned to the plains, many of the workers also became aware that the living conditions on the estates were far from ideal. Lovatt noticed the extremely poor state of health during the initial phases of plantation development. 'During the monsoon, when the labourers huddled shivering in their family huts, when the wind whistled through every corner, bronchial diseases took their toll.' Infection of the bronchi was not the only affliction. Most human suffering in the plantation districts was caused by hookworm, dysentery, smallpox, plague, cholera and malaria, which were often fatal as medical knowledge was still fairly limited. For instance, it was not until the end of the nineteenth century that the physician R. Ross discovered the part that mosquitoes played in the transmission of malaria. Be that as it may, medical breakthroughs did not automatically improve the quality of life of the labourers: often the estates lacked even the most basic medical facilities. Even after Ross' discovery 'many more years passed before effective steps were taken to reduce this menace'. No wonder the labourers feared the estates, particularly during epidemics. As Lovatt concludes: 'All the estates along the Periyar were very unhealthy, and when there was a bad outbreak of malaria it was impossible to get any work done.'[115]

At the socio-religious level, most agricultural labourers realized that caste background would continue to play a crucial role on the estates. They feared the cruel treatment which could be meted out by higher castes and Europeans. Finally, many labourers were aware that there was a discrepancy between the wages offered and the actual payments. The plantation workers were often paid less than promised.

Unfortunately for the European planters in south India, there was not only a lack of 'push factors' which made most agricultural labourers, especially those who lived in Travancore, reluctant to accept plantation work. There was a surplus of competing 'pull factors' as well: managers of estates situated in Ceylon, Fiji, Malaya, Mauritius and other British colonies were also interested in recruiting south Indian labour. The densely-populated and frequently famine-struck districts of the Madras Presidency, like

[115] Lovatt 1972: 12–13, 32.

Coimbatore, Madurai, Salem and Tinnevelly, functioned as important recruiting centres for plantation labour. To give an indication of the volume of this labour migration, in 1895 as many as 123,975 workers from south India went to Ceylon to labour on estates there.[116]

4.2.3 The Establishment of 'Unfree' Labour

The Kangany System. The way in which the unwilling agricultural labourers were mobilized by the south Indian planters, those of Travancore included, was written down in fairly detailed form by the South of India Planters' Enquiry Committee in 1896. It reported:

In securing labour for their estates planters occasionally deal directly with coolies, but these instances are very rare, being limited to engagements made with coolies who have worked for several years on the same plantations or with coolies living close to plantations. But even in such cases the practice is not universal and it may be said that in recruiting imported labour, *maistries*, who answer to the garden sardars of Assam, are invariably employed as the agents between the planters and the coolies. The maistry enters into a contract with the planter and the contract is generally reduced to writing. Such contracts are signed by the maistry, rarely by the planter, and they are seldom, if ever, registered. In the case of old and trusted maistries, informal verbal agreements sometimes take the place of written contracts and in some places, for instance, parts of Travancore, a verbal agreement is supplemented by a promissory note executed by the maistry. Receiving a lump sum in advance, the maistry undertakes to supply a certain number of coolies for plantation work for a certain time. These engagements are made usually in March or April of each year [when the plantation labourers return to the plains], the time fixed for the commencement of work on the plantations being two or three months later [in June or July]. The period of the contract is usually nine months, sometimes less, and it never exceeds a year (. . .). The maistry then proceeds to the village or villages where he collects labour and generally returns to his employer's estate with his gang, though some maistries remain in the villages and send their gangs to the plantations in charge of their representatives as headmen. The average strength of a maistry's gang varies between 20 and 40 adult males,

[116] RSIPEC 1896: 11.

females and children; the average maximum is 100, but in some cases a maistry furnishes as many as 200 or 250 coolies.[117]

The recruiter, who apart from the designation 'maistry' was also called 'kangany' in Travancore, in his turn, 'makes advances to coolies, entering into contracts with them (sometimes in writing, but more often not)'.[118] The advance offered by the kangany was of crucial importance in many regards. It enabled those workers who were financially bonded to a particular landlord and/or moneylender to pay off their debts. Many of these labourers, who were often extremely poor, could not resist taking the advances offered by the kanganies. They had to take care of their immediate basic needs, like food, clothing and shelter. Others, but this group was less significant, used the advances for extraordinary expenditures like marriages and other important social events. Apart from its disadvantages for the workers, which will be analysed in the next sub-sections, the advance was the only practical guarantee of his employment for the labourer. At the same time, the amount gave some indication about the financial background of the new employer and, consequently, about his ability to pay an attractive wage.[119] As Table 4.13 shows, the planters, through their recruiters, were able to advance relatively large sums of money. In central Travancore, where the wage of four annas was customary for men (Table 4.11), the advance of ten rupees was the equivalent of forty days of work.

TABLE 4.13

Advances in Central Travancore and the Kannan Devan Hills in 1893[120]

District	*Men*	*Women*	*Unspecified*
Central Travancore	10 rupees	6 rupees	
Kannan Devan Hills	unknown	unknown	10 rupees

The importance of the kangany was not restricted to the provision of advances. They were experts at spinning the workers

[117] RSIPEC 1896: 12.
[118] RSIPEC 1896: 17.
[119] RSIPEC 1896: 18–20.
[120] UPASI Proceedings 1893.

rosy stories about plantation life which, at least partially, assuaged their fear of all kinds of terrifying diseases, inhuman working conditions, and low wages. Uma Devi, who was able to study the information collected in Travancore for the benefit of the South of India Planters' Enquiry Committee, pointed out the dilemma many labourers had to face: either to believe the plantation labourers who returned to the plains or to have faith in the kangany's picture of plantation life.[121] In addition, the recruiters often obtained most of their labour from among their own family and friends, thus ensuring that, apart from their financial bonds, the workers were also closely related to them in a social sense. The recruiters also guided their gang of labourers through the jungle to reach the estate for which they were destined.

It should be emphasized that the advance system was not a phenomenon newly introduced by the British planters. The system of advances was in fact well-rooted in society and was closely related to indigenous modes of production, particularly in agriculture. As discussed in the second chapter, in eighteenth century Travancore the cultivation of and trade in pepper was based on advances. To mention yet another example, in 1800, in his research on Mysore, Malabar and Canara, Buchanan found that it was a common custom for a farmer to advance money to an agricultural labourer on condition that the worker (and his family) labour for him on his lands for a definite term. As Buchanan observed, the wages were barely sufficient for the labourers' maintenance, so that additional advances were regularly made by the employer to meet wedding expenses and other needs. The labourer was not free to go until he repaid his debt, often leading to a state of servitude which became hereditary. In return for his services the worker was housed, fed and clothed.[122] Though hampered by an absence of detailed historical

[121] Uma Devi quoted Krishna Iyer, the Tahsildar of Changanacherry. He stated: 'The contractors or maistries before engaging a labourer give him an attractive description of a labourer's life on the planters, while still under the charm he takes advances and enters into an agreement. Going home, his neighbours, relatives, wife and children describe to him the dark side of the picture and dissuade him from leaving his home' (quoted in: Uma Devi 1989: 82).

[122] RSIPEC 1896: 18 and cf. Kumar 1965: Ch. 3 and 4. Unfortunately, these

information about the recruitment of landless labourers in south-west India, there are no reasons whatsoever to assume that an advance system for the recruitment of agricultural workers did not exist in pre-colonial and early colonial Travancore. It was probably a common practice for wealthy landlords to offer advances to low-caste people, including (ex-)slaves, to work on their land, particularly in times of labour shortage, thereby competing with the interests of other landlords.[123]

What was the social composition of those who were hired? The principal group to be recruited by the kanganies were low-caste labourers. Mateer states that in south Travancore 'Pulayars, Pariahs, Vedars, and other low castes began to obtain employment (. . .) and Ilavars and Shanars working with them in the mountains'.[124] In the other plantation districts low-caste workers also formed the majority, which is reflected in the figures of the Census of 1901 (see Table 4.14). In that year there was a preponderance of labourers of the Channan, Maravan, Pallan, Paraya(n) and Pulaya(n) castes imported into the Cardamom Hills, an area encompassing the plantation districts of central and north Travancore. As can be seen below, castes like Pallan, Parayan and Pulayan especially were considered extremely depressed communities. The recruitment of low-caste labourers was partly related to the idiosyncratic characteristics of south India. Most members of the higher castes were not prepared to lower themselves to perform manual tasks like weeding, picking and plucking.[125] Conversely, to members of the depressed castes, especially in periods of unemployment and/or food shortage, accepting plantation work sometimes offered relief.

Quite apart from any choice on the part of the workers, it was also a deliberate policy on the part of the planters to recruit labour from the lowest castes. The planting community derived

---

and other sources dealing with the issue of advances in pre-colonial and colonial south India do not disclose any information regarding the role of middlemen. For this reason, the history of the concepts 'kangany' or 'maistry' cannot be traced.

123 Cf. George 1987.

124 Mateer 1883: 236.

125 Cf. Mateer 1883: 236.

TABLE 4.14
Castes in the Cardamom Hills in 1901, Except for the Categories 'Hill Tribes', 'Native Christians' and 'Vellala' and Numbering 700 or More[126]

Name	*Number*	*Description According to the 1911 Census*
Channan	911	A Tamil caste, more or less the analogue of the Izhava in the Malayalam-speaking tracts. Agriculture and tapping of palm trees form the chief occupations.
Maravan	742	Originally a police and military caste. They are found chiefly in Madura and Tinnevelly, and in Travancore, in the outlying taluks of Nanchinat and Shencottah. Besides being village watchmen, some of the members of the caste have too often taken to village raiding and highway robbery, presumably under stress of want added to lack of educational guidance.
Pallan	3,315	The agricultural labourers of the Tamil country. It may be mentioned that Pallan derived from Pallam, a pit, so called, it appears, because they were standing on low ground when the castes were originally formed.
Parayan	4,103	An important but depressed caste of agriculture and industrial labourers – matting and wicker-work being their speciality – found throughout Travancore, but especially common in the more southern taluks.
Pulayan	1,188	According to certain accounts, they are merely the most depressed layer of the comprehensive industrial caste of Indian society, their function in the organism being agricultural labour. In many cases, they are hereditary attaches to agricultural holdings.

[126] *Census* 1901, part 3: 332–95; *Census* 1911, part 1: 257–69.

a great deal of benefit from employing these categories of labourers, who, at least to a much greater extent than workers from higher castes, were accustomed to unhealthy living conditions and were used to working under difficult circumstances. For this reason especially they were considered more suitable. They would complain less. Also, they were often completely illiterate. Their total lack of education implied *inter alia* a more respectful and obedient attitude towards their superiors. And, in contrast to those who observed the religious rules of purity more assiduously, low-caste labourers did not look down on the European planters for their eating (meat) and drinking (alcohol) habits.

A second feature of the labour force was that it was mainly workers from the Madras Presidency who came to the plantations in Travancore, though especially in the early phase of the planting industry the number of Travancorean labourers was not negligible. As noted before, Mateer states that many Travancoreans obtained employment on the coffee estates situated in south Travancore.[127] On the basis of unpublished sources, Kooiman arrives at the same conclusion.[128] Lovatt, focusing on Peermade during the coffee period, states that Harry Baker brought up labour from Kottayam. She believes that most of the workers came from Travancore in these years.[129]

Later the bulk of the labour force who were employed on the Travancore plantations came from outside the area. This is reflected in the various censuses. To take a few examples, in the Cardamom Hills 59 per cent of those interviewed in 1881 mentioned Tamil as their vernacular. In 1901, 75 per cent gave the same answer. Simultaneously, the percentage of inhabitants of the Cardamom Hills considering Malayalam their native language decreased from 36 to 19 (see also Table 4.15). It should, however, be noted that a Tamil-speaking person did not necessarily originate from the Madras Presidency. In the southern tip of Travancore too Tamil was the dominant language, so these figures serve primarily as an *indication* of the number of non-Travancorean labourers working on the Travancore estates.

127 Mateer 1883: 236.
128 Kooiman 1989: 137.
129 Lovatt 1972: 12.

Nevertheless, there should be no doubt that, especially in the last decades of the nineteenth century, most plantation labour originated from the Madras Presidency. In 1896 the South of India Planters' Enquiry Committee gave evidence of this when it reported that of the labourers working at the Travancore plantations 60 per cent were imported from British India, 20 per cent had come in from Mysore, while the rest was local, i.e. Travancore, labour.[130]

TABLE 4.15

Mother Tongue in the Cardamom Hills in 1881, 1891 and 1901[131]

Year	*Total*	*T*	*M*	*Te*	*H*	*K*	*S*	*Ma*	*A*	*E*	*OE*
1881	6,228	3,683	2,253	250	20	7	-	-	-	14	1
	(100%)	(59%)	(36%)	(4%)							
1891	11,413	6,688	4,377	160	44	34	3	-	3	67	37
	(100%)	(59%)	(38%)	(1%)						(1%)	
1901	21,589	16,165	4,194	832	66	52	29	11	5	222	12
	(100%)	(75%)	(19%)	(4%)						(1%)	

T= Tamil; M = Malayalam; Te = Telugu; H = Hindustani; K = Kanarese; S = Singhalese; Ma = Malay; A = Arabic; E = English; OE = Other European (includes German 1891: 1, French 1901: 1, Portuguese 1901: 11)

The gradually swelling proportion of plantation labourers arriving from the Madras Presidency on comparison to those who came up from the Travancore plains was related to the nature of the labour required. When coffee had been the most important plantation crop, the labour required was mainly seasonal. The coffee picking season lasted only three or four months, from August till November. This enabled many agricultural labourers, especially those employed in toddy-tapping in south Travancore (season March–August), to alternate their main occupation with that of plantation labourer.[132] Tea on

[130] RSIPEC 1896: 94.

[131] *Census* 1881: 89 and 265–7; *Census* 1891, part 2: 535–7 and 546–7; *Census* 1901, part 3: 260–76.

[132] Kooiman 1984: 195; cf. Mateer 1883: 225.

the other hand required more permanent care and, as we observed earlier, involved nine months of work: from June or July till March or April. This made it impossible for agricultural labourers to combine their original occupations with estate work. The figures demonstrate that Travancoreans preferred their work on the plains, whereas most labourers in the Madras Presidency had no option other than to accept plantation labour.

Indubitably, many planters preferred Tamil workers to Malayali ones. Tamils were easier to discipline, presumably as these workers had fewer alternative sources of employment and food and felt more alienated, and thus more vulnerable, in the Travancore highlands than their Malayali counterparts. A European planter who managed an estate in central Travancore, however, explained the planters' preference in unvarnished racist terms. He said that 'on the plantations on Peermade, where the Tamil and Malayali coolies work together, the contrast between the two races is most striking'. 'The Tamil', the planter explains, 'possess a stronger physique and are in consequence better workers, they are more intelligent, honest and regular at work; and more trustworthy as a rule than their Malayali brethren'. Nor did he stop here in his comparison of the 'two races'. He compared the 'Malayali brethren' to animals. His conclusion is shocking. In his view the Malayalis were not only inferior to Tamils, they were on a par with insects and had less value than camels. In what was probably a characteristic feature of the nineteenth century European planters, he presented his opinion in a self-confident, almost scientific way. He averred that 'the low-caste Malayali will eat like the lesser white ant (a destructive white insect which eats without coming to the surface); he will keep in his line [the labourers lived in so-called "coolie-lines"] and devour in a day or two a whole week's rice'. He asked himself: 'the wonder is how they survive the gorge'. Answering his own question, he went on to state that 'the periods of repletion and starvation from which they suffer seem to have produced an internal system resembling that of the camel – without the endurance of that useful animal'.[133]

[133] *Planting Opinion* 1896, no. 11: 19.

Leaving aside caste and ethnicity, religion was the third main feature which shaped the work force. The predominant religious group in the plantation districts was Hindu. In the Cardamom Hills 98 per cent of the population was labelled Hindu in 1881. The gradual increase in the number of Christian workers in the same region provides a great deal of food for thought. In the period 1881–1901 alone the number of Christians rose from 81 to 3,286. In relative terms the gain in weight of the Christian share is even more impressive. This rose in the space of twenty years time from 1 to 15 per cent of the total population (see Table 4.16). In south Travancore many plantation workers were also Christian.[134]

TABLE 4.16
Religion in the Cardamom Hills in 1881, 1891 and 1901[135]

Year	*Total*	*Hindus*	*Christians*	*Mohammedans*	*Tribal*	*Buddhists*
1881	6,228 (100%)	6,113 (98%)	81 (1%)	34 (1%)		
1891	11,413 (100%)	10,235 (90%)	833 (7%)	345 (3%)		
1901	21,589 (100%)	16,026 (74%)	3,286 (15%)	347 (2%)	1,704 (8%)	226 (1%)

Many low-caste labourers, either before, during or after the period in which they were employed on an estate, were attracted by the Christian message. We have already seen that especially in south Travancore LMS missionaries converted many low-caste Travancoreans to Christianity.[136] The Church Missionary Society (CMS) was active in the region of Peermade/Vandiperiyar. Lovatt mentions that in 1887 Rev. Henry Baker, noting that most labour was enlisted outside Travancore, proposed

[134] Kooiman 1989: 138.
[135] *Census* 1881: 89 and 162; *Census* 1891, part 1: 402–3; *Census* 1931, part 1: 342–4; and cf. *Census* 1901, part 1: 118–22. In 1901, the Hindus numbered 17,730 according to the 1901 *Census*, which was divided into 16,026 Hindus and 1,704 Tribals according to the 1931 *Census*.
[136] Kooiman 1989.

establishing a Tamil Cooly Mission in Peermade. Acting upon his advice, in 1888 the CMS engaged Rev. Nallatambi, a Tamil clergyman who had previously worked in the Nilgiris, a plantation district situated in the Madras Presidency.[137] In 1890, Rev. S. Vethamanickam of the American Madura Mission Church became the pioneer evangelist in north Travancore.[138] The following year, the LMS also sent an Indian clergyman, named Samuel, to the plantations in the Kannan Devan Hills.[139] It needs no great stretch of imagination to understand why the Christian missionaries found such a fruitful field for conversion among low-caste Hindus, in particular those who worked as plantation labourers. Those who were despised, rejected and humiliated were taught about God's universal love for all mankind. Deliverance from evil, the call to start a new, more dignified life, and the assurance that all human beings possessed their own, inalienable worth as creatures of God, suited the aspirations of those low-caste people who cherished the hope of a better future.[140]

Although some plantation owners encouraged or at least did not object to the missionary activities (like the (ex-)missionaries and the converted Shanars in south Travancore, and Robert and George Baker in central Travancore), many planters did all that they could to prevent the recruitment of low-caste, Christianized labourers. For instance, a European manager, who called himself 'a much disgusted planter', had no doubt that 'the Native Christian is usually a scamp, lazy and insolent'. He also explains why 'a Native Christian is mostly below the average of his heathen friends'. He accuses them of drinking too much and putting themselves on a par with the planters, both undesirable features which had been put in their heads by the missionaries. He also states that 'the missionaries do not seem to be able to get hold of the higher castes and a good thing, too, in some ways'. He points out that a person of high caste is forbidden, like a Muslim, to touch alcohol. Once they become Christians they start to drink but know

[137] Lovatt 1972: 12.
[138] Martin 1930: 35.
[139] Mackay 1953: 313.
[140] Kooiman 1989: 83.

neither how to hold their liquor nor when to stop; 'they don't and seemingly never will'. 'You may, speaking roughly, say all Native Christians drink, and too much too.' A problem of an even more fundamental nature for the planters, 'his religion put him on an equality as it were with his master'. The manager claimed that 'the converts hear the mission people talking of planters as an unregenerate and unconverted lot of infidels, and morally pat themselves on the back and think they are better than we are'. This, he said, engendered reciprocal dislike and provided more than enough grounds for not employing Christian labour. In the peroration to his diatribe he concludes that 'Now all other things being equal, I'd rather have Christian coolies, but as they are not equal, give me the heathen who goes and worships his swami under the big tree and believes in him. I want no "Kistin boy", not by any means'. Apparently, he was trying to represent the view of many more planters for he concluded thus: 'And so say all of us'.[141]

Gender was a fourth major factor within the plantation population. Men as well as women were recruited to labour on the estates. The figures for the Cardamom Hills indicate that out of every five plantation workers three were males and two were females (see Table 4.17).

TABLE 4.17
Sex Ratio Cardamom Hills in 1881, 1891 and 1901[142]

Year	*Total*		*Males*		*Females*	
1881	6,228	(100%)	3,345	(54%)	2,883	(46%)
1891	11,413	(100%)	6,714	(59%)	4,699	(41%)
1901	21,589	(100%)	12,738	(59%)	8,851	(41%)

[141] *Planting Opinion* 1897, no. 11: 351–2. Also in other plantation districts the role of the missionaries was not unambiguous. In the case of Assam, for instance, (some) missionaries were aiding the planters directly by encouraging labour migration to the 'Christian' tea estates. At the same time, however, (other) evangelists publicly criticized the various coercive and other 'non-Christian' strategies and policies of the labour recruiters, the planters and the colonial government (Chattopadhyay 1987: Chapter 6).

[142] *Census* 1881: 89 and 162; *Census* 1891, part 2: 535; *Census* 1931, part 2: 4–5.

Unfortunately, there is no direct information available which explains why male labourers in particular accepted plantation work. However, it is not difficult to understand that for a woman carrying a child, the journey to the estates – by foot, uphill, and through thickly forested jungle – was almost impossible. Women also stayed behind to look after the children, for whom, many families realized, there were no advantages whatsoever in leaving for the plantations.

This did not stop the planters, with the help of their kanganies, from trying to recruit whole families. The Peermade planter, whom we have already met, mentioned as additional reason for his preference for Tamils over Malayalis: 'They come up with whole families at a time and remain on the estates for a year'.[143] The recruitment of whole families had many advantages for the planters. Apart from reducing the shortage of labour, the total amount of advances and wages could be cut down considerably. Women and children were offered a relatively small advance (see also Table 4.13) and were paid less (see also Table 4.11), although their labour was as indispensable to the planters as that of male workers. The planters'

TABLE 4.18

Language/Religion and Sex Ratio in the Cardamom Hills in 1881, 1891 and 1901[144]

Year	*Total*	*Tamil*		*Malayalam*		*Muslims*		*Christians*	
		M.	*F.*	*M.*	*F.*	*M.*	*F.*	*M.*	*F.*
1881	6,228 (100%)	1,935 (53%)	1,748 (47%)	1,238 (55%)	1,015 (45%)	3,251 (56%)	2,862 (44%)	64 (79%)	17 (21%)
1891	11,413 (100%)	3,835 (57%)	2,853 (43%)	2,687 (61%)	1,690 (39%)	5,877 (51%)	4,358 (49%)	616 (74%)	217 (26%)
1901	21,589 (100%)	9,098 (56%)	7,067 (44%)	2,857 (68%)	1,337 (32%)	9,231 (58%)	6,795 (42%)	2,207 (67%)	1,079 (33%)

143 *Planting Opinion* 1896, no. 11: 19.

144 *Census* 1881: 89, 162 and 265; *Census* 1891, part 1: 402–3, part 2: 535–7; *Census* 1901, part 3: 14–17 and 260–76.

preference for family labour to that of individual workers contributed to their choice for Tamil over Malayali labour and for Hindu over Christian workers. Tamil and Hindu men, to a greater extent than their Malayali and Christian counterparts, were accompanied by their wives, sons and daughters.[145] This is reflected in Table 4.18.

Finally, age was the fifth factor which influenced the nature of the work-force. Labourers between the ages of fifteen and forty were especially attracted to the estates, as can be seen in Table 4.19. In 1891 and 1901 a comparatively large number of people in that age group inhabited the Cardamom Hills.

TABLE 4.19

Age in the Cardamom Hills and in the Overall Travancore State in 1891 and 1901[146]

Age Group	*1891*		*1901*	
	C. Hills	*Travancore*	*C. Hills*	*Travancore*
0–5	9.4%	12.6%	8.1%	13.2%
0–15	27.6%	36.2%	33.3%	38.2%
15–40	55.7%	42.6%	54.3%	42.2%
40–60	13.7%	16.3%	11.4%	15.4%
60–	3.0%	4.9%	1.1%	4.2%

The Criminal Breach of Contract Act (1865). After they accepted work on a plantation, the mobilized labourers found themselves once more immobilized, enmeshed in a new set of bonds. Although freed from landlords and moneylenders, they were enslaved by their new employers. It should by now be obvious that the advances offered by the recruiters were often used immediately by the workers to pay off extant debts, to meet basic needs, and to defray expenses related to family affairs. From a practical point of view, the advances had to be accepted to pay for the journey to the plantations, particularly for food.

145 This phenomenon is fairly difficult to explain. Probably differences in economic circumstances, cultural values and geographical distances affected the choice of relatively large numbers of Tamil and Hindu women to leave for the Travancore estate regions.

146 *Census* 1901, part 1: 143 and 147.

Thus, even before having reached the estate, the contracted labourers were indebted to the kanganies, who, as seen above, were in turn bound through contracts and debts to their superiors: the planters.

Thus already indebted anew, the financial situation of most labourers did not improve during their period of contract. Most kanganies were in fact jobbers; apart from the recruitment of workers they supervised 'their' gang of labourers on the estates. This meant that the kanganies, who as well as their fixed pay as overseers received a commission on their gangs' earnings at the end of the year, derived a great deal of benefit from keeping their workers indebted. As long as the plantation labourers had to redeem their debts, the kanganies could force them to work longer hours than initially agreed, coerce sick workers to fulfil their daily tasks in the fields, and detain labourers beyond the period of contract, all of which contributed to higher wages paid by the planters, of which the kanganies obtained their share. In order to reinforce the debt slavery, the workers received a weekly food issue on credit accompanied by small cash advances. Only after the workers were allowed to return home, which often took place only months after the expiry of their contracts, did they receive the balance of their wages, in most cases paid out by the kanganies themselves. At these moments, the jobbers frequently cheated the often totally illiterate labourers and kept part of the workers' wages for themselves. To obtain an overall picture, it is essential to be aware that the planters' control, through the kanganies, over the labourers' debts, was manipulated in accordance to the seasonal labour requirements. From June/July till March/April about four times as many workers were needed than during the off-season.[147]

To strengthen their grip on the advance system, the European planters in Travancore had successfully convinced the Dewan to introduce a labour law similar to the British Indian Workman's Breach of Contract Act (Act XIII of 1859). This led to the passing of the Criminal Breach of Contract Act (Regulation I of 1040), which was indeed largely modelled on the legislation in force in British India, in 1865. As a well-informed Travancore

[147] Griffiths 1967: 403; Soutar 1976: 22; Tharian George and Tharakan 1986: 210; Uma Devi 1989: 80–4; Ravi Raman 1991: 248.

bureaucrat-cum-historian remarked later, 'its object (. . .) was the protection of the employer against the workmen, of capital against labour'.[148] This is immediately obvious from an examination of the legislation itself. Right at the beginning, Article 1 stated that employers could report any breach of contract to any criminal court, 'and such court shall thereupon issue a summons or a warrant as it shall think proper'.[149] If the accused was found guilty of the allegations, either the money advanced was required to be repaid, or the terms of the contract had to be met, and if the individual concerned failed to do so, he/she could be sentenced to imprisonment, though not exceeding a term of three months (Section 2). However, the basic purpose of the act was not to punish, but to compel performance. Apart from defending the interests of the planters against those of the kanganies and the latter against those of the workers, the Criminal Breach of Contract Act did not afford the employees any form of legal protection against their employers.[150]

148 Velu Pillai, vol. 4, 1940: 340.

149 Article 1 laid down: 'When any artificer, workman, or labourer shall have received from any master or employer resident, or carrying on business in Travancore, an advance of money on account of any work which he shall have contracted in writing to perform or to get performed by any other artificers, workmen, or labourers, if such artificer, workman, or labourer shall wilfully and without lawful or reasonable excuse neglect or refuse to perform or get performed such work according to the terms of his contract, such master or employer or any such person as aforesaid, may complain to the nearest Zillah Criminal Court, or to the Criminal Court within the jurisdiction of which the contract was entered into or violated, and such court shall thereupon issue a summons or a warrant as it shall think proper, for bringing before it such artificer, workman, or labourer, and shall hear and determine the case'.

150 The Criminal Breach of Contract Act, in: *Travancore Government Gazette* 4 February 1865. In the eyes of the law, the positions of the labourer (employee/non-European British subject), the kangany (employer and employee/non-European British subject) and the planter (employer/mostly European British subject) were fundamentally unequal. For example, special provisions were made for the trial of European British subjects committing offences in Travancore. The Royal Proclamation of 28 May 1875 states that 'We do hereby divest Our ordinary Criminal Courts of every grade of jurisdiction to try offences committed by European British subjects, and do further declare that such jurisdiction shall (. . .) be exercised exclusively by Special magistrates, being European British subjects' (Section 1). Paradoxically enough, upon conviction the special magistrates were authorized to pass any sentence though

A law enforcement agency was required to track down, arrest and sentence absconding labourers. Therefore, the planters urged the Travancore government to install a police force and to appoint magistrates in the plantation districts. Their cry having been heeded, the labour legislation was fully backed up by the executive and judicial agencies of the state. In Peermade (central Travancore) a Superintendent and Magistrate of the Cardamom Hills was installed. The government did not have to seek far for appointees; solely Europeans closely connected with the plantation industry (like Maltby[151] and J.S. Sealy) were selected.[152] Even planters, like J.D. Munro, were allowed to combine their task as estate manager with that of judge. For a very short period, the European planters in the Kannan Devan Hills jealously watched the situation in central Travancore. As requested by the secretary of the North Travancore Land Planting and Agricultural Society,[153] a magistrate was installed in 1887 in Devicolam as well.[154]

---

not exceeding three months' imprisonment or fine up to one thousand British Rupees or both (Article 3 in: APT, vol. 1, 1935: 46-7). Implicitly, the breach of a contract by a labourer or by a kangany was judged equally severely as, for example, murder by an English planter.

151 Unfortunately, it was impossible to find out whether the Maltby mentioned in CSF. No. 1681 and CSF. No. 7408, the Superintendent and Magistrate of the Cardamom Hills in the late 1880s, was the same as or related to F.N. Maltby, the British Resident in the early 1860s. If it was indeed one and the same person, it shows the great importance the Madras government accorded to this specific function in Peermade by supporting the appointment of such a politically influential person to this particular spot.

152 CSF. No. 1681; CSF. No. 2368; CSF. No. 7408; CSF. No. 11746.

153 CSF. No. 1681. In 1887, the secretary of the North Travancore Land Planting and Agricultural Society wrote Dewan Rama Rao that 'the Cardamom Magistrate resides the greater part of the year at Peermade between which place and Devicolam there is no sort of communication. The Committee venture to suggest that two Cardamom Magistrates should be appointed, one residing always at Peermade (. . .) and one at Devicolam (. . .). The High Range Devicolam and the Unjeenaad Valley with its many villages have at present no magistrate or police. Crimes are frequent – last year by the murder of three people by Pulliars who were arrested after much difficulty – and at present the case of Krishna Iyer which is before Mr Maltby – might be cited as instances' (letter of the secretary of the North Travancore Land Planting and Agricultural Society to Dewan Rama Rao on 8 June 1887, in: CSF. No. 1681).

154 Martin 1931: 11.

Other Disciplinary Measures. Apart from the 'kangany system – indebtedness – Criminal Breach of Contract Act' configuration, labourer discipline on the estates was enforced by the Coffee Stealing Prevention Act of 1879 (Act II of 1054). In the regulation itself, the Travancore government indicated the initiators and main beneficiaries of the legislation first. 'The Coffee Planters in the country having complained of the losses they were subjected to from theft of their produce, a law was passed on the lines of the British Indian Act to better protect their interest'. This having been established, the purpose of the regulation was made more explicit. 'The main feature of the Act is that possession of Coffee, in the region of the Coffee Estates, by other than planters is regarded as *prima facie* evidence of illegal possession, unless the provisions of the Act have been conformed to or the Coffee is intended purely for home consumption. Even in the latter case, the purchase cannot be from a "labourer" in a Coffee Estate'.[155]

Even more pertinent than the legal enforcement was the control of labour achieved by a clear-cut division of authority and responsibilities. The management, which on large holdings included some assistant managers, supervised both staff and field workers. If the manager was not at the same time the proprietor of the plantation, the management was held responsible by the owner or, in the case of a company, by the board of directors. The staff embraced those who worked either as office clerks, tea makers, or administrative field supervisors. The field workers, including the kanganies, were engaged in clearing jungle, planting, picking and plucking, weeding, pruning, factory work, and the construction of roads and buildings.

Apart from their clearly delineated responsibilities, the three groups were mostly obviously distinguished in terms of places of origin, social backgrounds, and lifestyles.[156] It goes without saying that the management was mostly European, especially on

[155] TAR 1054: 1; Coffee Stealing Prevention Act dated 14 August 1879 (Act II of 1054), in APT Vol. 1, 1935: 50–5.

[156] For that reason, I have chosen not to split up the employees into the four categories – management, staff, sub-staff, and 'coolies' – as done by Bhowmik (cf. Bhowmik 1981: 79–84). From a sociological point of view especially, both sub-staff (labour recruiters-cum-supervisors) and 'coolies' (factory and field workers) can be seen as one category. Workers could 'easily' become kanganies/sirdars, but social mobility among (a) management, (b) staff and

European-owned estates. In staff functions well-educated Travancore Christians, either high caste (Syrian Christians) or low caste (Shanars), were usually employed. Even in those sporadic cases in which the management was not in the hands of a European, this position was held by a member of the Christian community. Jeffrey suggests that many Syrian Christians were employed in high positions in the Peermade district, though he gives no precise information about whether they worked as managers or belonged to the staff. The same author also states that a Nayar converted to Christianity managed an estate belonging to the Baker family.[157] Mateer observes that apart from becoming overseers and confidential clerks, some LMS-educated, low-caste Christians were appointed managers on European estates as well.[158] The foregoing discussion had shown that most kanganies and labourers were illiterate, low-caste Tamils. This points to the fact that the social hierarchy within south India was not merely coincidentally reflected on the European estates, but was intentionally reinforced by the European planters. To prevent the staff from siding with the field workers against the management, the European planters manipulated differences in place of birth, religion, caste and education to tighten their grip on the labourers.[159] This strategy was not only restricted to the European-owned estates. Apart from fellow Shanars, the local Shanar planters mostly employed illiterate, lower-caste labourers to work on their plantations.[160]

Lastly, in order to impose their rule as forcefully as possible the planters also tried to keep the estates, which were often situated miles from the nearest village, totally isolated from the outside world. This ensured that the possibilities of labourers and kanganies leaving the plantation without informing the planter were considerably restricted. To assist the planters in their goal of keeping the workers in captivity, the Travancore government refused to grant land in the immediate surroundings

---

(c) workers including kanganies/sirdars was much more sporadic and mostly totally out of the question. Regarding the division of labour into different categories, see also the Mathew versus Mani debate (1984).

[157] Jeffrey 1976: 126.

[158] Mateer 1883: 236.

[159] Cf. George Tharian and Tharakan 1985: 15.

[160] Cf. Mateer 1883: 236.

of the estates to non-planters.[161] In fact, giving no permission to local cultivators to start agriculture in the high ranges served two goals. Apart from the control of labour, land was reserved for future plantation production.

4.2.4 Uncaptured Contract Coolies

Though successful to a very large extent, the planters did not find the bondage of labour to the estates totally satisfactory. In some respects their hands were tied. If, after having received an advance, a kangany took no steps to recruit labour for the particular planter, the courts did not consider the kangany to be subject to the penal provisions of the Criminal Breach of Contract Act. As H.M. Knight openly complained during a planters' meeting: 'a man gets an advance for coolies from one planter, he goes back with a gang and starts work on another place in order to get a reputation as a good man who gets coolies without any advances. They know very well that we are powerless to touch them'.[162] Knight, however, was being economical with the truth. The kangany could be judged under civil law, but these suits often resulted in the demand that the particular kangany should pay off his debts. In contrast to the labourer, the kangany could not be legally forced to fulfil his task, i.e. to recruit labour. This was not much help to the planters who were often desperately short of labour.[163]

Despite their debts and contractual obligations, individual workers often fled the plantation districts. Krishna Iyer, the

[161] Sealy, the Superintendent and Magistrate of the Cardamom Hills, also received strict orders to this end. Revealingly, he had to correct the mistake after he had granted E. Yagappa Pillay and J. Wood of Kumili some acres of land for the cultivation of paddy. Sealy explained the Dewan: 'I received an application for a grant of land for Paddy Cultivation at Kumili and caused enquiries to be made and as there was no objection I forwarded it to the Dewan Peishcar recommending that the land may be granted. No reference was made to Government on the subject, because the village of Kumili is not in the Peermade District and the land applied for is 10 miles from the nearest estate' (letter of the Superintendent and Magistrate of the Cardamom Hills to the Dewan dated 9 March 1897, in: CSF. No. 11746).

[162] 5th Annual General Meeting of the UPASI, Bangalore, 8–11 August 1898, in: UPASI Proceedings.

[163] Griffiths 1967: 397–8.

Tahsildar of Changanacherry who has already been mentioned, stated in 1896: 'Many of the labourers while on the hills get sick and they naturally desire to be by the side of their relatives and find their way home'.[164] At the same time, Krishna Iyer noted that many other discontented labourers were not prepared to take the same risk. Apart from being scared of being caught, they feared loss of their wages should they leave the plantations clandestinely.

Kanganies and workers together escaping their particular estate represented a problem of a more serious nature for the planters. It often meant irrecoverable debts and an immediate labour shortage. In many cases, the kanganies left the plantation in order to seek higher wages and better living conditions on a nearby estate. It was by no means unknown for planters, seriously short of workers, to take the initiative in such a situation. they contacted, either personally but more often through others, influential kanganies, already bonded to fellow planters, to induce them to come over to their estate.[165] Although planters with relatively healthy estates and good wages were comparatively well off, the fact that kanganies and labourers could search for the best employer handicapped all planters by driving up the price for labour. In efforts to counteract this chicanery, district planting associations frequently discussed rules which were intended to prevent the poaching or crimping of workers.

The premature departure of kanganies with their labourers from Travancore territory before the expiry of their contracts represented perhaps the most important labour problem with which planters had to deal. No legal means existed by which the planter could secure their extradition in order to enforce their contractual obligations.[166] Most probably this is the reason

[164] Quoted in: Uma Devi 1989: 82. See also page 87, note 121.

[165] This quite frequently occurred in the Peermade district. Lovatt stated: 'Under these circumstances [a shortage of labour due to among others bolting labourers], every planter did what he could to hold on to his labour, and to gather in new recruits, even if this meant not enquiring too closely from where they came. Poaching labour, or crimping as it was called, was universal and the cause of much bad blood' (Lovatt 1972: 36).

[166] Knight also complained about this. He told his European colleagues in 1898: 'our difficulty is to get sufficient gangs of coolies. We have maistries who are up to all sort of tricks. They take money in Travancore and go to

A. Martin, a planter in the Kannan Devan Hills, hated kanganies the most. As he once said during a planters' meeting, speaking about contractors: 'I think "sordid, unfeeling, reprobate, degraded, spiritless, outcaste" accurately describes them'.[167]

Indeed, 'absconding' led to large sums of money being written off annually by the estates under the heading of 'bad debts'. Fleeing with the advance, it must be said, was easier in (south) India than overseas: indentured Indian labour abroad often had no chance of returning home.

4.3 Travancore: 'A Model State'

4.3.1 Government Income Minimized

Besides land acquisition and labour control, being able to cut down on costs was another prerequisite for success for the planters. A detailed study of the state revenues derived from and expenditures made on account of the plantation sector will shed more light on this.

To start with, in the Memorandum of March 1862 the land tax was fixed at twelve annas per acre (Article 1), 'but should the applicant so desire, land for planting Coffee will be granted free of assessment for five years, on condition that quarter of the land shall be cleared and planted within the first three years' (Section 6). The export duty on coffee was ratified at 5 per cent *ad valorem* (Clause 10).

In the mid-1870s, the planters successfully convinced the

British territory, and we never see them again. He knows the law quite as well as the planter and understands that as matters stand at present he cannot be caught' (5th Annual General Meeting of the UPASI, Bangalore, 8–11 August 1898, in: UPASI Proceedings).

[167] 11th Annual General Meeting of the UPASI, Bangalore, 22–4 August 1904, in: UPASI Proceedings. The difficulties of the Travancore planters with regard to the recruitment and control of workers were certainly not unique. Most planters situated in south India had to take on the same problems. Despite the similar nature of their labour problems, it was only in the 1890s that the planters united at a south India level (see next chapter). The lack of good communications between the various planting districts and the often very individualistic attitude of the pioneering planters contributed especially to preventing the establishment of a south Indian planters' lobby at an earlier stage (Mayne 1953: 1–4).

government that the level of the export duty was too high, especially when it was compared to other coffee-exporting countries. Consequently, the government altogether abolished the coffee export duty in May 1875, though on the condition that the land tax might be raised in the near future.[168] Shortly after this, the sircar presented the planters with a choice: either no export tax but an increase in the land tax to two rupees per acre, or the reinstatement of a moderate export duty of 2.5 per cent on coffee. The coffee leaf blight, which caused an enormous drop in production precisely at this time, simplified the planters' dilemma considerably. Logically they selected the latter option, which was made law in May 1877.[169]

The planters' liabilities were reduced once more in July 1877. As mentioned earlier, those parts of the estates which were used to accommodate people (homesteads) or cattle (farmsteads) acquired a different status under the Grass Land Rules then promulgated. Land tax on these particular plots was limited to four annas per acre (Article 20).

In early 1883, the Travancore Planters' Association requested the Travancore government to lower their fiscal obligations even more. The planters' representative argued that the coffee industry was in a very depressed state and that the land tax weighed heavily on their shrinking resources.[170] The planters' lobby won the day and the land tax for the Ashamboo Hills and Peermade for 1882 remained in abeyance.[171] The government also approached the planters asking whether they would agree to an arrangement by which the levying of land tax would be postponed for an unspecified period of time and by which the export duty would be restored to 5 per cent. The planters replied

168 Proclamation of 9 May 1875, in: TAD 1879: 76.

169 CSF. No. 1858. Proclamation of 12 May 1877, in: TAD 1879: 76.

170 He continued his plea thus: 'not a few Estates have given crops realising but one quarter of the cost of producing them, and I could mention one or two where the value of crops scarcely covered the cost of picking and curing them, allowing nothing for the maintenance of the Estates. If another such disastrous year occurs the abandonment of Estates will be very general. (. . .) It is the custom of His Highness' Government to remit tax in the case of Ryots who have had a bad crop, and we may well ask that the same consideration may be shown to us' (CSF. No. 1760).

171 CSF. No. 1797.

unanimously in favour of the government proposal, which was enacted in the Proclamation of 1 October 1883.[172]

The planters settled in the Kannan Devan Hills, which enjoyed a separate legal status because of the Pooniat Concessions, also managed to obtain a reduction in their financial obligations during the mid-1880s. Munro, one of the concessionaries, enlisted the aid of the British Resident in their plans. He wrote to him on 12 March 1885: 'I have been endeavouring to get a reduction of the Pooniat tax and from the Rajah's early correspondence hoped some reduction would be agreed to but regret to say have failed to obtain it'. Apparently, in the days which followed the British Resident used his influence to good effect in favour of these planters as the Dewan, V. Ramiengar, wrote to Munro only seventeen days later: 'all the lands included in the grant will be free of assessment for 6 years'.[173]

With the successful development of tea in the 1890s, the plantation sector entered a new phase. The prosperity of the planters did not go unnoticed by the government. The Dewan intended to introduce a land tax of four annas per acre, though the British Resident was in favour of a levy of two annas. The latter even reformulated the text of the draft legislation. 'The draft Proclamation forwarded by you appearing in some measure defective, I have prepared another draft which may possibly appear preferable to you'.[174] The Dewan could not disregard the view of the British Resident. The Proclamation of 18 May 1891 imposed a land tax of two annas per acre.[175]

Only four years later, the financial responsibilities of the planters were again slightly increased. The Government Notice of 12 August 1895 raised the land tax on *newly* sold lands to twelve annas per acre,[176] and three years later the Rules for the Sale of Waste Land dated July 1898 obliged *all* estates to pay twelve annas per acre. Countervailing this rise, no export duty on tea was ever imposed during these years. The planters convinced the government that the introduction of such a levy

[172] CSF. No. 1760.
[173] Papers regarding the Kannan Devan Hills, in: CDS. RF. 546.
[174] CSF. No. 1797.
[175] CSF. No. 11629.
[176] CSF. No. 11629.

would be 'ill-timed', offering among their reasons 'owing to present prices realised by sale being barely remunerative'.[177]

Analyzing the level of the land tax and export duty, two features immediately stand out. The most striking is that the fiscal obligations of the planters, especially when compared to those of other agriculturists, were extremely modest. The Dewan realized that this was somewhat unfair. He showed the British Resident in 1878 that:

> this assessment is very low compared to the tax on other trees. Thus an acre of coffee pays 12 annas while the same extent of land planted with coconuts would give the Government 21 1/2 rupees at 300 trees to an acre and a tax of 2 chuckrams on each tree. The produce of the coconut tree pays in addition an export duty of 5% whether as fruit, or oil and coir, while coffee pays only 2.5% on a tariff valuation determined at less than half the market value. The duty thus virtually amounts to only 1 per cent. Thus while an acre of coffee (taken to produce 3 cwts at 20 Rs an acre) pays Government only 2 Rs a year in tax and duty, that extent of coconut cultivation brings into the treasury no less than 30 Rs from the same sources, not taking into account the durable duty on coir and oil, but reckoning only the fruit (6,000 nuts a year on an acre) as exported at the tariff valuation of 30 Rs per 1,000.[178]

The Revenue Settlement Proclamation of 24 February 1886 provides another indication of the extremely low level of taxation of coffee and tea lands. In this year farmers working lands fully cultivated with coconut, jackfruit, areca, palmyra, tamarind, punnai and mango all paid at least one rupee per acre (see Table 4.20), which was a higher tax than was ever levied on lands cultivated with coffee or tea in the second half of the nineteenth century. As has just been referred to, the highest tax for plantation lands was only twelve annas per acre.

A second feature is that the fiscal obligations of the planters were always in step with the prevailing situation in the plantation sector. The planters, with the help of the British Resident, managed to reduce their financial liabilities in difficult times, while keeping a tight curb on their fiscal responsibilities in prosperous years. Of course, the reverse is also true. The Travancore

[177] CSF. No. 1858.
[178] CSF. No. 11800.

government increased its income from the plantation sector whenever the planters' financial position improved and whenever the British Resident approved the sircar's decision.

TABLE 4.20
Land Tax in the Non-Plantation Sector in 1886[179]

Tree	*Tax Per Productive Tree in Travancore Currency*	*Number of Trees Per Acre When Fully Planted*	*Revenue Per Acre When Fully Planted in Travancore Currency*	*Revenue Per Acre When Fully Planted in British Indian Currency*
Coconut	1 to 4 chs	100	100 to 400 chs	3 Rs, 9 as to 14 Rs, 5 as
Jackfruit	4 chs	50	200 chs	7 Rs
Areca	0 chs, 8c	800	400 chs	14 Rs
Palmyra	0 chs, 8c	500	250 chs	8 Rs, 13 as
Tamarind	2 chs	50	100 chs	3 Rs, 8 as
Punnai	0 chs, 12c	50	37 chs, 8c	1 Rs, 5 as
Mango	0 chs, 12c	50	37 chs, 8c	1 Rs, 5 as

These two characteristics – the limited amount of government income derived from the planters and the role of the British Resident in manipulating this – is also detectable in the level of the upset price of land sold at the government auctions. Initially, in December 1864 the upset price was fixed at one rupee per acre. Dissatisfied with this, Dewan Seshiah Sastri decided to increase the upset price from one to ten rupees and prepared a draft notification in early March 1874. Before submitting the proposed legislation to the Maharaja, Seshiah Sastri wrote to the British Resident G.A. Ballard that 'I should be obliged for the expression of your views on the subject'. Some days later Ballard replied that in his view 'the step would be a very unadvisable one'. In the same letter he argued that nobody would contest the fact that both directly and indirectly coffee cultivation had been highly beneficial to the state. In what was a veiled threat Ballard argued that 'any considerable

[179] RPT, vol. 1, 1927: 24–49.

enhancement of the upset price of waste lands would tend materially to check extension'. In April the Dewan explained his considerations to the Resident. The raising of the upset price was meant as a antidote to collusions between bidders. Producing figures, the Dewan also argued that he failed to see how his plan would have a repressive influence on coffee cultivation as the 'cost of bringing to maturity an acre of coffee is about Rs 200. The proposed upset price of Rs 10 is only 1/20th of it'. In his reply, the Resident again tried to win the Dewan over to his way of thinking but probably had more than an inkling that his arguments were too weak. Therefore he resorted to threats, writing to the Dewan: 'as the question is one of some public interest I am laying the correspondence before the Madras government'. The Maharaja was shocked by the action of the Resident and advised the Dewan: 'I think the question of raising the upset price of lands sold for coffee cultivation may be deferred for the present'. Convinced of being in the right, and not afraid of the eventual consequences of his actions (see also the last note of this chapter), Seshiah Sastri was finally able to have his proposal passed. On 2 June 1874 the upset price of future plantation lands sold at the government auctions was raised from one to ten rupees.[180] In the late 1890s, the upset prices ranged from ten to forty rupees according to the quality, location and altitude of the plots.[181]

The state revenues from the plantation sector are summarized in Table 4.21. It is interesting to note the discrepancies between the amount of land tax demanded by the government and the actual sum paid by the planters. In this respect, the figures referring to the government auctions are somewhat misleading. To take one example, although the sale of waste lands in 1874–5 amounted to 214,761 rupees, only 126,716 rupees were actually paid in the same year.[182]

Apart from the amounts of money realized through land tax, export duty and government auctions, the planters did not provide the government with many other sources of income.

[180] CSF. No. 1772.
[181] CSF. No. 11684.
[182] TAR 1050: 21.

Planters were allowed to ignore the import duties, in which Maltby, the Superintendent and Magistrate Cardamom Hills, played a fundamental role. He mentioned in a letter that 'by Huzur No PD 37 of 27th January 1882, I was ordered not to interfere with the . . . boxes and postal parcels coming to planters, so that although dutiable articles do pass through the Established *Chowreys*, it is often in such a way that the duty is not recoverable, (. . .) it should be left to the honour of planters (. . .) to declare any dutiable articles'.[183] Moreover, in 1836 all inland transit customs had been abolished and in 1865 Travancore had joined the British Indian system of free trade.[184]

TABLE 4.21

State Revenues from the Plantation Sector (1859–1900); Land Tax, Export Duty and Government Auctions

Year	*Land Tax in Rupees (Demanded Land Tax in Rupees)*	*Export Duty in Rupees (Total Export Value in Rupees)*	*Land Offered for Sale at Government Auctions in Acres (Total Quantity of Land Sold in Acres)*
1859/60		(8,397)	
1860/61		(22,612)	
1861/62		(37,370)	
1862/63		(51,336)	
1863/64		(59,644)	
1864/65		(80,221)	
1865/66		(120,787)	
1866/67		10,577 (211,542)	(3,783)
1867/68		12,156 (243,000)	(2,900)
1868/69		12,567 (251,300)	(700)
1869/70	2,020	17,183 (338,800)	(700)
1870/71	2,880	27,242 (544,800)	(2,065)
1871/72	7,103	29,852 (596,800)	3,800 (2,685)

[183] CSF. No. 7408.

[184] Velu Pillai, vol. 4, 1940: 428–31; Kooiman 1989: 119.

Year	*Land Tax in Rupees (Demanded Land Tax in Rupees)*	*Export Duty in Rupees (Total Export Value in Rupees)*	*Land Offered for Sale at Government Auctions in Acres (Total Quantity of Land Sold in Acres)*
1872/73		41,963 (839,000)	5,444
1873/74		29,723 (584,000)	85,254
1874/75		26,224 (524,000)	214,761 (7,817)
1875/76		non (786,696)	no auction
1876/77		20 (990,058)	no auction
1877/78		20,093 (803,700)	no auction
1878/79	7,708 (17,105)	9,378 (374,600)	no auction
1879/80	7,545 (18,773)	22,084 (833,100)	no auction
1880/81	7,868 (19,316)	14,984 (599,400)	no auction
1881/82	8,329 (25,630)	20,877 (835,200)	no auction
1882/83	4,505 (28,104)	9,728 (389,135)	no auction
1883/84	non	13,565 (225,704)	no auction
1884/85	non	22,305 (446,085)	no auction
1885/86	non	11,061 (221,225)	no auction
1886/87	non	5,431 (100,714)	no auction
1887/88	non	7,950 (159,007)	no auction
1888/89	non	6,545 (130,908)	no auction
1889/90	non	5,158 (103,286)	no auction
1890/91	non	2,102 (42,061)	no auction
1891/92		5,478 (109,508)	no auction
1892/93		3,577 (71,549)	no auction
1893/94		3,459 (69,191)	no auction
1894/95		5,645 (112,909)	no auction
1895/96		2,624 (52,477)	
1896/97		2,648 (52,954)	
1897/98		2,073 (41,467)	
1898/99		5,566 (111,325)	
1899/1900		3,562 (71,239)	

Source: TAR 1035-75.

4.3.2 *Maximum State Expenditures*

The planters may have been let off lightly but the Travancore government spent large sums of money to stimulate plantation development. The experimental Government Garden in Peermade, established in 1862 at the suggestion of the British Resident, cost roughly 3,500 rupees annually.[185] The expenditures of the Survey Department were also considerable. The costs of surveying a tract of land which would be put up for auction amounted to more than one rupee an acre.[186] Resurveys, intended to discourage, or more often to recover, encroachments on government lands, cost twelve annas per acre.[187] The policemen and magistrates stationed in the plantation districts at the request of the planters were also a heavy burden on the state treasury.[188] Nor were these the only public services. At the end of the nineteenth century the Travancore government financed a hospital in the Ashamboo Hills,[189] a telegraph connection with Peermade,[190] and exhibitions of plantation products in Paris, France, and in St Louis, United States of America.[191] The most impressive outlays of all were the huge amounts the state spent on roads to connect the plantations with the coastal areas of Travancore. Financial aid was granted to planters who were willing to construct and maintain roads within the plantation districts, while the Public Works Department connected the south and central planting regions with the western parts of Travancore.[192]

The state expenditures fluctuated. This is revealed in regard

[185] TAR 1040: 54; cf. Griffiths 1967: 158 and Kooiman 1989: 120.
[186] CSF. No. 1772.
[187] CSF. No. 11716.
[188] CSF. No. 1772; IOL&R: MSS Eur D 1204/8.PT.2.
[189] CSF. No. 3928.
[190] CSF. No. 6598.
[191] CSF. No. 10133.
[192] The western outlet road of the Kannan Devan Hills was only completed in the early 1930s. Only after that were the products of this planting district shipped from Cochin. Before 1930, all produce from the northern district, as well as major portions of the south and central regions, were transported overseas through Tuticorin, a harbour situated in British territory, to the international auctions of Colombo and London (CSF. No. 1779; Langley 1962: 83).

to the treatment of those planters who undertook the building and maintenance of roads. At the beginning of the 1870s, the total costs of construction were refunded by the government, but by the end of the 1870s and the beginning of the 1880s, the government recompensed only half of the planters' expenses. At the end of the nineteenth century, apart from the construction costs the maintenance expenditures were also reimbursed.[193]

TABLE 4.22
Total Expenditures Public Works Department and Travancore State (in rupees), and Ratio Total Expenditures Public Works Department/Travancore State, 1863-1900[194]

Year	*PWD*	*Travancore*	*PWD/Travancore*
1863/64	200,688	3999,997	5.0%
1864/65	269,068	4047,734	6.6%
1865/66	333,346	4307,644	7.7%
1866/67	446,201	4359,177	10.2%
1867/68	282,576	4481,231	6.3%
1868/69	366,620	4535,088	8.1%
1869/70	751,743	5440,167	13.8%
1870/71	847,396	5310,187	16.0%
1871/72	1052,509	5493,682	19.2%
1872/73	711,281	5278,060	13.5%
1873/74	590,111	5321,292	11.1%
1874/75	531,707	5323,540	10.0%
1875/76	493,296	5594,814	8.8%
1876/77	515,681	5540,285	9.3%
1877/78	684,765	5615,211	12.2%
1878/79	791,437	5952,888	13.3%
1879/80	755,755	6027,802	12.5%

[193] CSF. No. 2441; CSF. No. 3444; CSF. No. 3468; CSF. No. 12423; CSF. No. 12446.
[194] TAR 1039–75.

Year	*PWD*	*Travancore*	*PWD/Travancore*
1880/81	815,574	6377,675	12.8%
1881/82	476,958	6235,328	7.6%
1882/83	552,628	5920,127	9.3%
1883/84	619,542	6013,122	10.3%
1884/85	606,599	6490,960	9.3%
1885/86	545,983	6462,541	8.4%
1886/87	541,530	6669,158	8.1%
1887/88	609,924	6992,458	8.7%
1888/89	773,870	7003,048	11.1%
1889/90	822,132	7333,891	11.2%
1890/91	767,150	7232,786	10.6%
1891/92	888,533	8715,053	10.2%
1892/93	949,185	7970,748	11.9%
1893/94	823,456	8317,273	9.9%
1894/95	1005,259	8265,528	12.2%
1895/96	890,071	8150,183	10.9%
1896/97	1363,244	8916,723	15.3%
1897/98	1687,840	9507,631	17.8%
1898/99	1407,601	8995,946	15.6%
1899/1900	1218,579	9624,202	12.7%

The same trend can be noticed regarding the outlay of the Public Works Department (see also Table 4.22). In the 1860s and 1870s, a mounting sum, in absolute and relative terms, was spent by the department. In the 1880s, the expenditures dropped considerably. In the last decade of the nineteenth century, the outlay of the Public Works Department again rose to a much higher level. It should be mentioned that at least until the turn of the century the Public Works Department spent a relatively large part of its budget on projects favouring plantation development. As already mentioned in Section 4.2.2, the establishment of the Public Works Department in 1860 was directly connected with the emergence of European coffee plantations.

Two major conclusions can be drawn in this context. To begin with, government expenditure was closely related to state revenue. In periods in which the plantation sector prospered, the state received a relatively high income from land tax, export duty and the sale of lands, while, at the same time, the administration also spent relatively large amounts on items such as the construction of roads intended to encourage plantation development. Conversely, in difficult times for the planters, the government both earned and spent less. Clearly, the planters contributed directly to government measures which stimulated plantation development. This was also mentioned in a report by Rees, the British Resident. He asked himself whether 'the tariff valuation of coffee should not be raised to something approaching actual value, and whether export duty might not be levied on tea if funds run short for road developments'.[195] At no time, and now we arrive at the second finding, did the state profit financially from plantation development. Taking into account the enormous expenses required for roads, the expenditures outweighed the revenues. Up to 1872 the outlay of the Peermade-Kottayam road alone amounted to 249,768 rupees.[196]

4.3.3 Travancore Catches Up

As explained in the previous chapter, the emergence of a European plantation sector in Travancore was a relatively late event. When it did start in the late 1850s, European plantations developed fast. Until 1876-7 the export of coffee rose quickly (Table 4.1). The rapid development is also revealed when the growth in the cultivation of tea in the Princely State is compared with that in the Madras Presidency or south India. Table 4.23 shows the explosive development of tea production in Travancore as compared to that in the Madras Presidency. A comparison of Tables 4.23 and 4.24 shows that Travancore became the most important tea cultivating region of south India during the last decades of the nineteenth century.

[195] R/2/879/41.
[196] TAR 1044-7. Unfortunately, I was unable to indicate the sircar's revenues from and expenditures to the plantation sector in later years. The information concerning the twentieth century was of less value than that regarding the nineteenth century, partially because of the disappearance of categories like 'planters roads' in the *Administration Reports*.

TABLE 4.23

Tea Cultivation in Travancore and the Madras Presidency (1885-1904); Average Area (in acres) and Average Production (in pounds)[197]

Period	Average Area		Average Production	
	Travancore	Madras	Travancore	Madras
1885-89	3,352*	5,558	342,718*	935,369
1890-94	7,186*	5,385	1,735,803*	1,095,590
1895-99	15,682*	7,016	3,505,938*	538,762
1900-04	24,931	7,263	6,107,429	1,171,594

* including Cochin state.

TABLE 4.24

Tea Cultivation in South India (1885-1900); Area (in acres) and Production (in pounds)[198]

Year	Area	Production
1885	6,370	939,000
1890	10,912	1,841,000
1895	16,276	4,394,000
1900	31,308	5,104,000

The development of plantations – relatively fast in comparison with the other south Indian regions – was not entirely favourable for the Travancore administration. Without any doubt, the government's policy towards the European planters contributed to a re-evaluation of the princely state by the British. During the latter part of the nineteenth century, Travancore was frequently labelled 'the Model State' by the British administrators. Official recognition was given in 1866 when the colonial authorities honoured the Raja, and his eventual successors, with the title of 'Maharaja' as a gesture of gratitude. Likewise, in the same year, Madhava Rao, the first Travancore Dewan who supported the emergence of a European plantation sector

197 *Indian Trade Journal* 1921: supplement 10.

198 *UPASI Planting Directory* 1979: 877.

convincingly, was knighted.[199] Apart from these political 'successes', the Travancore government did not achieve many of its other goals. The data in this chapter have shown that plantation development was not financially advantageous to the state treasury; there was a preference for labourers from the Madras Presidency on the estates; and only some Travancoreans opened up plantations.

It was a different story for the Europeans. The comparatively fast development of coffee and tea cultivation in Travancore does indicate that during the second half of the nineteenth century many European entrepreneurs preferred to open up an estate in the princely state to launching an enterprise in another south Indian region. The decision to do so was not taken lightly but because a level-headed weighing up of the pros and cons showed there was good reason to do so. As in directly-ruled areas, in Travancore the administration favoured plantation development. There is ample evidence that the Travancore government played a major role in the planters' efforts to acquire land, control labour and reduce costs. Travancore was more suitable to the planters than the Madras Presidency, particularly in the field of the fraudulent appropriation of government lands. In addition, in Travancore the detrimental Madras Planters' Act of 1903 was not imposed, an issue which will be discussed at length in the next chapter. Clearly, the planters in Travancore, like their counterparts in the Madras Presidency, profited from the colonial regime as the intervention of the British Resident

[199] In 1862, Resident Maltby wrote about Dewan Madhava Rao: 'I have never yet met with any Native of India who has obtained so thorough a mastery over our language, or so intimate a knowledge and application of the modern views of Englishmen in matters of political economy and Government' (quoted in: Jeffrey 1975: 276). In contrast, the English officials were not particularly happy with Seshiah Sastri, the Dewan who tried to counteract some of the planters' extra-legal strategies, like collusion at government auctions, buying of land for speculation, fraudulent encroachments on neighbouring land and destruction of the natural equilibrium. Consequently, the Resident was requested by the Madras administration on 5 April 1877 to 'inform His Highness the Maharaja that His Grace the Governor in Council in no way desires to press on him the renewal of an engagement which is about to expire (. . .); and that His Grace has no objection to offer to the selection of the Dewan Peishkar, Nanoo Pillay, for the post of Dewan' (R/2/879/8). In the same year, Seshiah Sastri was replaced by Nanoo Pillay.

was crucial in many respects. Unlike their fellow entrepreneurs in the directly-ruled areas, the European planters in Travancore did not suffer the disadvantages of the colonial state. In comparison with the Madras administration, the Travancore government was 'weak', namely less able to defend state properties, and remained 'out of range', that is less affected by the increasingly influential (British) public debate regarding labour conditions on estates. These were the reasons due to which the planters opted for Travancore. Here, in comparison with directly-ruled territories, the British Planters' Raj was more complete and remained unaffected longer.

Five

The Rise of a Syrian Christian Planting Elite: Early 1900s–*c.* 1947

The people of Kottayam, or at least a section of it, had waxed rich and prosperous. This was the result of the discovery that the High Range not far away was ideal for the cultivation of rubber. (. . .) The scramble for rubber was confined to the Syrian Christian community. The Nairs looked on with lordly disdain

K.P.S. Menon discussing his visit to his home town in 1922[1]

5.1 A Changing Political Arena

5.1.1 The Dominant Dewan

Perhaps the narrative of 'a king gradually losing his kingdom and the birth of a crown prince' is the best metaphor possible to characterize plantation development in early twentieth century Travancore.

The British planters' Raj did indeed dissolve. The European planters were bereft of most of their prerogatives. They gradually lost their grip on Travancore politics, were less effectual at excluding 'outside', Indian competitors, and lacked the strength to oppose the developments which would finally threaten the 'original' labour regime on the estates. Of these changes, the abolition of the Criminal Breach of Contract Act and the rise of labour unions in the plantation districts were the most important.

While this was taking place, a very powerful indigenous – predominantly Syrian Christian – planting group gradually made its presence felt. The Syrian Christian community expanded quickly in terms of acreage, some Syrians even emigrating to Malabar and other regions outside Travancore to open

[1] Menon 1981: 66. K.P.S. Menon was a successful bureaucrat-cum-diplomat and the first Foreign Secretary of the independent state of India.

up estates. Gaining in confidence, the indigenous planters' lobby successfully protested against the privileges enjoyed by 'foreigners' regarding the allocation of land and access to the world market. Given the group's prominent position in post-colonial south-west India, the rise of the Syrian Christian planters in early twentieth century did greatly resemble 'the ascending of an heir to the throne'.

Just what did lead to the British planters' gradual loss of their prominent place within the Travancore political system during the first half of the twentieth century? Having close and influential connections within the centre of power in the early 1900s, a few decades later they found themselves relegated to the side-lines of the political arena.

The most salient reason for this was the altered positions of the leading political figures. It was a fact of history that during the first half of the nineteenth century the British Resident dominated the Travancore political scene to an overwhelming extent. After the appointment of Madhava Rao as Dewan (1858), however, the relations between British representatives, Travancore monarchs, and Dewans had been much more 'balanced'.[2] In the first decade of the twentieth century another major shift occurred which, from the British planters' point of view, was to exercise a negative influence on the political situation.

Accusing the Travancore government of 'misrule' and the Madras Presidency of 'laxity of supervision' following various palace intrigues and an escalation in communal rivalry in the state, the *Viceroy*, Lord Curzon, changed the colonial policy towards Travancore in 1904. Except for some brief acting assignments, the colonial power appointed highly-anglicized bureaucrats from British India to the post of Dewan, men who had no long-standing personal ties either with the royal family or with other local interest groups. It was now insisted upon – and this would have especially serious repercussions for the planters – that these 'imposed outsiders' were obliged to correspond directly with the Governor of Madras, and, after 1923,

[2] Or as F.N. Maltby, British Resident in the early 1860s, wrote: 'measures are sometimes directed by His Highness to be initiated and sometimes suggested by the Resident, but more frequently brought by the Dewan himself' (quoted in: Jeffrey 1975: 269).

with the Political Secretary of the government of India.[3] This, of course, was intended to ensure that the Dewans remained loyal to the ideas and policies of the British administration. Ominously for the planters, it also meant that the position of the British Resident (called Agent to the Governor General, Madras States, after October 1923 and Resident for the Madras States after April 1937), the planters' longtime political ally, was reduced to that of a spy, a reporting spectator.

Curzon's measure, though drastic, did not seal the British Residents' fate – and thus the European planters' destiny – immediately. At first the latter continued to determine the careers of the Dewans. In this respect, the dewanship of S. Gopalachariar – the shortest of the twentieth century! – is most enlightening. The Dewan, appointed to office in 1906, decided to make a tour through Travancore, including a visit to Mundakayam. Being informed of Gopalachariar's plans, the Mundakayam Rubber Planters' Association prepared an address to be presented to him but the Dewan refused to attend a ceremony organized by the European planters. Disappointed by Gopalachariar's attitude but nonetheless eager to express their wishes, the members of the Mundakayam Rubber Planters' Association finally succeeded in handing over a petition. It summarized the planters' urgent needs, including the building of roads, a police station, a court house, a hospital, and a travellers' bungalow. It also asked for financial support for these purposes. Gopalachariar was not disposed to accede to the planters' requests, with the exception of the travellers' bungalow. In particular with regard to the maintenance of roads – illegally constructed through government properties and 'solely benefitting the planters' – the Dewan persisted in his critical opinions. Even after the intervention of the British Resident, Gopalachariar stuck to his guns. His inflexible attitude spelled the end of his public career. In 1907 Dewan Gopalachariar was dismissed from office.[4]

[3] This 1923 transfer of power from the Madras administration to the government of India barely changed the political landscape of Travancore (Jeffrey 1975: 263 and 273–5; Rangaswami 1981: 61).

[4] Letter of the MRPA to the British Resident dated 24 January 1907; memorandum of the MRPA; letter of the Dewan to the MRPA dated 8 March 1907, in: R/2/880/83; Rangaswami 1981: 57. The European planters experienced

Two years later the new official colonial policy of non-interference in the internal affairs of the princely states, as proclaimed by Lord Minto in his well-known 1909 Udaipur speech, did change the position of the British Resident for good. The European planters were deprived of their 'natural' political ally, but this did not mean that the indigenous cultivators had gained a political friend, namely the Dewan. For example, Sir C.P. Ramaswami Aiyer, who was the very epitome of the 'Dominant Dewan', was feared by all communities, particularly the Syrian Christians because of his autocratic, anti-Christian attitude. In contrast to his predecessors, in September 1936 Ramaswami Aiyer was officially appointed by the Maharaja, a decision initially sharply disapproved of by the Political Department.[5]

5.1.2 Constitutional Reforms

The most conclusive cause of the decline of the British planters' domination of Travancore state was the widening and deepening of the political arena. Powerful developments both within and outside the state were the catalysts which set this process in motion. The far-reaching effects of these upheavals meant that the leading political figures increasingly had to take into account the opinions and wishes of the members of the Travancore Legislative Council (established in 1888), the Sri Mulam Popular Assembly (inaugurated in 1904), the Travancore Sri Mulam Assembly and the Travancore Sri Chitra State Council (both since 1933).

The function of the Travancore Legislative Council was to support the Dewan in the legislative process. It was stated that he should 'have the benefit of discussing with and taking the opinion of responsible officers'.[6] Apart from its majority of bureaucrats, the Council consisted of non-officials (including

far less difficulties with the other, longer-terms Dewans, including C.P. Ramaswami Aiyer (see for example: Minutes MRPA dated 8 March 1910; *Addresses to the Dewan of Travancore with the Replies of the Dewan* 1914: 132–6 and 140–1; *Addresses to Dewan Bahadur M. Krishnan Nair, Dewan of Travancore* 1920: 208–10; Mackay 1984).

[5] Matthen 1951; Jeffrey 1975; 273–5; Rangaswami 1981: 4; Ouwerkerk 1994: 78; Kooiman 1995: 29; interview: Marthanda Varma.

[6] Quoted in: Sugeetha 1981: 12.

planters), who were all nominated and appointed by the government. At the outset, the members possessed only extremely limited powers in their dealings with the Dewan. No member was allowed to introduce bills affecting the public revenues of the state without the prior consent of the Dewan. It is a reflection of the balance of power in its early years that the Council met in the office of the Dewan. Only in the 1920s was the Travancore Legislative Council accorded some wider powers. When this finally happened, certain interest groups, like landlords, merchants and planters, were given the right to *elect* their own representatives.[7]

The Sri Mulam Popular Assembly was also primarily an advisory body to the Dewan, though there was a slight, but significant, difference in the formulation of its aims. The Assembly was also meant to afford 'an opportunity to the people of expressing direct to the Government their wants and wishes'.[8] Be that as it may, it possessed neither legislative nor executive powers, nor was it widely representative. The Assembly met just once a year and only on that occasion was each member allowed to pose questions to the Dewan. Unfortunately for its members, open discussion was not permitted. The questions, two at the most, had to be submitted previously, and, if approved by the government, were answered by the Dewan on behalf of his administration. Any further discussion was out of the question. For many years, the Assembly consisted of government nominated members (23) and elected representatives of taluks (43), municipalities (19), traders (7), *jenmies* (4) and planters (4).[9]

To smooth the transformation of the Assembly into a legislative body and to regulate the division of tasks between the Assembly and the Council more precisely, on 1 January 1933 the Sri Mulam Popular Assembly was re-styled the Travancore Sri Mulam Assembly (Lower House) and the Travancore Legislative Council was replaced by the Travancore Sri Chitra State Council (Upper House). After that, both the Assembly and the Council, as separate chambers of a bicameral body, were invested

[7] Sugeetha 1981: 12–16 and 19–24.
[8] Quoted in: Sugeetha 1981: 17.
[9] Sugeetha 1981: 17–19.

with extensive powers to bring their influence to bear on legislation and state budgets. This move did not as yet herald a real democracy. Some members were still nominated by the government, particular seats continued to be reserved for certain interest groups (including the plantation sector), and the franchise, although to a lesser extent than in earlier days, remained subject to certain qualifications like the possession of landed property and education.[10] The members had no constitutional right to dismiss the Dewan and/or the Department Heads from office in case of insurmountable disagreement regarding government policies. Formally, the Dewan remained solely responsible to the Maharaja. As indicated above, in actual practice he was accountable to the colonial authorities as well.

5.1.3 Communalism, Nationalism and Communism

To the regret of the British, the volume of the outcry against foreigners working in the Travancore bureaucracy, against the colonial regime, and against labour exploitation grew ever louder, both within as well as outside these constitutional bodies.

At first, communal groups entered the political field. At the outset, they fought the same battle: to reserve jobs in the state's bureaucracy for Travancoreans. As observed in preceding chapters, there was a tendency to appoint foreigners, British as well as non-Malayali Brahmins, to influential posts within the Travancore administration, a strategy which dated from the days of Marthanda and Rama Varma. In 1891 the pro-Travancorean, anti-foreigner movement reached its zenith when a joint petition, signed by more than 10,000 'subjects of His Highness', was presented to the Travancore monarch. This so-called Malayali Memorial expressed in no uncertain terms the idea that when it came to the distribution of government jobs the claims of Malayalis, i.e. Namboodiris, Nayars, Syrian Christians, other Christians and Ezhavas, should be given priority over those of non-Malayalis. Although Dewan Rama Rao in particular remained sceptical about the proposed change in the communal composition of the bureaucracy, the Travancore government was not in a position to overlook the wishes of the memorialists

[10] Sugeetha 1981: 25–6.

altogether. With the exception of the office of Dewan, non-Malayalis were gradually replaced by Travancoreans in the government administration.[11]

Narrowing down the term Travancoreans to more specific categories, we find that a preponderance of Nayars was appointed in the sircar's service, which meant that in the first third of the twentieth century the Nayars were able to dominate the political sphere to an overwhelming degree, both in the bureaucracy as well as in the various Councils and Assemblies. It has already been mentioned in passing that enfranchisement was connected to certain property and educational qualifications and on both counts the Nayar community enjoyed the status of most privileged caste. The not unnatural upshot was that the non-Nayar sections of Travancore society were extremely frustrated by the fruit borne by the Malayali Memorial. Caste organizations led the way in challenging the predominance of the Nayars in the legislative bodies and the civil service. Of these, the Sree Narayana Dharma Paripalana Yogam (SNDPY), the influential political lobby of the Ezhavas, deserves a special mention as the first formal caste organization to be formed in Travancore. It was formed in 1903.

In January 1933, the Syrian Christians, Ezhavas and Muslims even went so far as to form a joint organization, the Samyukta Rashtriya Samiti (Joint Political Congress), which demanded universal adult franchise and proportional representation in the legislature. To isolate the Nayars politically, the leaders of the movement boycotted the legislature, organized mass meetings and exhorted the voters to abstain from voting. The Abstention Movement, as it came to be called, eventually won the day. After more than two and a half years of protests, the government slightly modified the rules regarding the franchise, thereby raising the enfranchised percentage of the population to 10–12 per cent, and allowed a small number of reserved seats for Ezhavas, Muslims, and Latin Catholics. Equally important, the composition of the bureaucracy, which had hitherto been clearly biased in favour of the Nayar community, increasingly reflected the political strength of the various communities in Travancore. Thus, a growing number of Syrian

[11] Lewandowski 1980: 36; Bhaskaran 1981.

Christians, Ezhavas, and Muslims found their way into government service.[12]

The nationalist movement only became a serious contender in Travancore politics after the (second) formation of the Travancore State Congress in 1938, though it was fairly weak compared to its power in the provinces of British India. The Travancore State Congress itself was predominantly concerned with the 'internal', not the 'national', power struggle. Initially, the demand for the resignation of the authoritarian Dewan C.P. Ramaswami Aiyer and for the introduction of 'Responsible Government' reunited Nayars and other communities on a common platform. However, a Memorandum submitted to the Maharaja setting out the Dewan's unfitness for office and the subsequent civil disobedience campaign were disapproved of by M.K. Gandhi and other influential leaders of the Indian National Congress (INC). They insisted that their party's policy of non-interference in the internal affairs of princely states be observed. Furthermore, the Dewan convinced the Nayar community, whose position was already under attack, that any further constitutional changes would lead to Syrian Christian ascendancy. In turn, the Nayars' enthusiasm for the Mahatma's prohibition campaign alienated the Ezhavas, many of whom were toddy-tappers by profession, from the movement. In later years, the discouraging attitude of the INC, the often stringent measures introduced by Ramaswami Aiyer, not to mention communal rivalries, weakened the strength of the Travancore State Congress.[13]

There are some indications that supporters of communist ideologies, who would become a major political force after Indian Independence, were already active in early twentieth century Travancore. A short biography of Marx published in Trivandrum in 1912 is evidence of this and workers in Alleppey organized themselves as far back as 1920. Interestingly, those who would become the most influential communist leaders, like

[12] Namboodiripad 1968: 150; Koshy 1972: Ch. 4; Lewandowski 1980: 36–7; Sugeetha 1981: 26; Vijayakumar 1981: 262; Rangaswami 1981: 130–2; Nossiter 1982: 80; Raj and Tharakan 1983: 43; Kooiman 1995: Ch. 8 and 9 and Appendix 1.

[13] Namboodiripad 1968: 160–6; Vijayakumar 1981: 262; Rangaswami 1981: 180–92 and 206–22; Nossiter 1982: 78–82. Unfortunately, no information was obtained about the *first* formation of the Travancore State Congress.

E.M.S. Namboodiripad and A.K. Gopalan, started their political careers within the nationalist movement. In 1935, these leftist-oriented Congressmen, though remaining within the fold of the parent organization, formed the Congress Socialist Party. This gave them a double advantage. They could remain 'independent of the bourgeois leadership of the national movement'[14] but at the same time use the party's organization to arrange meetings, spread pamphlets, journals and books, and to build up labour unions. In December 1939 the political leaders of the left wing went even one step further. They transformed the Congress Socialist Party into the Communist Party.[15]

These political transformations influenced plantation development considerably, while changes within the plantation sector itself, particularly the emergence of a Syrian Christian planting segment, transformed the political landscape of Travancore fundamentally.

5.2 A Fierce Competition for Land

5.2.1 Rubber Surpassing Tea

Carrying on with the impetus it had gained in the 1890s, the tea sector developed extremely rapidly in the early twentieth century. Starting with about 25,000 acres planted with tea in the first decade of the twentieth century (see Table 4.3), the total acreage increased to 50,296 by 1925 and 76,659 by 1934. Good prices on the world tea market, with the exception of a brief depression in 1920,[16] contributed greatly to this growth. The prospect was no longer as rosy in the early 1930s because reduced consumption of tea in the period following the Crisis (1929) and the growing competition from tea plantations in Sumatra and Java lowered prices considerably. After some earlier attempts had failed, the most influential British and Dutch producers, supported by the governments of India, Ceylon, and the Netherlands

[14] Namboodiripad 1968: 152.

[15] Namboodiripad 1968: 149–83; Nossiter 1982: 65, 71, 75, 83–4. P. Krishna Pillai, who was actively involved in the foundation of the Congress Socialist Party in 1935, is widely considered the 'founder' of the Communist Party in south-west India/Kerala (cf. Nossiter 1982: 71 and 384).

[16] Harler 1956: 247; Griffiths 1967: 170, 178–9.

Indies, agreed to adjust exports to tally with world demand and to restrict the extension of areas under tea in February 1933. This so-called International Tea Agreement[17] also had implications for south-west India. Since then, the acreage under tea in Travancore has remained fairly stable. In 1948 the area cultivated with tea covered 78,700 acres (see Table 5.1).

TABLE 5.1
Acreages Under Tea and Rubber in Travancore in the Years 1925, 1934 and 1948[18]

Year	*Tea*		*Rubber*		*Total*	
1925	50,296	(57%)	37,498	(43%)	87,794	(100%)
1934	76,659	(61%)	49,553	(39%)	126,212	(100%)
1948	78,700	(42%)	110,012	(58%)	188,712	(100%)

Much more spectacular was the entry on to the scene and the development of rubber estates. In less than half a century after the first rubber tree was planted on Travancore soil, the area under rubber had already surpassed the tea acreage (see Table 5.1). Leaving aside some experimental efforts,[19] rubber cultivation really started in Travancore in 1902, when some European planters, J.J. Murphy, J.A. Hunter, K.E. Nicoll and C.M.F. Ross, opened up Periyar Estate at Thattakad near Alwaye.[20] The following year, other large-scale agro-industries were begun around Mundakayam, a village west of Vandiperiyar and Peermade at the foot of the Western Ghats.[21] The previous experience of rubber

17 Harler 1956: 248-9; Griffiths 1967: Ch. 15 and Ch. 36.

18 *UPASI Planting Directory* 1925: 88–114; *UPASI Planting Directory* 1934: 126–76; *UPASI Planting Directory* 1948: 107–42 and 144. According to Panicker et al. (1978: 9); the 10-year averages of tea and rubber in relation to the total value of commodities exported from Travancore were 18 and 8 per cent respectively (1910–20), 27 and 6 per cent respectively (1920–30) and 23 and 5 per cent respectively (1930–40). Considering the rapid increase in the acreage planted with rubber, the figures provided by Panicker et al. are somewhat unrealistic, particularly since almost all rubber produced in the state was exported.

19 See for example: Mahadevan Nadar 1980: 220.

20 Clay 1911: 18; Ghosh 1928: 136; *UPASI Planting Directory* 1948: 4.

21 *UPASI Planting Directory* 1960: 439; cf. Clay 1911: 23; Hatch 1939: 107; Haridasan 1975: 119.

planting gained by some of the Europeans in Ceylon and Malaya was of enormous benefit to the emergence and subsequent spread of the cultivation of rubber in Travancore.[22] Mundakayam even took the palm as the leading centre of rubber estates in India, with an acreage of 5,493 in 1907[23] and 9,652 in 1910.[24] In the 1920s, the period in which the Stevenson Restriction Scheme was enforced, the development of rubber continued unhindered. This international agreement, which had been introduced to cut back the supply of rubber on the world market, was not applicable to India. It took the Great Depression to end the period of growth. Worse affected than tea, prices dropped drastically, leaving the rubber growers to incur heavy losses for many years. The financial situation of the Central Travancore Rubber Company is illuminating. It deteriorated so badly that the pay-out of dividends on ordinary shares went into abeyance for eight years, from 1929 till 1936.[25] Rubber-producing nations did not stand still and, as in the case of tea, a world-wide solution was sought. The ultimate outcome also resembled the International Tea Agreement in many respects. The International Rubber Regulation Agreement (May 1934) was meant to restrict the production and planting of rubber. The conquest of various rubber-producing South-east Asian countries by Japan (1941–2) totally reversed the situation for south-west India. To provide the allies with sufficient quantities of rubber, all kinds of measures, such as fixed minimum prices, were adopted in 1942 to maximize production and to encourage the planting of rubber trees. The rubber growers reacted positively; in the years that followed the increase in acreage under rubber was enormous.[26]

To go back half a century, the introduction of rubber altered plantation development in many respects. It led to the opening up of two other planting districts. The Periyar Estate at Thattakad alluded to earlier marked the beginning of plantation development in north Travancore. Despite its early start, the district never really became very important. Table 5.2 reveals

[22] Malankara 1961: 2.

[23] Proudlock 1908: 47.

[24] Clay 1911: 23; minutes MRPA 13 January 1912.

[25] Figures obtained from various *Tea Share Manuals*, in: MSS Eur F 174.

[26] Plantation Enquiry Commission – Rubber 1956: 1–2; Unny and Jacob 1972: 1–2; Haridasan 1975: 119; Haridasan 1978: 63–6.

that in 1925 and 1934, 5,584 acres and 6,376 acres respectively were planted with rubber, which meant that the influence of the West Coast Planters' Association was kept to a minimum. By contrast, Mundakayam became the most important rubber district in south-west India with an acreage of 16,875 in 1925 and 23,068 in 1934. The Mundakayam Rubber Planters' Association, established in 1905, and after the introduction of tea in the district transformed in 1912 into the Mundakayam Planters' Association, was extremely influential in rubber development. This district organization was the driving force behind the formation of the UPASI Rubber Scientific Department. The researchers, stationed in Mundakayam, were highly successful in their attempts to check diseases of trees and to develop higher yielding varieties.[27] The district of south Travancore experienced a revival because of the introduction of rubber. Besides being the third ranking tea region, it became the second ranking rubber area. The cultivation of rubber was only practised on a very small scale in central Travancore and the Kannan Devan Hills because the high altitude and cold climate here were not conducive to the large-scale introduction of rubber.

TABLE 5.2
The Plantation Sector in Travancore in 1925 and 1934 (acreages)[28]

District	*1925*			*1934*		
	Tea	*Rubber*	*Total*	*Tea*	*Rubber*	*Total*
Kannan Devan	22,982	25	23,007	33,655	65	33,720
	(46%)	(0%)	(26%)	(44%)	(0%)	(27%)
Central Trav.	18,949	-	18,949	27,016	5	27,021
	(38%)	(0%)	(22%)	(35%)	(0%)	(21%)
South Trav.	6,917	15,014	21,931	10,685	20,039	30,724
	(14%)	(40%)	(25%)	(14%)	(40%)	(24%)
Mundakayam	1,448	16,875	18,323	5,301	23,068	28,369
	(3%)	(45%)	(21%)	(7%)	(47%)	(22%)
North Trav.	-	5,584	5,584	2	6,376	6,378
	(0%)	(15%)	(6%)	(0%)	(13%)	(5%)
Total	*50,296*	*37,498*	*87,794*	*76,659*	*49,553*	*126,212*
	(100%)	*(100%)*	*(100%)*	*(100%)*	*(100%)*	*(100%)*

27 Ghosh 1928: 149; *UPASI Planting Directory* 1960: 440.

28 *UPASI Planting Directory* 1925: 88–114; *UPASI Planting Directory* 1934: 126–76.

As well as a geographical change of emphasis, the rise of a rubber sector implied a change in the pattern of ownership. In the last chapter it was established that tea was produced mainly by limited liability companies. There was a clear logic behind this. Tea leaves needed processing in a nearby factory before being transported. These large investments in machinery required, in their turn, large investments in land. To maximize the profitability of such a factory, an input of sufficient quantities of tea leaves was essential. All this adds up to the fact that tea cultivation remained primarily the domain of large undertakings during the first half of the twentieth century (see also Table 5.3).

TABLE 5.3

Size of the Tea and Rubber Undertakings in Travancore in 1925, 1934 and 1948[29]

Size of the Holding	*1925*		*1934*		*1948*	
	Tea	*Rubber*	*Tea*	*Rubber*	*Tea*	*Rubber*
Less than 10 Acres	–	–	6 (3%)	3 (2%)	6 (3%)	2 (1%)
Between 10 and 100 Acres	9 (8%)	24 (28%)	36 (20%)	93 (49%)	46 (24%)	224 (61%)
More Than 100 Acres	103 (92%)	62 (72%)	134 (76%)	93 (49%)	137 (72%)	140 (38%)
Total	*112 (100%)*	*86 (100%)*	*176 (100%)*	*189 (100%)*	*189 (100%)*	*366 (100%)*

Rubber was clearly a different matter. After the cultivation of rubber had been initiated by large undertakings (more than a hundred acres), smaller estates (between ten and a hundred acres)

[29] *UPASI Planting Directory* 1925: 88–114; *UPASI Planting Directory* 1934: 126–76; *UPASI Planting Directory* 1948: 107–42 and 144. In 1948, with the exception of eight cases, the UPASI enumeration did not include the category smallholdings: 447 acres in the case of tea and 39,246 acres (!) in the case of rubber (*UPASI Planting Directory* 1948: 142). Especially with regard to rubber, the number of holdings covering less than 10 acres was therefore much higher in reality. Taking these holdings on an average of 5 acres, this category included almost 8,000 rubber cultivators in 1948.

and smallholdings (less than ten acres) began to predominate in the field.[30] Table 5.3 shows that at the beginning large estates did indeed dominate rubber cultivation. In 1925, 62 out of 86 estates belonged to this category. Less than a decade later, the number of large and other estates was equal. In 1948 the smallholders, who profited greatly from the prices fixed by the government, far outnumbered the estates. The reason smaller estates and smallholdings were also able to profit from the introduction of rubber was that no large investments were necessary to process the latex collected from the rubber trees. The removal of dirt, sand and other impurities, coagulation, sheeting, and smoking were in fact simple tasks which the rubber peasants themselves were quite able to undertake. The lack of a special smoke house, used by large undertakings to dry the rubber sheets, could be compensated for by 'sun drying' and 'kitchen smoking', although these methods resulted in an inferior quality rubber compared to that produced on the large estates in particular, which possessed the facilities essential for producing high quality rubber.[31] This deficiency was largely overcome by a favourable price for which it was offered. Unencumbered by costly machinery and crushing overheads, small growers were able to offer the traders cheaper rubber sheets.[32]

5.2.2 The Rise of Syrian Christian Planters

The possibility of starting an estate without the immediate need to procure expensive processing equipment encouraged many Travancoreans to open up plantations. P.J. John, who started

[30] Shortly after Indian Independence the official definition of an estate was a holding above 100 acres. Later, the acreage was brought down to 50 acres (Haridasan 1980: 173). Because most tea and rubber holdings with an acreage of 10 acres or more require the use of wage labour, my concepts differ from the official ones. I would like to add, however, that these definitions are also arbitrary, particularly since the labour input differs per crop: on the average tea estates employ 1.22 workers per acre whereas rubber plantations employ 0.37 of a labourer per acre (RCL 1931 Evidence Vol. 11: 383). See also Section 4.1.4 with regard to coffee smallholdings and plantations.

[31] Unny and Jacob 1972: Ch. 8.

[32] Interview: M. Mathew.

the Kaliyar rubber estate in 1905, was the pioneer in this.[33] The rapid rise of local rubber growers is illustrated by Table 5.4. In 1925 the indigenous cultivators owned about 39 per cent of the planted area. By 1934, the Travancoreans already outnumbered the British entrepreneurs. In that year the local growers possessed some 61 per cent of the acreage under rubber. Their share increased even further to about 85 per cent in 1948.

Despite the large investments required, an indigenous tea segment began to appear, though of lesser significance compared to rubber. The same table shows that in 1925 the British planting associations represented 88 per cent of the total area under tea. Nine years later, their share had declined to 82 per cent. In 1948 the sterling companies owned 69 per cent of the fields planted with tea. Although British interests actually dominated to a greater extent than is suggested by these figures,[34] the table clearly

[33] According to most sources, P.J. John (Aiymanam) was the Indian pioneer planter (Malankara 1961: 2; cf. Haridasan 1975: 119). Menon, however, states that P.T. Thomas (Palampadam) was the first Travancorean to open up a rubber estate (Menon 1981: 66). Certainly, P.T. Thomas, like P.J. John a Syrian Christian, was the founder of Tropical Plantations Ltd, one of the earliest indigenous plantation companies in Travancore (interview: Abraham Kurien and Abraham Eapen). Dominic Thommen (Karipparambil) and Chacko Abraham (Kallivayalil) were yet other Syrian Christian planting pioneers (interviews: Hannibal Dominic and Michael A. Kallivayalil).

[34] For example, some British planters did not join the British planting association in their district (see the figures with regard to 1925 and 1934) and some British entrepreneurs were themselves the sole proprietors of their estates (consult the data concerning 1948). Moreover, the emergence and gradual growth of the indigenous segment remained within a framework dominated by foreign interests. For the Travancorean tea planters in particular, conducting business was virtually impossible without making use of the services of either Aspinwall and Company, Darragh Smail and Company, Harrisons and Crossfield, James Finlay and Company, Madura Company, Peirce Leslie and Company or Volkart Brothers. These so-called coastal firms, of which most had started their activities in the coffee era, had monopolized the export of tea and functioned as insurer, banker, producer, shipper, broker and distributor. To diminish the financial risks involved in their business, the coastal firms claimed the right to interfere, through their visiting agents (also indicated as 'managing agents'), in the internal affairs of the estates under contract. Therefore, the planting policies and labour strategies of most Travancorean plantations were influenced to a very great extent by the British trading companies. But it should not be assumed that the relations between the indigenous planters and the coastal firms, though unequal in nature, were

TABLE 5.4

A Rough Indication of the Growing Strength of the Indigenous Plantation Segments in Travancore (in acres)[35]

Year	*Feature*	*Rubber*		*Tea*	
		Area	*Estates*	*Area*	*Estates*
1925	British PAs	22,803 (61%)	33 (38%)	44,271 (88%)	83 (74%)
	Non-Members	14,695 (39%)	53 (62%)	6,025 (12%)	29 (26%)
	Total	*37,498 (100%)*	*86 (100%)*	*50,296 (100%)*	*112 (100%)*
1934	British PAs	19,315 (39%)	20 (11%)	62,498 (82%)	90 (51%)
	Non-Members	30,238 (61%)	169 (89%)	14,161 (18%)	86 (49%)
	Total	*49,553 (100%)*	*189 (100%)*	*76,659 (100%)*	*176 (100%)*
1948	Sterling Cos	16,634 (15%)	13 (4%)	54,126 (69%)	72 (38%)
	Others	93,378 (85%)	353 (96%)	24,574 (31%)	117 (62%)
	Total	*110,012 (100%)*	*366 (100%)*	*78,700 (100%)*	*189 (100%)*

PA = Planting Association; Co = Company

solely disadvantageous to the former. Apart from the necessary inputs, at a time when a home market for tea was almost virtually non-existent the trading concerns provided the planters an opportunity to market their produce abroad. In the relatively low capital-intensive rubber sector, the role of the coastal firms was of secondary importance. The gradually growing demand for raw rubber from within Travancore and within British India played a key role as well (*UPASI Planting Directory* 1925: 88–114; Banking Enquiry Committee 1930: 74; Economic Depression Enquiry Committee 1931: 71 and 73; *UPASI Planting Directory* 1934: 126–75; Malankara 1961: 4–5; Langley 1962; Ganapathy 1978; Jeffery 1982: 247; Thomas Isaac and Tharakan 1986: 7; Tharian George and Tharakan 1986: 221–2; Janson 1990; Dharmarajan 1991: 27–8; Mahadevan 1991: 165 and 174; Ravi Raman 1991: 245; interviews: V. Haridasan; Sethuram; Sundar; Tharian George; K.T. Thomas and T.J. Joseph; A. Venugopal).

[35] *UPASI Planting Directory* 1925: 88–114; *UPASI Planting Directory* 1934: 126–76;

indicates the growing significance of Travancorean planters in the tea sector during the first half of the twentieth century. Often after having achieved success in the rubber agro-industry, the indigenous entrepreneurs, including P.J. John,[36] were prepared to invest their capital in the opening up of tea estates.

TABLE 5.5
Community Background of Indigenous Proprietors of Privately-Owned Rubber Estates in Travancore in 1925, 1934 and 1948 (total area in acres and number of estates)[37]

Community	*1925*		*1934*		*1948*	
	Area	*Est.*	*Area*	*Est.*	*Area*	*Est.*
Syrian	1,858 (47%)	18 (64%)	8,693 (76%)	95 (75%)	11,403 (55%)	152 (59%)
Chr.	1,839 (46%)	8 (29%)	1,967 (17%)	22 (17%)	4,545 (22%)	40 (16%)
Brahmin	130 (3%)	1 (4%)	225 (2%)	1 (1%)	850 (4%)	10 (4%)
Nayar	165 (4%)	1 (4%)	220 (2%)	3 (2%)	930 (5%)	20 (8%)
Muslim	-	-	78 (1%)	3 (2%)	1,123 (5%)	19 (7%)
O,U,U	-	-	330 (3%)	3 (2%)	1,776 (9%)	15 (6%)
Total	*3,992 (100%)*	*28 (100%)*	*11,513 (100%)*	*127 (100%)*	*20,627 (100%)*	*256 (100%)*

Est. = Estates; Chr. = Christian (which can either be a Syrian Christian or newly-converted Christian, most of whom originated from castes considered lower in the existing hierarchy); O,U,U = Others, Uncertain, Unknown

---

UPASI Planting Directory 1948: 107–42 and 144. According to the Economic Depression Enquiry Committee, in 1931 the indigenous planters owned 17,000 acres of rubber (total 61,986) and 4,000 acres of tea (total 74,618), representing 27 per cent and 4 per cent of the rubber and tea sectors respectively (Economic Depression Enquiry Committee 1931: 69 and 74).

[36] *UPASI Planting Directory* 1925: 92.

[37] *UPASI Planting Directory* 1925: 88–114; *UPASI Planting Directory* 1934: 126–76; *UPASI Planting Directory* 1948: 107–42 and 144. See also Appendix 6. It should

If we accept the names of estate owners as a valuable indication of the caste background of the indigenous plantation segment, one major feature overrides all others: Syrian Christians – bearing names like Chacko, Cherian, Devasia, Eipen, Ittiyavirah, Koshy, Kurian, Mathai, Oommen, Ouseph, Thommen, Varghese and Varkey – took the lead in opening up rubber estates. All other communities, Nayars included, were far behind both in terms of planted acreage and in the number of estates owned (see Table 5.5). P.J. John, mentioned twice above, was also a Syrian Christian. Syrian Christians also predominated in the Travancorean tea segment. Tamil Brahmins came second, owning about one-fifth of the acreage under tea (consult Table 5.6).

Apart from the predominance of this category as proprietors of privately-owned estates, many, if not most, Travancorean rubber and tea companies were also owned by Syrian Christians. The enterprising P.J. John also played a crucial role in floating the earliest Travancorean rubber company. The background and career of this pioneer are worth recounting. His father, the Syrian Christian Marthomite[38] P. John Kurien, was educated by CMS missionaries, after which he entered government service and later, assisted by his sons, turned his attention to various economic ventures.[39] After his father's death, P.J. John was employed by

---

be noted that all tables dealing with the community background of indigenous planters as presented in this book are unique. Particularly in view of political considerations (the Indian constitution prohibits the usage of caste categories), the Rubber Board is not allowed to publish its data regarding the caste identity of the agro-entrepreneurs. The tables given in this book come very close to the *unofficial* estimates and figures of the Rubber Board (interview: V. Haridasan). I would like to thank Dr J.L. Chiriyankandath, Dr V. Haridasan, Dr Raman Mahadevan, Motty Philipose and Nannu Antony for their information regarding the community background of local planters.

[38] European religious influences divided the Syrian Christian community into three major sections. First, Portuguese Catholic missionaries stimulated the foundation of the Syrian Catholic Church, which acknowledged the pope as its religious leader. Later, the influence of British Protestant missionaries caused the emergence of the independent, so-called Mar Thoma Church. In contrast to the Syrian Catholics and the Marthomites, the Jacobites (nowadays divided into Jacobite and Orthodox factions) established connections anew with the Jacobite patriarchs in western Asia (Jeffrey 1976: 17; Kooiman 1993: 288; interview: P.K. Michael Tharakan).

[39] Chiriyankandath 1985: 114.

TABLE 5.6
Community Background of Indigenous Proprietors of Privately-Owned Tea Estates in Travancore in 1925, 1934 and 1948 (total area in acres and number of estates)[40]

Community	*1925*		*1934*		*1948*	
	Area	*Est.*	*Area*	*Est.*	*Area*	*Est.*
Syrian	804 (62%)	4 (57%)	2,127 (38%)	31 (55%)	2,246 (34%)	28 (43%)
Chr.	200 (15%)	1 (14%)	1,111 (20%)	12 (21%)	1,649 (25%)	22 (34%)
Brahmin	240 (19%)	1 (14%)	1,544 (27%)	6 (11%)	1,436 (22%)	7 (11%)
Nayar	-	-	-	-	9 (0%)	1 (2%)
Muslim	-	-	107 (2%)	1 (2%)	472 (7%)	2 (3%)
O,U,U	48 (4%)	1 (14%)	765 (14%)	6 (11%)	741 (11%)	5 (8%)
Total	*1,292 (100%)*	*7 (100%)*	*5,654 (100%)*	*56 (100%)*	*6,553 (100%)*	*65 (100%)*

Est. = Estates; Chr. = Christian (which can either be a Syrian Christian or newly-converted Christian, most of whom originated from castes considered lower in the existing hierarchy); O,U,U = Others, Uncertain, Unknown

Brander,[41] a European planter who had started an estate in 1903. Only two years later, P.J. John, with the aid of his brothers, opened up the Kaliyar Estate in north Travancore. Within three years of planting, he sold the enterprise to Malayalam Plantations, a large British company. He had no plans to remain idle

[40] *UPASI Planting Directory* 1925: 88–114; *UPASI Planting Directory* 1934: 126–76; *UPASI Planting Directory* 1948: 107–42 and 144. See also Appendix 7.

[41] In the late 1910s or early 1920s, Brander was killed by a local cultivator. Epitomizing the attitude of many European planters, he thought he could encroach on the latter's land and get away with it. Ultimately, Brander's properties were auctioned off and bought by the father of K.T. Thomas, the secretary of the Central Travancore Agricultural Association (interview: K.T. Thomas and T.J. Joseph).

and in 1910 his dream of starting a large rubber plantation under corporate management was realized. In that year, he in conjunction with other promoters like Rev. C.P. Philipose Kassessa and K.C. Mammen Mappillai, established the Malankara Rubber and Produce Company.[42] The Malankara Estate, situated like the Kaliyar Estate in north Travancore, possessed 1,475 acres under rubber in 1925, 1,455 acres under rubber in 1934 and 1,735 acres under rubber in 1948.[43] Besides the Malankara Rubber and Produce Company, other examples of early planting companies in which Syrian Christian interests loomed large were Tropical Plantations Limited with estates in central Travancore and Mundakayam[44] and the Mar Thoma Rubber Company based in north Travancore.[45]

The prevalence of Syrian Christians in the indigenous plantation segment, which has been recognized by many other scholars,[46] is often explained in terms of the community's positive work ethic, which stands in fairly stark contrast especially when compared to that of the Nayars. Syrian Christians were 'more entrepreneurial'[47] and far more 'dynamic and flexible',[48] in contrast to the Nayars who looked down on the economic activities of the Syrians 'with lordly disdain'.[49] Although a difference of attitude towards labour ethos as such might serve as a valid explanation, my interest is drawn especially to the various factors underlying the different attitudes towards work and economic enterprise.

[42] Malankara 1961: 2; Haridasan 1975: 119. P.J. John's son, P. Kurien John, also a rubber planter, was first appointed as the Controller of the Rubber Regulation Committee in India (1934) and later as the Rubber Production Commissioner of the Indian Rubber Production and Marketing Board (1942) (proceedings TSCSC 17 July 1942; Malankara 1961: 5–7).

[43] *UPASI Planting Directory* 1925: 106; *UPASI Planting Directory* 1934: 160–1; *UPASI Planting Directory* 1948: 122.

[44] *UPASI Planting Directory* 1925: 92–3 and 104–5; interview: Abraham Kurien and Abraham Eapen.

[45] *UPASI Planting Directory* 1925: 106–7.

[46] See for example: Varghese 1970: 116 and 119; Jeffrey 1976: 127 and 202; Tharakan 1976: 44; Lewandowski 1980: 35; Chiriyankandath 1985: 113; Mahadevan 1991: 170–2.

[47] Interview: N.E. Kurien and N.K. Elias.

[48] Joseph 1988: 132.

[49] Menon 1981: 67.

Before starting an enterprise, many Syrian Christians obtained the necessary skills, like a knowledge of planting and processing, from European planters.[50] For some Travancoreans, applying for a job on a European-owned estate may have been the first step, consciously taken, to achieve their own ultimate goal: opening up a plantation of their own. In the earliest days of rubber planting, however, the initiative was firmly in the hands of the Europeans. To create an intermediate group above the predominantly low-caste, Hindu workers from the Madras Presidency, European planters preferred to employ Christian Malayalis, especially those with a high social status (namely Syrian Christians), as managers and clerks. Only after having saved some of the income earned and having obtained considerable planting experience did the possibility of starting a planting enterprise for themselves dawn on these Syrian Christians.

Articles published in the *Malayala Manorama*, a Malayalam newspaper with a predominantly Syrian Christian readership, encouraged them to put their ideas into practice.[51] In December 1905, for instance, the *Malayala Manorama*, in an editorial, advocated:

> For a long time we have been hearing of the great interest taken by the Europeans in this country in rubber cultivation. But we are surprised to note that none of our people has taken any step or shown any interest in the matter. When one realises the enormous profit this cultivation will bring to us, our lethargy and lack of enterprise will be condemned by all patriotic people.

Indeed, for indigenous agriculturalists the opening up of rubber estates was considered to be a patriotic, perhaps even anti-colonialist, nationalist duty. As 'the finest land suitable for rubber cultivation is plenty in Travancore', the editor argued, 'it is high time for the people of the state to realise the

[50] Lewandowski 1980: 35; Chiriyankandath 1985: 113; interview: K.T. Thomas and T.J. Joseph.

[51] In this study only some of the articles published in the *Malayala Manorama* are dealt with. According to most scholars, the newspaper has had a wide impact on south-west Indian society (Abraham 1971: 42). It should be mentioned, however, that the *Nazrani Deepika*, another Syrian Christian-led, Kottayam-based Malayalam daily, was started in 1887, i.e. one year before the *Malayala Manorama* (1888), and presented similar views. The *Pouradhwani* was also pro-planting, i.e. pro-Syrian Christian planters (Tharakan 1984a: 1924).

opportunity and make a concerted effort even by promoting some joint stock companies to enter the field'. He ended his plea: 'We draw the immediate attention of all people who have received education not to miss this opportunity'.[52]

With regard to the acquisition and exploitation of land, Syrian Christians also had a great number of advantages. The regions suitable for the cultivation of rubber, like north Travancore and Mundakayam, were already the homes of large Syrian communities.[53] In contrast to the Nayars household organization (to be looked into below), Syrian Christian family structures were patriarchal, with every adult male having an equal claim to his father's property, consequently providing enough incentive to individual initiative.[54] Apart from the built-in advantages of their highly suitable geographical situation and the inherent strengths of their inheritance system, the Syrian Christians also profited from the gradual break-down of the Nayar *taravad* system.

Historically, the taravad, or extended household, was a largely self-sufficient, prosperous unit. The land was cultivated mainly by bonded, low-caste labour, while the Nayars themselves managed food production, settled local and regional disputes, and supervised marriages and other public ceremonies. A fundamental feature of the Nayar taravad was the matrilineal system of the transmission of property. Closely related to the inheritance customs of the Namboodiri Brahmins, a patrilineal caste which ensured that its property would not be subdivided by allowing only the eldest male to marry within his community, Nayar

[52] Quoted in: Haridasan 1992: 34. To give two other examples, in February 1907 and August 1909 respectively the journal tried to counteract the reluctance of the Travancorean entrepreneurs by informing them that: 'From the available calculation of the profitability of rubber cultivation, we like to assure the public that at present prices rubber cultivation is more paying than gold mines. It is a great pity that *our* farmers are so disinterested about such a profitable venture' (italics mine, quoted in: Haridasan 1992: 34–5). 'If the Europeans can conduct this lucrative cultivation in such an organised manner, all are wondering why people of *our* country are not entering this field' (emphasis added, quoted in: Haridasan 1992: 35). Like the statement of December 1905, both accounts emphasize the need for Travancoreans/Syrian Christians to start rubber enterprises.

[53] Varghese 1970: 119; Chiriyankandath 1985: 113.

[54] Tharakan 1984b: 1960.

women had hypergamous unions with both Brahmins and Kshatriyas. Complementing these social rules, the children took the caste of their mother and family property was transmitted through the women of the household and could not be divided. Certain vicissitudes in the nineteenth and twentieth centuries undermined the prosperity of these taravads. The abolition of slavery certainly played a role, but of greater importance were legal changes regarding landed property. Once land became a commodity, i.e. an exchangeable, divisible unit which could be bought and sold for cash, tensions within the household, which sometimes encompassed more than a hundred people, grew more pronounced, especially since some were in desperate need of financial means to pay for education or to support their prestigious lifestyle. The Nair Regulation of 1924, which allowed the partitioning of matrilineal families, accelerated the division and sale of taravad properties. Syrian Christians in particular profited from the process of Nayar land alienation.[55] Many Syrian planters started their estates on land previously owned by Nayar taravads.[56] They also bought up lands formerly belonging to the Vanchipuzha and Pooniat Chiefs.[57] A third source of property was that of European planters who abandoned their estates during the Depression and sold these properties to Syrian Christians.[58]

Syrian Christians were also blessed with better access to financial resources. In contrast to other communities which were primarily dependent on so-called Hundi merchants or other Tamil Brahmins for the supply of capital, joint funds known as *chitties* or *kuris* were common among Syrians. In these, the generation of capital was mostly village-based, with deposits being accepted from small savers and credit given to small peasants and traders. Making use of a joint fund was much more attractive than entering into business with Tamil moneylenders who charged interest of up to 70 per cent. Quite apart from their lending activities, the chitties also provided popular

55 Varghese 1970: 103; Tharakan 1976: 44; Chiriyankandath 1985: 110; Joseph 1988: 136.

56 Interviews: Abraham Kurien and Abraham Eapen; N.E. Kurien and N.K. Elias; K.T. Thomas and T.J. Joseph.

57 Interviews: Hannibal Dominic, Michael A. Kallivayalil, Michael Tharakan.

58 Interview: Michael Tharakan.

opportunities for investment. The willingness of Syrian Christians to lend capital was also seized upon for the opening up of rubber estates. For instance, Rev. C.P. Philipose Kassessa, the promoter of the Malankara Rubber and Produce Company, successfully employed his influence as leader of the Evangelical Association of the Mar Thoma Church to request the faithful to subscribe to the company's capital.

The Syrian Christian laws of inheritance, which stipulated that land was a definite share with sole proprietary rights, also stood the Syrians in good stead in their applications for banking loans, giving them an advantage over members of other communities. In this context it needs mentioning that the other financier of the Malankara Rubber and Produce Company, K.C. Mammen Mappillai, also has a fascinating background. He was a prominent member of the Kandathil family. These Jacobites, who had a century-old tradition of running chitties, founded the Travancore Bank (1893), the first indigenous credit institution organized on the lines of a modern commercial bank, and later established the Travancore National Bank (1912), which expanded to 40 branches spread over the whole of British India and Ceylon (1930). In 1937, the Travancore National Bank and the equally successful Quilon Bank, managed by another Syrian Christian family, amalgamated into the Travancore National and Quilon Bank. Figures illustrate its significance: in India the new bank ranked first in the number of branch offices and third in the total volume of trade. Indubitably, the growing significance of Syrian Christian planters and the expansion of Syrian Christian bankers went hand in hand. Including among them many family members who wanted to open up and run estates, the bankers' customers were primarily Syrian Christians. The prosperity of the indigenous plantation segment contributed, in its turn, to the growth of commercial banking in Travancore.[59]

Unfortunately, not much can be said about the backgrounds of the Brahmin and Muslim planting communities. There is simply not much historical material available on these topics,

[59] Banking Enquiry Committee 1930: Ch. 3, 12 and 13; Malankara 1961: 3; Tharakan 1976: 25; Oommen 1977; Mahadevan Nadar 1980: 223; Chiriyankandath 1985: 112–14; Mahadevan 1991: 170–2.

except for the fact that most Brahmins originally came from the Madras Presidency.[60]

Within the Shanar planting group, one individual stood head and shoulders above all the others: A.V. Thomas.[61] Before opening up land for tea, this Christian Shanar had built up experience in two closely-related fields. For a while he had worked for J.A. Richardson, the prominent European planter, in the function of estate staff. Later he was employed by the European trading firm Darragh Smail and Company. Deeming that he had obtained sufficient knowledge about both the production of and trade in plantation products, Thomas opened up Pasupara Estate in 1927, gradually building up an enormous enterprise encompassing various production units and wide trading networks.[62] Making a long-term historical comparison, A.V. Thomas' career resembled that of P.D. Devasahayam: both were notable exceptions within the Shanar community (see also the Sections 4.1.4 and 7.1.1). As successor to D.C. Kothari (1950–1), A.V. Thomas became the second *Indian* President of the United Planters' Association of Southern India (1953–4).

5.2.3 Syrian Christian Emigration

The explosive development of plantation crops in Travancore was the antecedent of the migration of Syrian Christian rubber planters from Travancore to Malabar from the 1930s. In fact, they were 'pushed out' because large tracts were already occupied by Europeans who had profited from their influence on state politics, while the efforts among indigenous cultivators to occupy government lands also exacerbated the shortage of land. The combined result of foreign and indigenous agricultural expansion meant that almost the entire cultivable area in Travancore was brought into production by the end of the 1920s.[63] There were still the reserved forests but the government, motivated by both economy and ecology, was reluctant to open portions of these for rubber production. The sircar derived a

60 Cf. Mahadevan 1991: 174.

61 Both in Table 5.5 and Table 5.6, A.V. Thomas was included in the category 'Christian'.

62 Muthiah 1993: 262, 480; interview: Michael A. Kallivayalil.

63 Joseph 1988: 95.

considerable income from jungle products – like valuable timber – and feared damage to the natural assets of the state.[64]

Quite apart from these compelling reasons, pull factors were also involved in the migration. Large areas of Malabar had remain uncultivated. Reference was made earlier to the fact that landlords had been primarily interested in the intensification of existing modes of exploitation, consequently tenants had been hindered in their efforts to introduce new crops and to open up new regions. Furthermore, by the mid-twentieth century many Malabar Nayar families, like most of their Travancore counterparts, found themselves confronted with serious financial difficulties and were willing to sell off parts of their properties relatively cheaply.[65]

The migration of Travancorean rubber cultivators to Malabar is to be seen in the figures provided in Table 5.7. Initially, the acreage under rubber in the Madras Presidency remained more or less stable: 10,080 in 1925 and 10,975 in 1934. However, by 1948 the area planted with rubber amounted to 24,696 acres. To be more precise, in the plantation district of Malabar 9,016 acres, 9,228 acres and 21,851 acres were planted with rubber in 1925, 1934 and 1948 respectively.[66]

Considering princely Travancore's position in relation to other, including the directly-ruled, south Indian regions, Travancore remained the outright leader in the field of plantation crops during the first half of the twentieth century. In 1925, 1934 and 1948, 52.4 per cent, 49.4 per cent and 49.2 per cent respectively of the total acreages under tea in south India were situated in Travancore. In the case of rubber, the state's position was even more impressive. Its share in the total area planted with rubber in south India increased from 63 per cent in 1925 to 67 per cent in 1934 and 72 per cent in 1948 (figures calculated from Table 5.7).

[64] KSA Bundle No. 7: 1101/21; TLRM, vol. 2, Suppl. 3, 1935: 132, 137, 142–3, 285–6, 315 and 538; TLRM, vol. 2, Suppl. 4, 1943: 55–6 and 199; TLRM, vol. 2, Suppl. 5, 1950: 172–4, 326–7.

[65] Varghese 1970: 121; Tharakan 1976: 7, 44 and 49–50; Joseph 1988: 111 and 113; Kurup 1988: 151–2.

[66] *UPASI Planting Directory* 1925: 114; *UPASI Planting Directory* 1934: 176; *UPASI Planting Directory* 1948: 144.

TABLE 5.7
Rubber and Tea Cultivation in South India in the Years 1925, 1934 and 1948 (acreages)[67]

Region	*1925*		*1934*		*1948*	
	Rubber	*Tea*	*Rubber*	*Tea*	*Rubber*	*Tea*
Madras Presidency	10,080 (17%)	45,210 (47%)	10,975 (15%)	73,355 (47%)	24,696 (16%)	74,756 (47%)
Coorg	1,757 (3%)	316 (0%)	3,038 (4%)	416 (0%)	3,140 (2%)	415 (0%)
Total British India	*11,837 (20%)*	*45,526 (47%)*	*14,013 (19%)*	*73,771 (47%)*	*27,836 (18%)*	*75,171 (47%)*
Cochin	9,348 (16%)	50 (0%)	10,507 (14%)	555 (0%)	14,061 (9%)	1,683 (1%)
Mysore	1,269 (2%)	192 (0%)	345 (0%)	4,330 (3%)	581 (0%)	4,310 (3%)
Travancore	37,498 (63%)	50,296 (52%)	49,553 (67%)	76,659 (49%)	110,012 (72%)	78,700 (49%)
Total Indian States	*48,115 (80%)*	*50,538 (53%)*	*60,405 (81%)*	*81,544 (53%)*	*124,654 (82%)*	*84,693 (53%)*
Total South India	*59,952 (100%)*	*96,064 (100%)*	*74,418 (100%)*	*155,315 (100%)*	*152,490 (100%)*	*159,864 (100%)*

1. The Madras Presidency encompasses the districts Anamalais (Coimbatore), Madura, Malabar, Malabar-Wynaad, Nilgiris, Nilgiri-Wynaad, Shevaroys, Tinnevelly and Vizgapatam.
2. In 1948, the name 'British India' changed into 'Indian Union'.

Two factors explain the relatively quick development of the plantation crops, especially rubber. Most importantly, within the regions suitable for rubber cultivation Travancorean peasants were in an especially advantageous position to respond to new opportunities. In the second chapter there is a discussion of the strategies adopted by Marthanda and Rama Varma – including that of expropriating influential landlords – which had made the state the owner of the bulk of the agricultural land. In the

[67] *UPASI Planting Directory* 1925: 114; *UPASI Planting Directory* 1934: 176; *UPASI Planting Directory* 1948: 144.

mid-nineteenth century, the tenants of these sircar lands acquired full ownership rights, including that of conveyance of their properties (so-called Pattom Proclamation of 1865). This measure, and other regulations which strengthened the position of peasants holding tenures from private landlords, converted Travancore into an area of predominantly small owner-cultivators and secure tenants. This situation contrasted sharply with that in the Malabar district of the Madras Presidency, where the landlords owned the major part of the cultivable lands. Moreover, with the support of the British colonial regime these janmkars had succeeded in reinforcing their powers, leaving most cultivators to struggle as best as they could under an oppressive land tenure system.[68] It is therefore hardly surprising that when the opportunity of opening up rubber holdings arose Travancorean peasants especially profited. Tenants in Travancore required no permission to cultivate the land with rubber. There was great flexibility as the owner-cultivators were able to sell their properties if they wanted to finance their migration to areas suitable for rubber production. As already mentioned, among the Travancorean cultivators the Syrian Christians were the main beneficiaries of the introduction of rubber.

The second reason for the rapid rise of plantation crops was that many British entrepreneurs were attracted to Travancore in the early twentieth century. They ensconced themselves in a highly satisfactory position regarding the allotment of land until the mid-1920s and the recruitment of labour until the mid-1930s. Travancore's policy regarding land will be discussed in the next section. Compared to their counterparts in all other parts of southern India, the planters situated in Travancore benefited the longest from labour legislation which was slanted in their direction. The Criminal Breach of Contract Act was abolished in Travancore only in 1935 (consult Section 5.3.3).

5.2.4 *The Privileges of British Planters Openly Contested*

The growing weight of the indigenous planters was also given voice in politics. Guided by some influential Syrian Christian entrepreneurs, the Travancorean planters entered the public

[68] Tharakan 1976: 42; Chiriyankandath 1985: 109; Joseph 1988: 33–8.

domain and gradually hammered themselves into a major political force, thus ending the political hegemony of the British planters. The European lobby, which had been overwhelming during the second part of the nineteenth century, was gradually drained of its strength.

The *Malayala Manorama*, owned by the Kandathil family, led the way in the protests against the enormous influence of the European planters who, through their planting associations and the British representatives, were highly successful in their efforts to manipulate the Travancore administration and restrict indigenous competition. This situation, a relic of the nineteenth century, was still largely intact in the early twentieth century. There are numerous examples to show that the European planters, aided by the British Resident, acquired land relatively easily. The correspondence between aspirant rubber planters and the Travancore administration, the Superintendent and Magistrate Cardamom Hills, and the Conservator of Forests, both of the last two being Europeans, betrays unequivocally that the latter favoured applications from their fellow countrymen. In their efforts to establish themselves, indigenous entrepreneurs encountered many difficulties regarding the allocation of government waste land and the registration of land in their names.[69]

More explicitly, the *Malayala Manorama* openly questioned whether European entrepreneurs were entitled to fertile land in Travancore. This was demonstrated nicely in December 1905 when the editor regretfully noted that: 'Europeans are moving heaven and earth to corner all *our* available land'.[70] The notion

[69] TLRM, vol. 2, 1915: 256–60 and 336–7; and cf. KSA Bundle No. 7: 1174/21; Tharakan 1976: 50; Pandian 1990: 79–80 and 82–4. In contrast to both coffee and tea (Rules for the Sale of Waste Lands on the Travancore Hills for Coffee or Tea Cultivation, dated 28 November 1913, in: TLRM, vol. 2, 1915: 1035–42), in the case of rubber there was no particular law regarding the allotment of government lands and taxation. Areas cultivated with rubber were covered by the rule regarding cherikal, *puduval*, and *sirkar thanathu*, *tharisu* and *poramboke* lands, all these concepts representing different forms of tenure (TLRM, vol. 2, 1915: 256–60, 336–7, 407, 928; TLRM, vol. 2, Suppl. 3, 1935: 441, 628–9; TLRM, vol. 2, Suppl. 4, 1943: 10; *Addresses to Dewan Bahadur M. Krishnan Nair, Dewan of Travancore* 1920: 207).

[70] Emphasis added, quoted in: Haridasan 1992: 34.

that land in Travancore should be solely or primarily cultivated by indigenous proprietors had hitherto never been openly expressed. Just over a year later, in February 1907 the newspaper claimed that the Travancore state should put the prosperity of its own citizens first and foremost, and not feel obliged to take care of the well-being of foreigners. In fairly blunt terms, the *Malayala Manorama* accused the Travancore administration of negligence in defending and promoting the interests of Travancorean cultivators. The newspaper propagated that 'it is (. . .) unpardonable and shameful that *our Government* is doing nothing to encourage *our farmers* or assist them to undertake this profitable venture [i.e. rubber planting]'.[71]

Gradually other people, members of the Travancore Legislative Council and the Sri Mulam Popular Assembly in particular, joined the *Malayala Manorama* in disputing the privileges of the European planting community. A member of the Assembly attacked the government's attitude towards plantation development, for 'capital, labour and everything is foreign'. He informed his colleagues that 'neither the state nor the people are benefitted'.[72] The protests in the constitutional bodies culminated in the period 1926–7, when a British concern intended to set up tea cultivation on a huge scale. The press claimed the firm had its eye on 200,000 acres.[73]

The political upheaval started while the Dewan, M.E. Watts, was on leave in England.[74] Meeting with representatives of Brooke Bond and Company, he discussed the further development of tea cultivation in Travancore. In their conversations, the Dewan suggested developing a plan whereby small owners might share in the advantages of tea production. Brooke Bond and Co., though finding the proposal 'contrary to the policy of the ordinary producing company, who only spend capital on their own interests', reacted positively to the Dewan's remarks, averring that it was more than just a producer but was greatly

[71] Italics mine, quoted in: Haridasan 1992: 34–5.

[72] Minutes MPA 24 June 1916.

[73] Fortnightly Reports on the Political Situation in the Madras States for the Year 1927, in: R/1/1/1644.

[74] It is worth noting that their Watts himself or some of his family members owned coffee estates in Peermade (R/2/884/147).

interested in the distribution and sale of tea in India. The British firm planned to create units of 1,000 acres, comprising 600 acres for the company and 400 acres for smallholders. According to Brooke Bond, with at least 25 acres planted with tea the latter group especially would profit from the proposed scheme. Apart from lending assistance in acquiring capital, ensuring the right type of seed, and learning the requisite planting skills, the company would permit the use of a fully-equipped modern tea factory. At the marketing stage, the small growers would be able either to sell their tea at public auction in London to obtain the world market price, paying the costs for sorting, packing, insurance, transport including shipping, brokerage and agency commission to Brooke Bond and Company, or dispose of their produce on the spot, getting the local value from the British concern.[75]

Having arranged matters with Brooke Bond, Watts decided to listen to the opinion of the Economic Development Board. When the Board met in November 1926 its President, N. Kunjan Pillai, waxed enthusiastic about the positive features of the proposed scheme. In his view, 'the large majority of our people could not take up tea cultivation for want of sufficient capital and technical knowledge'. For that reason, he argued, 'Messrs Brooke Bond and Co., Ltd, a rich and influential company of world-wide reputation, have offered to find the capital, provide technical knowledge and open factories for the benefit of smallholders on terms favourable to the latter'.[76]

The other members of the Economic Development Board, however, were much more cautious, preferring to reserve their judgements about the firm's intentions and the implications of the scheme. Although not against the plan as such, they demanded more detailed information, especially with a view

[75] Confidential letter of G. Brooke, Chairman of Brooke Bond and Company, to Watts, Dewan, dated 15 July 1926, in: R/2/884/163.

[76] Somewhat later in his speech, N. Kunjan Pillai argued on similar lines: 'In the past, foreign capital had come into the country. (. . .) There is, however, one vital difference between this and the present offer. All the development that had already taken place with the help of foreign capital had benefited only the foreigner, whereas if this scheme is worked out it will benefit the local people and particularly the middle classes who have no capital at their command'.

to finding out whether the industry could not be developed without the aid of foreign capital. Pertinently, they held the opinion that consideration of the scheme should not stand in the way of the registry of lands by Travancoreans for the cultivation of tea. Sensibly, the members also requested that a few Travancoreans with practical experience in the tea industry, like P.J. John, should be invited to attend the next meeting on the subject. Dr K. Madhavan Pillai, however, wanted no further discussions. In contrast to his fellow board members, he could not accept the principle of inviting foreign capital to develop industries. Profits would drain away from Travancore, he argued.[77]

Later that month, in a meeting of the Travancore Legislative Council, K.P. Nilakanta Pillai expressed astonishment that the Brooke Bond scheme, 'a definite matter of urgent public importance', was not openly being discussed. Instead, 'the Government are taking confidential steps in the matter and are carrying on secret negotiations with the company', causing 'great consternation in the minds of the public'. Watts, in his turn, denied the clandestine character of the negotiations and subsequently informed the Council about the scheme and the discussions in the Economic Development Board. Watts' reaction to Dr K. Madhavan Pillai's objection that land should not be registered in the name of foreigners was most enlightening. Reflecting the (anti-)nationalist discourse of those times, the Dewan stated: 'I can assure him that there is no proposal to register lands to *foreigners*. (. . .) It is an *English* company that has made these overtures to Travancore'.[78]

[77] Proceedings of the Economic Development Board dated 13 November 1926, in: R/2/884/163.

[78] Italics mine. Proceedings of the Travancore Legislative Council dated 24 November 1926, in: R/2/884/163. The full text included: 'Happily, Travancore was grafted on to the British Empire some hundred years and more ago and the same political tree cannot produce two varieties of citizenship. People of Travancore are members of the British Empire the world over and are fellow-citizens thus with other members of the British Empire. Therefore, those fellow-citizens from overseas should not be excluded in the rights of citizenship in Travancore any more than Travancoreans should be excluded from sharing in the rights of citizenship all over the British Empire. So British subjects are not foreigners in the sense Dr Madhavan Pillai contemplates'.

Though usually remaining in the background in the twentieth century, the royal family also entered the debate. Setu Parvati Bai, the Junior Maharani, wrote to C.W.E. Cotton, the Agent to the Governor General, Madras States, that 'the withdrawal of such a large extent of cultivable waste land from the total area of normal agricultural extension would be detrimental to the tenantry of Travancore'. She went on to say that it might well stir up needless discontent amongst those elements of the population who were articulate, probably conjuring with the idea of the Syrian Christians. She pointed out that so many people had taken up the cultivation of rubber and tea that the demand for land was growing by the day. Clearly, the Junior Maharani, the head of the regency administration during the minority of the Maharaja, was reluctant to take a decision which would tie the hands of the future government and which could create an unbridgeable gap between rulers and subjects. For this reason above all, she strongly emphasized that 'the proposal should neither have been mooted nor pursued during the minority of the Maharaja and during a Regency administration'.[79]

To the great annoyance of Dewan Watts, during the twenty-third session of the Sri Mulam Popular Assembly even more negative opinions about the Brooke Bond scheme were voiced. Narayana Pillai Kochukunju Pillai is a case in point. He feared the destruction of natural resources, the desecration of an important Hindu temple, scarcity of labour and a shortage of land. R. Nilakanta Pillai, another member of the Assembly, stated that it would be 'a suicidal policy if the Government made a permanent alienation of one-twentieth part of the State in favour of a foreign company', while in Travancore itself 'an educated and intelligent set of people, steadily advancing in economic organization and enterprise, were on the look-out for land for extension of cultivation'. Pointing to the rapid development of indigenous rubber and tea estates, the following speaker,

[79] Letter Setu Parvati Bai, Maharani, to Cotton, the Agent to the Governor General, Madras States, dated 19 February 1927, in: R/2/884/163. Even before the Brooke Bond scheme was mooted, the relationship between Setu Parvati Bai and Watts was already fairly fraught. In 1925, the junior Maharani had protested against his appointment as Dewan. She informed the British government that 'only a Caste-Hindu should hold this appointment' (R/2/884/115; cf. R/2/884/147).

S. Subbaraya Aiyer, underscored that 'the native planter did not stand in need of tutelage' by a European plantation company. During the discussions, Watts lost his temper many times. Petulantly, he told Narayana Pillai Kochukunju Pillai that 'you have just read out a long written essay full of facts and fancies of your own creation'. Watts went even one step further in his efforts to publicly humiliate the Travancorean. He accused Narayana Pillai Kochukunju Pillai of coming from a place where people indulged 'in necromancy by which they see visions and call up phantoms and spirits'.[80] At the conclusion of the twenty-third session of the Sri Mulam Popular Assembly, a Memorial, signed by a great number of the members stating once again their grievances against the Brooke Bond plan, was handed over to the Dewan.[81]

While this was going on, Cotton, the British agent, adopted a policy of 'wait and see'. This strategy was the logical outcome of a dilemma. Balancing on a knife-edge, his job was to promote British entrepreneurial interests, but he also had to prevent political instability. A less problematic task of the British representative agent was to brief the government of India regularly about current events. His assessment of the anti-Brooke Bond movement provides extremely interesting information. In his view, 'those who have interests already in tea and rubber, chiefly Syrian Christians, continue hostile, but the voting of the Nairs may turn the scale when the scheme is submitted for the approval of the Legislative Council in July'.[82]

[80] Proceedings of the Sri Mulam Popular Assembly dated 21 and 23 February 1927, in: R/2/884/163.

[81] Memorial presented to M.E. Watts, Dewan, in: R/2/884/163.

[82] In another letter to his superiors, Cotton reconfirmed that Syrian Christians were the major driving force behind the movement. Nayars pursued other objects. 'On the 1st May, when Mr Watts was about to leave his house to call upon me, he was surprised to find a deputation of Nair gentlemen waiting to see him including Messrs A. Narayana Pillai, K.P. Nilakanta Pillai, Dr M.N. Pillai and other signatories to last year's memorial against the Recency administration. They asked him to believe that they were now convinced that he was the Nairs best friend and assured him that they would support him in and outside the Council and even help him to get the Brooke Bond scheme through if he was convinced it was in the country interest' (Fortnightly Reports on the Political Situation in the Madras States for the Year 1927, in: R/1/1/1644).

This was not to be: the Brooke Bond scheme was never submitted to the Legislative Council for approval. The British company decided to withdraw the proposal. The argumentation of the board of directors of Brooke Bond and Company shows that the protests of members of the Economic Development Board, the Travancore Legislative Council and the Sri Mulam Popular Assembly, the warning letter of the Maharani, and the lobby of the Syrian Christians won the day. Brooke Bond and Co. informed Watts that 'the great natural difficulties to be overcome in road-making in what, to begin with, would prove an unhealthy area, is an obstacle which might be overcome, but we feel that an active opposition on the part of the Travancoreans themselves, even of a somewhat unintelligent minority, would prove an insuperable obstacle to the scheme'.[83]

Perhaps the most significant accomplishment of the anti-Brooke Bond movement was that from the late 1920s in the registration of lands suitable for the cultivation of plantation crops preference was given to subjects of the Travancore state.[84]

In the mid-1930s, the implementation of the International Tea Agreement and the International Rubber Regulation Agreement triggered off strong protests from the indigenous planters. As alluded to above, both supra-national regulations were introduced to restrict the export and cultivation of tea and rubber respectively in order to boost the international market price of these crops. In Travancore, the large, European-owned plantation companies – by no accident from the same stable as the initiators of the programmes! – were most in favour of a rapid implementation of the schemes. They approved the Travancore Tea Cultivation Restriction Bill and the Travancore Rubber Control Bill. The European companies, in a position to influence colonial policy, were certain that they would be granted permission to transport their products abroad.[85] Logically, many Travancoreans were against the introduction of the bills. The indigenous

[83] Letter of Brooke Bond and Company to Watts, Dewan, dated 9 June 1927, in: R/2/884/163.

[84] Dis. No. 807/29/Rev., from Chief Secretary to Government to the Land Revenue and Income Tax Commissioner dated 16 May 1929, in: TLRM, vol. 2, Suppl. 3, 1935: 75, and cf. TLRM, vol. 2, Suppl. 5, 1950: 383.

[85] R/1/1/2238.

planters feared that they would be denied access to the profitable world market and local entrepreneurs, planning to open up an estate, were afraid that they would not be given permission to act accordingly.

The attitude of the European planters, their planting associations and the colonial government, in particular with regard to the implementation of the programmes, galvanized the institutionalization of the Indian planting lobby. The South Travancore Planters' Association (initially named the Travancore Planters' Association), the Central Travancore Planters' Association, the Kannan Devan Planters' Association, the West Coast Planters' Association and the Mundakayam Planters' Association, jointly represented in the Combined Travancore Planters' Association[86] and the United Planters' Association of Southern India, were no longer the sole formal planters organizations. The Indian Tea Planters' Association represented the interests of the indigenous tea entrepreneurs with estates in central Travancore and Mundakayam. The Kanjirapalli Planters' Association and the Travancore Rubber Planters' Association promoted the interests of the local rubber cultivators.[87]

The latter organizations especially were leading lights in the struggle against the introduction of the rubber restriction scheme. In 1934 at least four protest meetings were called: one in May, the All Travancore Rubber Cultivators' Conference held in Palai and presided over by K.C. Chandy, and the others in June and July, held in Kottayam and presided over by P.J. Varghese, Thomas Dominic and Karimpanal Thomas respectively. The object of all four meetings was to make clear, both to the Travancorean cultivators and to the government officials, that the rubber export quota granted to India was far too limited. Therefore, resolutions were passed which requested the Travancore, Cochin and British Indian governments 'not to accept the Rubber Restriction Scheme or pass laws in support of it unless the harmful conditions therein are removed'.[88]

Conversely, the relatively weak lobby of the local tea planters

[86] Minutes MPA 24 June 1916; *UPASI Planting Directory* 1925: 139.

[87] *UPASI Planting Directory* 1934: 135, 143, 159, 163 and 175; Weekly Secret Bulletin 1934 No. 1: § 201; Hatch 1939: 104–5.

[88] Weekly Secret Bulletin 1934 No. 1: § 155, 167, 201 and 220; and cf. proceedings TSMA 28 July 1934; proceedings TSCSC 15 August 1934.

contributed to the prompt acceptance of the Travancore Tea Cultivation Restriction Bill by the legislature, the bill becoming law in 1934.[89]

Predictably, it was the large tea estates especially, owned mostly by European companies, which profited most from the new situation. R.V. Thomas, a member of the Travancore Sri Chitra State Council, claimed that the legislation caused 'great hardship on the tea estate owners in Travancore'. The scheme was particularly detrimental to the small tea estates. 'Their grievances are not heard by the [Indian Tea Licensing] committee', Thomas said, while 'those who are owning large acreages are granted quotas'.[90] Also illustrative of the subordinate position of the indigenous planters is the fact that neither they personally nor their planting association were extended membership in the Indian Tea Market Expansion Board. This was all the more frustrating since the Travancorean planters, like their European counterparts, were obliged to pay a tea cess, which was introduced to finance the expansion of the South Asian tea market.[91]

The anti-restriction movement organized by the indigenous rubber cultivators was able to exert more leverage. Its first victory was when the Travancore government agreed to absorb some of the superfluous rubber by opening up a rubber factory. At the opening ceremony in August 1935, the Maharaja once more reiterated the purpose of the Travancore Rubber Factory in no uncertain terms. He informed his audience that 'as a result of the agreement', meaning the International Rubber Regulation Agreement, 'it would not be possible to export the whole quantity of rubber produced in Travancore to other countries, and it will therefore be necessary to devise other methods to utilise as much of the excess rubber as possible'. Therefore, he continued, 'the starting of a rubber factory at this juncture is peculiarly opportune'.[92] The Travancore government also played its part by refusing to join the scheme unless the quota was raised. Fortunately for the rubber cultivators, the International

[89] Regulation II of 1109, dated 6 March 1934, in: RPT, vol. 7: 614–17; later changed into Act XV of 1119, dated 5 March 1944, in: RPT, vol. 14: 122–38.
[90] Proceedings TSCSC 9 August 1937.
[91] TLRM, vol. 2, Suppl. 4, 1943: 524–5 and cf. 538–9, 678.
[92] Varma 1938: 103.

Restriction Committee complied with their wishes, to some extent at least.[93] Matters were helped along even better when in certain respects the original Travancore Rubber Control Bill was altered in favour of the local cultivators, first by a Select Committee and later by the Travancore Sri Mulam Assembly,[94] before it was finally passed by the Travancore Sri Chitra State Council.[95] In 1940, the legislation was replaced by another Travancore Rubber Control Act, amending the previous legislation on some minor points.[96]

To complete the picture, within the indigenous plantation segment the large estates especially profited from the growing demand for rubber, the enlarged export quota and the amendments to provisions in the legislation. Taking the opinion of the Plantation Enquiry Commission as a guide, one important reason was that many smallholders were not 'alert and assertive'.[97] But to put the record straight, other sources point to the combined efforts made by the small cultivators to combat the privileges of the large planters. It was not uncommon for their representatives in the legislature to make statements like 'it will cause a great deal of hardship to the ordinary poor agriculturists'.[98] Voices were raised in 1937 and again in 1939 when the owners of small estates in Meenachil protested about the partiality shown by the rubber quota assessors to the proprietors of large plantations.[99] Yet despite the pains they took, in comparison with the influential lobby of the large estates the efforts of the small rubber cultivators proved bootless.[100]

93 Proceedings TSCSC 9 December 1936.

94 Proceedings TSMA 19 and 20 November 1936.

95 Proceedings TSCSC 9 December 1936; Regulation II of 1112, dated 25 December 1936, in: RPT, vol. 9: 25–35.

96 Regulation XXI of 1115, dated 16 February 1940, in: RPT, vol. 11: 814–31.

97 Plantation Enquiry Commission – Rubber 1956: 99.

98 Proceedings TSMA 19 November 1936; and cf. proceedings TSMA 20 November 1936 and 24 June 1939.

99 Weekly Secret Bulletin 1937 No. 4: § 105; proceedings TSCSC 11 August 1939.

100 Indian membership in the 'European' UPASI was again allowed since 1937 (Muthiah 1993: 473–8).

5.3 Changing Labour Conditions

5.3.1 Indian Methods of Labour Management

Control of production costs was another field in which European supremacy was contested. In general, the Indian planters were operating much more cheaply than their European counterparts. Some Travancorean rubber planters informed the Economic Depression Enquiry Committee that a realization of three and a half annas per pound was sufficient to run the estates without loss. This contrasted with the statement issued by the British planters that four annas a pound would just about keep the rubber estates going. In the tea sector, indigenous estates were producing tea at six annas per pound in the pre-Depression years and at four and a half to five annas per pound in the subsequent period. During the same time span, the European tea estates reduced the costs from seven annas per pound to six annas per pound. The greater difficulties encountered by the large British estates especially during the Depression are indicative of their relatively expensive mode of production.[101]

These figures may seem surprising given the fact that the European planters had definite advantages over the Travancorean entrepreneurs. One such advantage was that most foreign planters had large tracts of highly fertile land at their disposal. Taking into account the economies of scale and the quality of the land, it would be reasonable to expect that the British estates would produce more cheaply per unit than the frequently smaller-sized and worse-situated indigenous plantations. Another of their plus points was that the Criminal Breach of Contract Act was especially designed to protect the interests of the European entrepreneurs. This particular piece of labour legislation, which was indeed primarily, if not solely, used by the European community, was another factor which might give the impression that the foreign entrepreneurs were able to produce more cheaply than the indigenous planters. Undeniably, the members of the United Planters' Association of Southern India, which remained an overwhelmingly European organization, were aided by its Labour Department. The latter, an initiative of

[101] Economic Depression Enquiry Committee 1931: 72–4.

James Finlay and Company, took in hand the large-scale recruitment of labourers from the Madras Presidency as well as the round-up of runaway plantation workers.[102]

Despite these conditions which weighted the odds in favour of the Europeans, the Travancorean planters were able to run their estates at a lower cost. To a certain extent, this was related to the quantity and quality of the capital goods available on the estates. Lacking adequate financial resources, most indigenous entrepreneurs were unable to procure the latest equipment to process either tea-leaves or latex, leaving the European planters to produce better products for which they obtained higher prices. Yet all the while, they were also under a greater financial pressure.

Going some of the way towards an answer, this explanation is still not enough. In weighing up this matter, the indigenous methods of labour control must on no account be overlooked. Like their European counterparts, indigenous planters allotted

TABLE 5.8
Community and Plantation Managements in Travancore in 1921[103]

Community	*Directors of Companies*		*Proprietary Planters*		*Managers*	
	Tea	*Rubber*	*Tea*	*Rubber*	*Tea*	*Rubber*
Europeans/ Anglo-Indians	49	23	15	5	63	27
Indian Christians	-	-	2	10	5	14
Brahmins	-	-	-	1	1	1
Konkanis	-	-	-	-	-	1
Nayars	-	-	-	-	-	1
Indians Unspecified	3	5	-	-	-	-

[102] Martin's Memorandum on Rates of Coolies' Pay, in: IOL&R: MSS Eur D 1204; minutes KDPA 6 June 1923; RCL 1931 Evidence Vol. 7, part 2: 199; 25th–30th and 32nd Annual General Meeting of the UPASI, Madras (1918) and Bangalore (other years), 19–28 August 1918, 25–28 August 1919, 23–26 August 1920, 22–25 August 1921, 21–24 August 1922, 20–24 August 1923, 17–20 August 1925, in: UPASI Proceedings, Martin 1931: 21–23; Soutar 1976: 23; Ravi Raman 1991: 248; interviews: A Manuel; Sundar.

[103] *Census* 1921, part 2: 120–1. It should be noted that some managers did

particular tasks to particular categories of people. Members of high castes generally tended to be found in the higher echelons of the plantation hierarchy. Table 5.8 provides some more detailed information about estate management. Apart from Europeans and Syrian Christians, Brahmins, Konkanis and Nayars also worked as plantation managers, though it is difficult to establish whether they were employed on foreign or indigenous estates. The Konkani caste, not yet discussed, also enjoyed a relatively high status, slightly below that of the Brahmins.[104]

Conversely, low-caste Tamil Hindus were preferred as field workers, a category including kanganies. To keep the plantation hierarchy safe, both the Indian and European planters working in the Mundakayam district, where Malayali labour was close at hand, still imported large numbers of Tamil labourers.[105] Table 5.9, which indicates the community background of the work force, supports the view that members of the castes considered lowest were preferred as field workers. This group included the Parayas, Sakkiliyans ('leather-workers'), Pulayas and Kuravans ('agricultural labourers').[106] One step above them were the low and middle castes, like the Channan/Shanar, Ezhava, Maravan,

not co-operate with the Census officials: in actual practice a much larger number of managers and workers was employed on the Travancore plantations. Given the total area planted with tea (50,296 acres) and rubber (37,498 acres) in 1925 (see Table 5.1), and basing ourselves on the assumption that tea estates employ one worker per acre and rubber plantations one worker per three acres (see page 123, note 30), the Travancore plantation sector amounted to approximately 63,000 estate workers in the early 1920s.

[104] *Census* 1911, part 1: 261.

[105] Especially after the 1930s, the recruiting practices of the planters were not to the liking of the administrators of the Travancore state. The *Census* of 1931 for example stated: 'As against 76,699 foreign labourers working in estates, the Travancorean labourers employed therein number only 25,791. It is a legitimate question to ask why the estates import foreign labour when there is a large volume of unemployment in the country. If the estates would recruit all their labour from the State itself, more than 75,000 persons could find work, and to that extent the number of the unemployed here would go down' (*Census* 1931, vol. 1: 87). 'It would be in the interest of the State if a progressive reduction is made in the number of coolies imported into the estates', the author of the *Census* of 1941 informed (*Census* 1941, part 1: 152).

[106] *Census* 1911, part 1: 262 and 267.

Kammalan ('artisans'), Muslim and Vellala ('farmers') communities.[107] It is noteworthy that only a very limited number of Brahmins and Europeans/Anglo-Indians were categorized as 'estate workers'. The fairly large number of unskilled Nayars working at plantations is food for thought. It reflects the great financial problems with which the members of that community often had to cope. It also suggests that while members of low castes were unable to become managers, some people with a high community status were forced, by economic circumstances, to take on the work of a field labourer. The category 'Indian Christians' gives rise to some confusion. In Table 5.8 this group

TABLE 5.9
Community and the Estate Work-force in Travancore in 1921[108]

Community	*Skilled Workers*	*Unskilled Workers*		*Total*
	All Estates	*Tea*	*Rubber*	
Paraya	1,587	1,455	650	3,692
Indian Christian	616	1,649	549	2,814
Channan/Shanar	940	926	188	2,054
Sakkiliyan	955	500	172	1,627
Ezhava	96	800	626	1,522
Maravan	267	305	262	834
Nayar	46	422	336	804
Pulaya	–	296	349	645
Kuravan	145	200	262	607
Kammalan	107	174	286	567
Muslim	7	293	127	427
Vellala	27	224	115	366
European/ Anglo-Indian	20	–	–	20
Brahmin	1	2	–	3
Other	173	181	154	508

[107] *Census* 1911, part 1: 260, 268–9.
[108] *Census* 1921, part 2: 122–3 and 130–1. See also page 145, note 104.

indicates especially, and probably exclusively, Syrian Christians, but in Table 5.9 the same category hints at the great number of newly-converted, low-caste Christians.[109]

In contrast to the situation prevailing on the British estates, the management structure of the local estates was less comprehensive and less static. In the previous chapter it emerged that most European plantations had been absorbed by large companies, and consequently the distance between the top and bottom of the hierarchy had grown enormously. The structure of the plantations situated in Travancore and owned by James Finlay and Company is a good example: shareholders (mostly British) – board of directors (Glasgow) – managing agents (Calcutta) – general manager (Munnar) – managers (estate level) – assistant managers – staff members – kanganies – field labourers. Indicative of the limited autonomy allowed to the managers, the managing agent represented 'the most powerful man throughout the (. . .) industry'[110] and the head office in Glasgow was 'like God'.[111] Antithetically, the management structure of most indigenous estates was fairly simple and dynamic. Normally, all executive persons connected with the enterprise were actually to be found at least in south-west India if not on the estate itself. Another of their strengths lay in the fact that in most companies, the persons or families owning the majority of the shares simultaneously functioned as directors and managers.[112] Because of the relatively small size, on many privately-owned plantations there was no need for assistant managers. A further advantage was that the indigenous management structures were more direct. To cut

[109] Interviews: Tharian George; V. Haridasan; V.J. Joseph and V.J. Shaju; Abraham Kurien and Abraham Eappen; Suresh; P. Madhavan; K.T. Thomas and T.J. Joseph; Michael A. Kallivayalil. Cf. Minutes MRPA 7 January 1911; Clay 1911: 24; *Census* 1931, vol. 1: 83–4; Lewandowski 1980: 35; Clough 1984; Tape 1 Side 1; Mackay 1984; Tape 1 Side 2; Tewson 1984, part 3: 2–3; Chiriyakandath 1985: 113. Indicating once again that particular tasks were allotted to particular groups, in the Kannan Devan Hills the house servants – probably regarded as an intermediate category in between (high-caste, Christian, Malayali) staff and (low-caste, Hindu, Tamil) field labourers – were their Hindu Malayalis or Christian Tamils (Clough 1984: Tape 1 Side 1).

[110] Mackay 1984: Tape 1 Side 1.

[111] Clough 1984: Tape 1 Side 1.

[112] Malankara 1961: 5 and 10; interview: Abraham Kurien and Abraham Eapen.

a long story short, the local enterprises possessed two overwhelming benefits: they needed less personnel and they were in a better position to adapt to the constantly changing reality.

Less hampered by their background, the indigenous planters had fewer difficulties with labour recruitment. They were not afraid to attract workers with anti-British sentiments and, even more importantly, the Travancorean planters were in closer contact with the kanganies. This was first and foremost closely related to the more compact plantation hierarchy of most local enterprises. Of almost equal importance, there was no language barrier between the management, the staff, and the field workers. Even if the latter category spoke Tamil, most Mayalali managers had little difficulty in carrying on a conversation. Thirdly, most indigenous entrepreneurs had had previous experiences of the employment of a temporary or permanent labour force. In sharp contrast to most European managers, because they had been born and raised in the countryside of Travancore, the indigenous planters were more in touch with the labour conditions in south-west India in general and on their estates in particular.[113]

An important external factor was that most journalists, politicians and government officials did not pay much attention to the living and working conditions on the Travancorean-owned estates. Like the South of India Planters' Enquiry Committee (1896), the Royal Commission on Labour (1931) focused on the labour policies of the European planters.[114] The findings – including inadequate education for the workers' children, child labour, low wages, almost total absence of health care and overcrowding of 'cooly lines' – implied that at least some measures had to be taken by the British planters to improve labour conditions. The European entrepreneurs feared that should they not do so the British-Indian administration would impose retaliatory policies which would turn out to be more costly than their own measures. The nationalist movement[115]

[113] Cf. Malankara 1961: 11; Chiriyankandath 1985: 111–12.

[114] RCL 1931: 399–402, Evidence Vol. 7, part 1: 154–64 and part 2: 193–218, vol. 11: 383–5.

[115] It should be noted that the party loosely affiliated with the nationalist movement, the Travancore State Congress, was dominated by Syrian Christians, the same group which had great interests in the plantation sector (see also Sections 5.1.3 and 6.1.3).

had no interest whatsoever in accusing the Indian planters of similar abuses, and the colonial administration was too busy counteracting the grievances against British rule to be interested in the labour conditions on indigenously-owned estates located in a princely state. All this lack of attention meant that there was no immediate need for the Travancorean planters to improve the conditions of the workers which were at least as appalling as those on the European estates.[116]

5.3.2 *The End of the 'Good Old Days'*

Although threats against the British planters' prominence in politics, the allocation of land, and finance, began in the early 1900s, European methods of labour control came under serious attack only in the 1930s and 1940s. The Criminal Breach of Contract Act was abolished in 1935 (see Section 5.3.3) and labour unions emerged in the late 1930s (see Section 5.3.4). Therefore, at least during the first three, perhaps even four, decades of the twentieth century the relationship between British management and Indian labour was that between king and subjects, between a master and his servants, or, to phrase it in terms more appropriate to the Asian context, between sahib and coolies.

It is therefore hardly surprising that, reflecting on their lives in Travancore after repatriation, the British planters remembered their years in India as the *Good Old Days*. The nostalgia was heightened as most tyro planters arrived in Travancore without much self-esteem. In their home countries they had little or no status. At the risk of making an understatement, most planter-to-be were neither brilliant pupils nor rich and influential citizens back home. In colonial India, the colour of their skin, combined with qualities like enthusiasm, perseverance, and leadership, guaranteed personal success in the plantation business. Richard R. Tewson (born in 1915) left school in England 'with no particular regrets and few academic distinctions'. In the Kannan Devan Hills, despite being 'new' and 'inexperienced', he immediately became 'monarch of 450 acres of tea and sahib and arbiter of a labour force of some 500 souls'.[117] In subsequent

[116] Cf. RCL 1931 Evidence Vol. 7, part 2: 213; Thomas 1969: 1.
[117] Tewson 1981, part 1: 1 and 8.

years, the acreage and workmen under Tewson's control gradually expanded.[118] According to another European planter, W.S.S. Mackay (born in 1904), the planting fraternity quite rightly gained the reputation of being blessed with 'more brawn than brain'.[119]

This is not to say that the management's position remained fixed and immutable, particularly when compared to the late nineteenth century. Aware that their political and economic power in India was waning, the British planters realized that preserving their prestige and status had become of crucial importance in maintaining the existing *modus operandi.* The boards of directors of the various European plantation companies played significant roles in this regard. By strictly regulating the behaviour of their European personnel, they formalized, intensified, and justified the racial character of the plantation hierarchy.

Worried about the morals of their employees, the directors of James Finlay and Company tried to change the planters' sexual mores. Concubinage with Indian women was discouraged in various ways. At first, the coming out to India of European females was stimulated. Meetings with European girls were organized and allowances were given to those planters who were planning to bring their fiancées or wives into the Kannan Devan Hills district.[120] In later years attitudes toughened and having an Indian mistress could even spell the end of a planter's career. Apart from maltreating labourers 'too violently', a relationship with an Indian woman was the only other major reason for dismissal.[121]

Sensing that their position was being threatened by Indian competitors, most European women supported the policy of the company directors wholeheartedly. Generally sure in their moral rectitude, the wives, especially those of senior planters, publicly condemned European-Asian or European-Eurasian relationships. Being married to an Eurasian himself, Mackay was irritated by the undisguised attitude of superiority flaunted by these European ladies. As he recalled: 'Liaison with Indian girls

[118] Tewson 1984, part 1: 8 and part 3: 3.
[119] Mackay 1984: E1.
[120] Minutes MRPA 8 March 1910; Martin 1931: 9; Clough 1984: Tape 1 Side 1; Tewson 1984, part 1: 10.
[121] Clough 1984: Tape 1 Side 1.

had become taboo and some of "les grande mesdames" paraded themselves in the Club with an air of conscious rectitude I was later to find quite nauseating'. The European women were also strongly in favour of keeping the 'European' offspring separated from the 'Eurasian' children. One illustration which underlines the fact that the latter category was treated as inferior, indeed as outcaste, was their preferred invisibility: they could not be seen in public and were sent out of the district for their education and employment.[122]

Tragically for the planters' mistresses and their children, they were not respected by high-caste Indians either.[123] Probably the most salient reason for this denigration was the low-caste origin of most Indian concubines, but the privileges enjoyed by some of these Indian women and their children, in terms of housing, education and/or employment, was also resented as unfair and improper.

Despite the company's official policy backed up to the hilt by the European ladies and the Indian staff, many planters were not prepared to have the terms of their personal lives dictated. Mackay wrote scathingly condemning the ignorance of the wives of influential planters: 'few of whom, if indeed any, were aware of their own husband's coffee coloured progeny at school in the plains'. Mackay also had information that venereal disease was rampant in the district, especially at the time when many planters had returned to the Kannan Devan Hills after having served their country as soldiers during the First World War. In one particular year, 'no less than 11 cases of V.D. [venereal disease] among Europeans had been privately treated in the Munnar General Hospital'. Such was the fear of losing prestige, especially with the other members of the European community, and of the severe sanctions exacted for breaking the moral code, that – as Mackay stated further – 'One boy was so ashamed, he failed to report to the company's medical officer until too late. He became seriously ill and when his case came before his superiors at headquarters, his contract was not renewed'.[124]

[122] Mackay 1984: C8.

[123] Mackay 1984: Tape 1 Side 1.

[124] Mackay 1984: C8–9; Mackay 1984: Tape 1 Side 2. With regard to the regulation of sexual relations in colonial societies, and in plantation societies in colonial Asia in particular, see: Stoler 1990.

The boards of directors also chose to enlarge the social distance between the management and the staff, and thus between the European planting community and the high-caste Indians. Europeans were discouraged from socializing with Indians, a regulation which brought a lot of loneliness in its wake, especially among the women who were mostly confined to the bungalow during the working week. As Mrs Clough noted: 'it wasn't a good idea, it was thought, to be familiar with your own office or field staff – and that probably from a managerial point of view was right – but there was nobody else'. The European children were not allowed to mix with their Indian contemporaries either.[125]

The strict separation was not solely the policy of the company directors. The Indians themselves were often disgusted by the drinking (alcohol) and eating (meat) habits of the Europeans. The religious scruples of the higher castes prevented most staff members and their relatives from becoming intimate with the managers' families. Reminiscently, Clough noted: 'If a Brahmin ever called on my father [who was the manager of an estate owned by James Finlay and Company], he always talked to him on the verandah' for 'that would entail less exercises to recover from the pollution from going into (. . .) a European house'.[126]

Besides providing regulations regarding women and staff, the directors taught the managers how to deal with the field workers. In fact, the management was trained to regard the plantation as a military organization and to act accordingly. Consequently, plantation life resembled a soldier's life in many ways, thus justifying the use of E.T. Thompson's concept of estates as 'military agriculture'.[127] According to Mackay:

Our daily routine seldom varied. At 6.00 a.m. we had a quick cup of smoky tea and dashed off to the muster ground half a mile from the bungalow. The head conductor (. . .), an educated Syrian Christian (. . .), was always there before us, and had already sounded a great brass gong, signalling the headmen to turn out their workers from the lines. There was also a special man whose job was to (. . .) blow

[125] Clough 1984: Tape 1 Side 1; cf. Mackay 1984: Tape 1 Side 2.
[126] Clough 1984: Tape 1 Side 1.
[127] Cf. Wolf 1982: 315.

on an instrument like the fog siren of a ship in distress, or the warning boom from a lighthouse. The labour was mustered in gangs like a military parade and dispatched under junior headmen to various works. The women were sent for plucking leaf, the men to pruning, road work or one of many tasks calling for considerable physical effort. After the early parade, the 'muster chit' or labour distribution statement had to be signed and dispatched to the P.D. to reach him not later than 8.00 a.m. We then had breakfast and started our day of supervising and checking work loads and tasks. The green leaf plucked by the women was weighed up at noon, and again at night, and dispatched on a wire chute down to a point close to the factory.[128]

Tewson told much the same story, adding some interesting information about child labour.

Morning muster was at 6.30, when work was allotted to men, women and children (weeders), some who seemed small enough to run to their mothers at break time for a few minutes at the nipple!

Evening muster took an hour or so, really the equivalent of beating the retreat in the Army. All the work force went back to their lines (houses), shaking the wet out of their blankets [which they used during the field work to protect themselves against the rain and cold] ready for the evening meal and bed, wrapped in the same blankets, poor devils.[129]

To prevent the danger of the labourers regarding the managers simply as exploitative dictators, the directors tried to create the image of a 'strict but righteous' planter. In other words, they attempted to fashion a paternalistic figure, a person who gave as well as took, who had certain rights but also particular responsibilities, and, simultaneously, one who stood above the whole situation. To give concrete form to their ideas, the directors required the management to preserve law and order. Exactly as Tewson's statement quoted above indicates: simultaneously becoming *'arbiter'* and *'monarch'*.[130] The following story shows that Mackay combined both privileges as well. When a murder was committed on an estate on which Mackay was acting manager in 1930, he sent for the local Indian Police

[128] Mackay 1984: C15–16. P.D. refers to *Peria Durai* or 'big gentleman', i.e. the manager. The assistant manager was called *Sinna Durai* or 'little gentleman'.

[129] Tewson 1984, part 1: 8.

[130] Cf. Clough 1984: Tape 2 Side 1.

Inspector. The policeman made enquiries and reported to Mackay that he 'suspected foul play'. On being asked what they should do with the body, the Inspector replied 'As master pleases!!' 'A sum of Rs 20 settled the matter and I heard no more of it', Mackay recollected.[131] The managers' tasks at the estate level, also in this particular field, were supervised by the general manager. As Mackay stated, 'Law and Order were largely in the General Manager's hands'. 'The High Range was a state within a state, and the General Manager was King'.[132] That the management was requested not to behave 'too violently' towards their subordinates should also be seen in this specific context. Such behaviour would invalidate the image of the 'righteous planter'.

Although some knowledge of the workers was essential to act as both judge and ruler, incongruously, at the same time the planters, as well as their wives and children,[133] were prevented from becoming too personally involved with the labourers. Learning Tamil, beyond what was absolutely necessary, was discouraged. 'We were not expected to take too close an interest in the labour who worked for us.' As Mackay explains 'the official reason': 'popularity with labour meant time wasted in listening to their complaints, and often pathetic little domestic worries'. For the management, 'the job was to get as much work out of them as possible'. Focusing on the issue of language, Mackay went on to say that there was no encouragement to speak the language

[131] In the interview, Mackay gave the subsequent, almost similar, version of the murder case. 'I tell you what happened to my first murder. (. . .) There was a murder! So I sent for the inspector of police. (. . .) The inspector arrived (. . .) I knew him quite well and we had a talk and he said: "Sir, there has been some foul play". I said: "Yes inspector, that's fine, the man is dead, he obviously has been murdered – What am I to do?" He said: "As master pleases". So for 20 rupees the matter was settled and then we never heard anymore about it'. Subsequently the interviewer asked Mackay why he did not want a proper investigation? 'Why should one be involved in trouble if it can be avoided, and, after all, it might as well happen all damn day in India', Mackay answered (Mackay 1984: Tape 2 Side 1).

[132] Mackay 1984: C2. In a similar phrase, Mackay stated: 'We were a state within a state; we were law and order; we were everything . . . up to the second war' (Mackay 1984: Tape 2 Side 1).

[133] Clough 1984: Tape 1 Side 1.

fluently. All that was required was enough to make them understand how the work was to be done.[134]

Therefore, the most obvious conclusion is that, in contrast to the nineteenth century in which most planters were proprietors-cum-managers, during the twentieth century the majority of the planters – salaried managers – were not free to determine their own attitudes and behaviour towards their staff and field workers. It was an epoch in which most planters themselves fell victim to stringent labour strategies imposed from above. The new leaders of the plantation industry, the directors of the various planting companies, disciplined the management as part of their strategy to control the labour force.

Given such an enclave turned in on itself it is no wonder that the first half of the twentieth century witnessed the heyday of European Clubs. Having long working weeks – from 6 a.m. till 6 p.m. daily with the exception of Sundays – and socially totally isolated from staff and field workers as well as from other Europeans – the managers were not supposed to leave their estates during the working week[135] – the planters sought their diversions at the Club. The men gathered to hear the latest news from fellow planters, to play games[136] and to drink, while their wives and children busied themselves arranging the library and organizing social events like amateur theatricals or Christmas celebrations. Without any doubt, the Club functioned as an outlet for the tensions generated by a plantation regime of which the planters were both begetters and victims. It must be said that the planters' sources seem to indicate that the amusement enjoyed at the weekend and during holidays far outweighed the other, often desolate, days.[137]

Before ending this particular sub-section, one further observation should be made. It was certainly no coincidence that most European planters obtained some practical military training. As Mackay commented about the European planters in the

[134] Mackay 1984: C7.

[135] Clough 1984: Tape 1 Side 1; Mackay 1984: C10.

[136] The sports included billiards, cricket, football, rifle-shooting, rugby and tennis. The Club at Munnar even had the disposal over an 18-hole golf-course.

[137] Martin 1931: 39–41; Clough 1984: Tape 1 Side 1; Mackay 1984: C10–11; Tewson 1984, part 1: 10.

Kannan Devan Hills district: 'all of us were expected to join the Southern Provinces Mounted Rifles'.[138] The original idea of auxiliary corps came from a leading figure in the plantation industry, A.W. Macrae, the director of the coastal firm Peirce Leslie and Company. With the unanimous support of the European planting community and sanctioned and financed by the British-Indian government (1904), in 1905 cavalry units were established in the various plantation districts of southern India, including those in Travancore. The colonial officials saw the volunteer corps as part of the machinery for maintaining internal security in times of unrest or war. From the point of view of the European planters, these forces were especially useful in times of peace. Undoubtedly, the military divisions enhanced their prestige, supporting the belief that the *European* planters protected India against *foreign* intruders. More practically, the planters' experience with military exercises encouraged them to behave in a like manner during working weeks. 'I found myself adopting Army methods with my labour force', Tewson wrote.[139] Lastly, the regiments were also supposed to aid the local magistrates and police forces. Mackay says 'the raison d'etre of the entire A.F. [Auxiliary Forces] organization throughout India was not only internal security but military aid to the civil power and close liaison with the army, particularly in respect of isolated and scattered European communities'.[140] This view is confirmed by Tewson. The lessons in horsemanship and shooting were primarily intended to be 'in aid of Civil Power'.[141] It is perhaps superfluous to mention that in the plantation districts, one of the main aims of both magistrates and policemen was to enforce the Criminal Breach of Contract Act.

5.3.3 *Towards the Abolition of the Criminal Breach of Contract Act*

Though the various company policies changed the nature of plantation life, the planters' control over the field workers' debts remained the core factor in the recruitment and

[138] Mackay 1984: C17.
[139] Tewson 1984, part 3: 1.
[140] Mackay 1984: C17.
[141] Tewson 1984, part 1: 9.

disciplining of labourers.[142] In this respect, the Criminal Breach of Contract Act continued to play an important role. With the coming of the twentieth century, prosecution of 'absconders' was facilitated by taking the workers' thumb prints on arrival at the estates.[143] Undeterred, kanganies and labourers continued to flee the estates: sometimes trying to reach home; sometimes searching for better working conditions elsewhere in the plantation regions.[144]

Paradoxically, the first threat to the Criminal Breach of Contract Act was initiated accidentally by the planters themselves. The planters' miscalculation began thus. In March 1892 a petition, dealing with the planters' labour problems and signed by the chairmen and secretaries of the various district associations, was presented to the government of India. Failing to respond to the planters' high expectations, the administration was not prepared to pay serious attention to their difficulties. The response was dismissive of the Memorial as a letter from 'certain planters of the South of India' and stated that the labour problems were created by the planters themselves. Therefore, the government refused to take any action whatsoever.[145]

In order to strengthen their lobby, of which the aim was to bring changes in the labour situation, the planters formed the United Planters' Association of Southern India (UPASI) in August 1893. In the subsequent period this central organization of the planters kept up a fairly frequent barrage of addresses to

142 RCL 1931 Report: 400, Evidence Vol. 7, part 1: 155. The author of the *Census of Travancore* of 1921 clearly felt compassion for the workers when he wrote: 'In the case of labour recruited for work in the tea and rubber estates in the hills, the recruitment is made by kanganies who advance money to the coolies to meet their urgent wants, to purchase new clothes and to spend a little for drink also. During their stay in the estates, they pay off the debt and return home with little or no savings to be recruited again under similar conditions. The coolies themselves are in a chronic state of indebtedness. Their relation to the kangany is not one of love but is that of a debtor to a creditor' (*Census* 1921, part 1: 156-7).

143 17th Annual General Meeting of the UPASI, Bangalore, 1-5 August 1910, in: UPASI Proceedings; RCL 1931, vol. 7, part 2: 198; proceedings TSMA 24 January 1935.

144 Muthiah 1993: 301-7; see also: Baig and Henderson 1978: 19-20.

145 Mayne 1953: 4-5.

the governments of India and Madras about the Labour Question and how to solve it. In contrast to the administration of Madras, the government of India was opposed to the introduction of new legislation which would give the position of the planter any extra leverage over that of the kangany and the labourer.[146] In this epoch when especially in Western Europe more and more attention was gradually being paid to the interests of the employees, the Indian administration was more concerned with criticism from the British parliament than the Madras government.

Finally, in December 1895, following a short meeting with representatives of the UPASI, the Viceroy agreed that there was a case for setting up an enquiry into the difficulties experienced by the planters, although he considered the remedies proposed 'opposed to the whole spirit of modern legislation'.[147] Nonetheless, in March 1896 the South of India Planters' Enquiry Committee was installed.

The composition of the committee, the questions tabled and the nature of the evidence collected reveal all too clearly that the enquiry was meant to defend the interests of the employers only. The committee was composed of one representative of the UPASI and two civil servants. It was innocent of representatives of kanganies or labourers, who at that time were also not yet formally organized. Even more revealingly, the goal of the committee was to give an insight into: (a) the nature of the advance and kangany system; (b) the necessity for its continuance; (c) the problems experienced in its operation; and (d) the best methods of removing these difficulties. Putting two and two together, obviously finding a solution which would suit the planters best and which was at the same time acceptable to the governments of India and Madras was the core issue. This assertion is supported by the fact that most of the evidence was collected from the planters themselves: 197 proprietors and managers of estates and twelve secretaries of planting associations answered the questionnaire sent by the committee, and at least seventy officials gave their opinion. In contrast, only fourteen

[146] Griffiths 1967: 398–9.

[147] Quoted in: Griffiths 1967: 399.

kanganies and not a single plantation worker were heard. The outcome of the enquiry was therefore a foregone conclusion. The committee recognized the need for advances and kanganies and appreciated the planters' difficulties. The upshot was that it recommended the enactment of an amended labour law for the Madras Presidency. It appended the advice that the princely states, which apart from Travancore also included Mysore and Cochin, should be requested to adopt the new legislation as well.[148]

Although most planters were content with the results of the enquiry, the subsequent discussions between government officials and planter representatives about the new labour legislation were fraught with frustration. Nevertheless, the Madras Planters' Labour Bill went a considerable way towards solving the planters' difficulties. For example, kanganies who absented themselves from work or who failed to account for the advances received, were, like the labourers, liable to imprisonment (see Clause 24 of the final draft of the bill). To back this up, the bill introduced reciprocity in the matter of issue of warrants between the Madras Presidency and the princely states (Section 44). Try as they might, the planters could not prevent certain rules which protected the interests of workers and administrators being included. The labourers were given the right to sick pay (Article 14), the planters were obliged to provide their labourers with house accommodation, water-supply, sanitary arrangements and medical attention (Section 15) and certain officials were allowed to inspect the estates (Clause 16). After prolonged deliberations, the Madras Planters' Labour Bill, approved by the representative of the UPASI, was passed in early 1903.[149]

The new legislation was only to apply to the particular areas which might be notified (Article 1). However, at the tenth annual general meeting of the UPASI it was evident that no planting district in the Madras Presidency was really interested in implementing the act because of the new responsibilities it involved. Or as one planter stated: 'We asked for bread and they have given us a stone. (. . .) I say we should take Act XIII in

[148] RSIPEC 1896: 1–5 and 65.

[149] The Madras Planters' Labour Act (Act I of 1903), in: UPASI Proceedings.

preference of the new Act'.[150] Another planter made his point even more bluntly when he observed: 'the house to house inspection is the great block'. The butt of such carping criticism, the Madras government, not unnaturally felt betrayed by the planting community. In a private conversation about the introduction of the Madras Planters' Labour Bill, the Governor told Romilly, a Wynad planter and former Planting Member of the Madras Legislative Council: 'We have done all this for you, and if this is all the return we are going to get in the future, you will get precious little from us in the future'.[151] It was also about this time that the Indian administration began to show mounting opposition to the old legislation, Act XIII of 1859. Stokes, a representative of the government of India, pointed out the reasons. 'That Act was one which, under an unjust Judge, or a stupid Magistrate, might place the cooly in an actual position of servitude'. For Romilly there was therefore no option. He stated: 'I don't think the Government will force the Act on any District; but they might take away Act XIII *and then you will have free labour*' (emphasis added). Apart from Wynad, Nilgiris, another planting district likewise situated in the Madras Presidency, also came under the jurisdiction of the new legislation in 1904.[152]

The Indian government was not interested in the health and education figures of plantation workers in princely states,[153] so

[150] The planter's address began thus: 'These provisions are not what we asked for. What we did ask for was increased protection against loss by wilful breach of contract. That is what we asked for. We had Act XIII of 1859 and we simply asked that it might be amended so that contracts entered into under it might be enforced and proper punishments meted out for wilful and fraudulent breaches of contract' (10th Annual General Meeting of the UPASI, Bangalore, 3-7 August 1903, in: UPASI Proceedings).

[151] As reproduced by Romilly, the Governor commenced his talk by saying: 'You ask for an Act. You accept the Act through your Member of Council; and then before you ever give the Act a trial, you ask to have it amended. I will have nothing whatever to say to you in this matter of amendment until you come with a concrete case showing where the Act has caused a hardship. Then we may listen to you and may amend the Act. But to come up before the Act has ever been tested is absolutely illogical and absurd, and is practically putting the Madras Government in a false position' (10th Annual General Meeting of the UPASI, Bangalore, 3-7 August 1903, in: UPASI Proceedings).

[152] 10th Annual General Meeting of the UPASI, Bangalore, 3-7 August 1903, in: UPASI Proceedings.

[153] 11th Annual General Meeting of the UPASI, Bangalore, 22-24 August

the Travancore planters had more room for manoeuvre than their less fortunately placed colleagues. They had therefore a real option: either to introduce labour legislation along the lines of the Madras Planters' Labour Act or to keep the Travancore version of Act XIII of 1859. The latter, the Criminal Breach of Contract Act (Regulation I of 1040), was first amended in Regulation II of 1052 and later updated in Regulation II of 1080. To investigate the opinion of the planters on this particular subject, a Select Committee of the Travancore Legislative Council was installed by the Dewan. The Resident, G.T. Mackenzie, detested the proposed Planters' Labour Bill and ranged himself against the formation of such a committee. His position was not unique; the committee's report commenting on the planters' point of view stated that 'they were unanimously of opinion that the Bill was unnecessary and would destruct the progress of their industry'.[154]

The report needs some interpretation as it suggested more consensus of opinion among the Travancore planters than was the actual case during the deliberations. A.H. Mead, the representative of the planters of central Travancore, was initially in favour of the new legislation. H.M. Knight, representing the Cardamom Hills, also voted in favour of the proposed Labour Act, although he preferred Regulation II of 1080 if the planters could get an enactment that empowered extradition. Taking a contrary view, A. Martin, representing the Kannan Devan Hills, thought that 'the Act was totally unsuited to labour in south India, and introduced the thin end of the wedge of interference by Government with their labour'. He concluded thus: 'We do not want inspection'. It would seem that the view of Martin, one of the leading figures of the largest plantation company in Travancore, James Finlay and Company, carried great weight with the administration, because the Madras Planters' Labour Act was never introduced in Travancore.[155]

1904, in: UPASI Proceedings.

[154] The report stated further: 'and that the Breach of Contract Regulation II of 1080 recently passed would answer their purpose provided a clause was inserted securing to them the privilege of extraction of coolies who run away to British India after receiving advances of money'.

[155] R/2/903/441; 13th Annual General Meeting of the UPASI, Bangalore, 13-16 August 1906, in: UPASI Proceedings.

The Criminal Breach of Contract Act again became an issue with the so-called Slavery Case in the Travancore Highlands (1913). The facts laid before the *Sessions Judge*, P. Raman Tampi, were these. Mylen Chothan, plantation labourer, complained that Killian Panikan, another estate worker, had kidnapped his son, Kutten aged twelve, from his lawful guardianship. Originally, Mylen Chothan, with one son (Azhakan) and one nephew (Karumpan), had been recruited by a man called Chothi to work under the direction of a kangany named Onchi Thevan on Haileyburia Estate. Later, the labour gang, including a fourth member of the Chothan family – the boy Kutten – moved to Kozhikanom Estate owing to some differences between Onchi Thevan and the management of the Haileyburia Estate. Mylen Chothan, however, wished to return home with his children and nephew, and asked Onchi Thevan for permission to do so. Onchi Thevan stated that he first had to settle his accounts with Chothi. Acting on this information, Mylen Chothan left for Mundakayam, hoping to meet Chothi, but was unable to find him. Upon his return to Kozhikanom Estate, Onchi Thevan told Mylen Chothan that his three kinsmen had disappeared the previous day. After making fruitless searches on various plantations, Mylen Chothan finally discovered them, but this did not mean that he was able to take them home. Their new kangany, Abdullah Avukar, and the person who had enticed the three youngsters away, Killian Panikan, were unwilling to release the boys. In attempts to stifle his persuasive efforts, Mylen Chothan was even threatened with personal violence. In desperation, finally he turned his back on the estate, hurried up to the Magistrate's Court of Peermade and presented his grievances.

It was not the sentence handed down by the Sessions Judge that shocked the European plantation community. On 24 September P. Raman Thampi sentenced Killian Panikan to be strictly confined for a three month prison term but what really raised the ire of the planters were the remarks the Sessions Judge thought fit to utter when he passed sentence. Raman Thampi's words expressed as much compassion for the accused as for the victims, hereby implicitly condemning the conditions under which the field labourers worked and lived. He stated:

In passing sentence on the prisoner [Killian Panikan] I must not omit to consider the facts that he was a mere tool in the hands of designing persons. He is a victim of circumstances. He was working as a coolie for about three years in Kozhikanom Estate, and being ill-paid and under-fed, was anxious to go back to his native village. He had no money to pay his passage expenses. His masters would not willingly release him from bondage. (. . .) The temptation [receiving a commission by a kangany for bringing in three labourers] was too strong to be resisted. The boys (. . .) were working in Kozhikanom Estate and their father was absent. The boys themselves had ceased to be enamoured of highland life and were eager to return home. They were induced to believe that Chothan [respectively their father or uncle] had arranged for their clandestine departure from the Hills. They at once took the hint and without waiting to ask Thevan [their kangany on Kozhikanom Estate] for leave followed the prisoner who had suddenly presented himself before them as deliverer, friend and guide.

The planters had even more problems in swallowing the subsequent remarks of the judge. Without being explicit, Raman Thampi accused the European planters of creating a financial, judicial, and military framework in which labour bondage, especially between worker and kangany, was not only approved of but encouraged as well.

A careful consideration of the evidence in the case lends colour to the impression that slave traffic is carried on with helpless children and credulous adults to their doom. The boy Avutha [alias Kutten Chothan], for instance, was treated to a cup of coffee and sweet meats and decoyed from Manakumam [his home town] without the consent and permission of his mother and elderly brother. Once attached to an Estate as coolies, the defenceless creatures are deprived of all individuality and freedom and looked upon as chattels. (. . .) It is a pity that such conditions should prevail in modern days when we hear so much of philanthropic endeavours to elevate the depressed classes. The arms of the law are unfortunately not strong enough or long enough to reach the worst offenders. It is permissible to doubt whether the encouragement of the labour contract system by statute is not mainly responsible for this regrettable state of affairs.

Even more of an anathema to the planters, journalists took the issue outside the courtroom. Newspaper headlines and comments, closely following the words of the Sessions Judge, suggested that the plantation workers were in fact slaves.

Alarmed by the Sessions Judge's remarks and the response of the press, the managers of the estates concerned and the Secretary of the Central Travancore Planters' Association demanded that the British Resident launch a counter-offensive. Thus prodded, the latter informed the Dewan 'that the High Court should draw his attention to the rather lurid language in which the Sessions Judge has indulged'. Some days later it transpired that Krishnan Nair, the Chief Justice, had already taken action. He had pointed out the impropriety of his language to Raman Tampi and the utter absence of any justification for his generalizations. Afraid of being dismissed, the Sessions Judge acknowledged to the Chief Justice that he deserved the rebuke and that he promised to be more careful in future.[156]

This is an interesting intimation of undercurrents within the state but the real threats to the penal powers of the planters came from outside the state. Initially, such perils were successfully averted. For instance, the Workman's Breach of Contract Act (Act XIII of 1859) and the Madras Planters' Labour Act (Act I of 1903) were abolished in British India in 1925 and 1929 respectively. To prevent the Travancorean legislature from following suit and abolishing the Criminal Breach of Contract Act, the planting community and the British Resident kept their mouths shut about the legal changes in the neighbouring districts.[157] Moreover, in 1926 when the colonial government, requested to do so by the League of Nations, asked the Dewan whether slave trade, slavery or any similar condition existed in Travancore, Watts certainly misled the higher authorities when he answered that 'there is now no survival in the country or any trace of bondage or compulsory servitude'.[158]

The flood could not be stemmed and in the early 1930s the forces against the Criminal Breach of Contract Act took on a higher profile. First, the International Labour Organization – an institution formed and financed by the League of Nations – discovered that the information provided by Watts was incorrect

[156] All information regarding the so-called Slavery Case in the Travancore Highlands was obtained from file R/2/882/108, cf. *Planters' Chronicle* 1913: 579.

[157] 32nd, 33rd, 34th and 35th Annual General Meetings of the UPASI, Bangalore, on 17–20 August 1925, 23–26 August 1926, 22–25 August 1927 and 22–25 August 1928 respectively, in: UPASI Proceedings.

[158] KSA Bundle No. 224: 5531/1358.

(1930). Later, the planters established in the princely state of Mysore were deprived of their version of the Workman's Breach of Contract Act (1933). Having been acquainted with the opinion of the International Labour Organization and with the judicial changes in British India and Mysore, the Travancore legislature repealed the Criminal Breach of Contract Act (Regulation II of 1080) in January 1935.[159] The Travancore Breach of Contract (Repealing) Act ended the situation in which agreements between planter and kangany and between kangany and worker were subject to penal law.[160] This did not automatically imply that all contracts were now simply covered by civil law. Agreements entered into before the date of the repealing regulation remained subject to penal clauses until these contracts had expired.[161]

5.3.4 *The Emergence of Labour Unions*

Only during the last years of the British Raj did labour unions begin to make headway in the plantation regions.[162] Until the early or mid-1940s, depending on the degree of isolation to which the plantation labourers were subjected, acts of resistance by individuals or small groups continued to pose the most significant menace to the planters.

Of these, bolting and thefts of rubber and tea were most common,[163] but the workers were learning new strategies to resist the plantation regime as well. The Slavery Case in the Travancore Highlands already illustrated that labourers dared to go to the police in case of a serious felony by their employers. To provide yet another example, in early 1905 a (most probably Syrian Christian) employee named Koshi reported to the Peermade authorities that he had been tied up by F.G. Richardson, the

159 Proceedings TSMA 24 January 1935.

160 Regulation I of 1110, in: RPT, vol. 8: 1–2.

161 *Planters' Chronicle* 23 February 1935: 77.

162 This was relatively late: the first trade unions in India were established in industrial centres like Bombay, Calcutta and Madras in the late nineteenth and early twentieth centuries (Sharma 1963: Ch. 4 and 5; Karnik 1966: Ch. 1–3).

163 Minutes MRPA 2 July 1910, 27 August 1910 and 15 November 1910; *Addresses to Dewan Bahadur M. Krishnan Nair, Dewan of Travancore* 1920: 208–9; minutes KDPA 11 July 1925; Mackay 1984: F9.

well-known European pioneer planter, when the latter thought that Koshi had tried to poison him. After the combined intervention of Richardson's wife and son, other planters and the British Resident, Richardson was finally exempted from immediate prosecution on the grounds of 'insanity at the time of the trial'.[164] The same year, another worker, a kangany called Moses, accused a European planter named Lee[165] of homicide. Acting upon this information, the Inspector of Police posted in Devicolam arrested Lee and launched an investigation into the case. The findings, based on interviews and a *postmortem* examination, were shattering. Jacob, kangany by profession, had complained to Lee, manager of Yellapatty Estate, that he and his coolies were not being given enough food. Lee replied that they had no cause for dissatisfaction, blaming Jacob for the great number of complaining and absconding labourers. When Jacob denied this, Lee caught him and 'struck him with hands, horsewhip and boots'. About two weeks later, as a direct consequence of Lee's assault, Jacob died. After successful lobbying, the Kannan Devan Planters' Association and the British Resident were able to get the case transferred from Travancore to the Madras Presidency.[166] Although most probably Lee like Richardson escaped conviction, these were the first cases in which plantation workers had laid accusations of misconduct against planters. Hitherto, only the reverse had been the case.[167]

Apart from accusing the planters of violence, the labourers themselves began to show themselves to be less subservient and more aggressive. In 1930 a rubber-tapper threw a bucket of latex over a young planter named Seabrooke and then stabbed him

[164] R/2/880/69. The statement about Richardson's 'insanity' was most probably correct. Muthiah says that F.G. Richardson, once a key figure in the opening up of the Peermade/Vandiperiyar area, 'took up residence in a shack along a godforsaken road, went mad and died' (Muthiah 1993: 379).

[165] Fellow planters called Lee 'Bullocky Bill': he was in charge of the cattle farm in the Kannan Devan Hills (Muthiah 1993: 67).

[166] R/2/900/402.

[167] In late 1912 or early 1913, another European, a rubber planter named Hamond, was arrested on the accusation of a kangany. Unfortunately, no further information was found about the reason for Hamond's arrest and the subsequent law case (minutes MPA 1 February 1913; *Planters' Chronicle* 1913: 123 and 127).

in the back with a tapping knife. The British Resident claimed that this incident, as well as signs of unrest in other plantations, were closely related to the wage cuts introduced because of the Depression.[168] In 1933, W. Gillespie, manager of Isfield Estate, was stabbed with a hunting knife by Ponnaya Thevan. The latter, an engine driver employed on the plantation, was taking revenge for some earlier disciplinary action meted out to him by the planter. Gillespie had to be hospitalized for weeks.[169]

Policemen, who kept a close eye on the labour movement in Travancore, discovered that the establishment of labour unions in the plantation districts was being encouraged by activists based in the non-estate regions. The Travancore Labour Association, strongly rooted in the western parts of the state,[170] organized a meeting, intended for all plantation workers, in Mundakayam on 30 January 1938. In his opening speech, the President, M.S. Anurudhan, began by setting out the handicaps from which the labourers in general and the estate workers in particular were suffering. Having expatiated on their grievances, he advised the plantation labourers to get themselves organized. The meeting was fairly successful, going by the fact that a branch of the Travancore Labour Association was established in Mundakayam on the very same day.[171]

Although both at that particular and at subsequent gatherings resolutions were passed requesting the government to enquire into the disabilities suffered by the estate workers,[172] the first direct claims were only made four years later by a labour union, when the Estate Workers' Union demanded an increase in wages of at least 25 per cent in 1942. They also took the opportunity to protest about a tendency of certain estate owners to dismiss

[168] R/1/1/1889.

[169] R/1/1/2238.

[170] Chiriyankandath 1985: 116. Established on 31 March 1922, initially the Travancore Labour Association was 'Ezhava dominated', maintaining strong ties with the SNDPY, and 'Alleppey based', drawing its members from labourers and contractors in the coir industry at a time when European interests dominated that industry. It is interesting to note that trading firms which had a fixed interest in plantation crops, including Darragh Smail and Company, were active participants in this sector of the Travancore economy as well.

[171] Weekly Secret Bulletin 1938 No. 5: § 27.

[172] Travancore Police Abstract of Intelligence 1939, vol. 6: § 104.

local workmen who were members of their union. Reflecting the anti-communist strategies of the planters, the estate managers replaced the members of labour unions by importing outside labour.[173] Organized strikes were never called in the colonial period, though in some cases labour union members, without the consent of their leaders, downed tools as a measure of protest and/or merely to show their growing strength.[174]

The staff were also becoming more aware of their rights. The Estate Staff Association of South India (ESASI), formed at the instigation of the European planters in 1929, became a union only in 1946 (ESUSI). Though formally independent, the ESUSI pursued a very moderate policy. The planters' strategy of creating and maintaining a three-class division of plantation employees, each with its own social and religious characteristics, precluded the staff organization from linking up with the unions of the field workers. Quite apart from these obstacles, most staff members were expecting a promotion to (assistant) manager when the Europeans left India.[175]

The initial absence of formal trade unions on the estates surprised both British Indian and Travancorean officials. The planters were uniformly of the opinion that there was no reason for the establishment of labour unions or for the enactment of laws of which the purpose was to improve the working and living conditions of the workers. In 1930 H.L. Pinches, representing the planters in Travancore, replied to the Royal Commission on Labour that 'a trade union movement amongst our very illiterate labour would do more harm than good' and that 'we are quite willing to develop on welfare lines without legislation'.[176] Likewise, in 1937 the representative of the planting community, Eric Hall, informed the Travancore Sri Chitra State Council that 'it has been stated that there is no representative

[173] R/1/1/3786.

[174] Travancore Police Abstract of Intelligence 1939, vol. 6: § 223.

[175] Weekly Secret Bulletin 1938 No. 5: § 209; Mani 1984; Mathew 1984; Mackay 1984: Tape 2 Side 2. Most probably, some staff members acquired the position of assistant manager during the Second World War. At that time, many planters had joined the army, though at least some of their positions on the estates were taken in by elderly, retired British planters (Mackay 1984: G1).

[176] RCL 1931 Evidence Vol. 7, part 2: 217 and 209 and cf. 207, 216.

or advocate of labour in this House'. He continued his argument, 'I represent employers of about two lakhs of Indians, and I am here on behalf of those two lakhs of Indians'. His peroration stated: 'Sir, after working for nearly thirty years among labourers, I know the inside of their minds very well'.[177]

TABLE 5.10

Composition of the Population of the High Range Division in 1921 and 1931[178]

Feature		*Year*			
		1921		*1931*	
Sex	Male	31,783	(56%)	59,573	(57%)
	Female	25,138	(44%)	45,651	(43%)
Religion	Hindu	41,444	(73%)	79,685	(76%)
	Christian	11,929	(21%)	22,393	(21%)
	Muslim	1,667	(3%)	2,810	(3%)
	Tribal	1,877	(3%)	305	(0%)
	Buddhist	4	(0%)	20	(0%)
	Sikh	–		6	(0%)
	Jew	–		5	(0%)
Mother-Tongue	Tamil	40,585	(71%)	76,506	(73%)
	Malayalam	13,768	(24%)	23,178	(22%)
	Telugu	2,116	(4%)	3,981	(4%)
	Hindostani	177	(0%)	234	(0%)
	Kanarese	41	(0%)	539	(1%)
	Tulu	20	(0%)	311	(0%)
	Marathi	20	(0%)	19	(0%)
	Singhalese	10	(0%)	25	(0%)
	Arabic	6	(0%)	10	(0%)

177 Proceedings TSCSC 16 March 1937.

178 *Census* 1921, part 2: 32–3 and 46–9; *Census* 1931, part 1: 342–4, part 2: 4–5, 134–5 and 154–63. In fact, the 'High Range division' (1921 and 1931) closely corresponded to the 'Cardamom Hills', an earlier administrative unit used in 1875, 1881, 1891 and 1901 (see also Chapter 4). The 'Devicolam Division' (1911), however, clearly represented a very different area. For this reason, the figures for 1911 were not included in the various tables.

Feature		*Year*			
		1921		*1931*	
	Konkani	–		92	(0%)
	Punjabi	–		7	(0%)
	Sanskrit	–		6	(0%)
	English	166	(0%)	276	(0%)
	Other European	11	(0%)	24	(0%)
	Unknown	1	(0%)	16	(0%)
Caste (> 999 in 1931)	Channan	1,743	(3%)	5,823	(6%)
	Chetti	822	(1%)	1,564	(1%)
	Ezhava	1,713	(3%)	3,848	(4%)
	Kammalan	710	(1%)	1,142	(1%)
	Mannan	244	(0%)	1,023	(1%)
	Maravan	2,117	(4%)	5,805	(6%)
	Nayar	1,002	(2%)	1,857	(2%)
	Pallan	8,450	(15%)	22,517	(21%)
	Paraya	11,010	(19%)	23,983	(23%)
	Pulaya	2,218	(4%)	4,785	(5%)
	Sakkiliyan	2,679	(5%)	4,549	(4%)
	Vellala	2,864	(5%)	4,423	(4%)
Total		*56,921*	*(100%)*	*105,224*	*(100%)*

At first sight, the recruiting policy pursued by the European planters seems to have been immutable during the entire first half of the twentieth century. The planters were at least unwavering in their preference for family labour units which were composed of low-caste Hindu Tamils. This is once again borne out by the figures in Table 5.10, reflecting the composition of the population of the High Range *division*, an area inhabited mainly by plantation workers.[179]

One item which did alter, however, was the planters' policy regarding the length of the labourers' stay on the estates. To begin with, the field workers were recruited for a specified period,

[179] Cf. *Census* 1931, part 1: 283.

usually nine months. This was also noted by the British Resident as he wrote that 'the planter does not want his coolies permanently on the estates. He wants them to go back to their villages (. . .) and to return when work has to be done'.[180] In these early years, the size of the labour force was closely related to the fluctuations in green leaf crops and latex respectively, which were the natural outcomes of extreme variations between heat and cold, drought and monsoon rainfall.[181] It seems obvious that these methods of recruitment and of maintenance of plantation workers were developed to cut the direct labour costs to a minimum. An added advantage was that the migratory, often circular, movement of labour implied that the costs of providing for the non-working family members of the plantation workers were avoided. Young children (possible future estate workers) and old people (including ex-plantation labourers) stayed behind at home and were taken care of by the other members of the village communities. The recruitment of temporary, outside labour can also be observed in Table 5.11. In 1931 – seventy years (!) after the opening up of the region for plantation agriculture – only 15,393 inhabitants, or 15 per cent of the total population of the district, had been born in the High Range division.

TABLE 5.11
Birthplace in the High Range Division in 1921 and 1931[182]

Year	*Born in Travancore in:*			*Born Outside Travancore in*		Total
	Division	*Contiguous Divisions*	*Other Parts*	*Contiguous Parts*	*Other Parts*	
1921	13,032 (23%)	8,015 (14%)	1,538 (3%)	19,039 (33%)	15,297 (27%)	56,921 (100%)
1931	15,393 (15%)	14,595 (14%)	4,668 (5%)	20,403 (19%)	50,165 (48%)	105,224 (100%)

[180] Letter of the British Resident to the Madras government dated 23 December 1913, in: R/2/882/108.
[181] RCL 1931 Evidence Vol. 7, part 2: 197–8 and 207; Soutar 1976: 22; Mackay 1984: Tape 2 Side 2.
[182] *Census* 1921, part 1: 31; *Census* 1931, part 1: 88.

Very gradually, the estates shifted from what was basically a temporary to a fundamentally permanent work force. To some extent this mutation was brought about by the workers themselves. Some plantation workers, either because they had left behind personal or financial difficulties in their home villages or because they had no family left on the plains, preferred to stay on the estate.[183]

Of far greater significance was the change of heart of the planters, who eventually welcomed the formation of a stable work force, especially after the mid-1930s. In the first place, this was triggered off by the abolition of the Criminal Breach of Contract Act. The alteration to the judicial situation meant that the planters came across more difficulties in the recruitment and disciplining of labour. The kanganies and workers could no longer be threatened with imprisonment. To compensate for their weakened positions in dealing with the labour force, the planters changed their initial policy of contracting the field labourers temporarily for one intended to keep the workers chronically indebted. With this in the back of their heads, the managers started to pay the workers more regularly instead of after the expiry dates of the contracts. The wages, especially when paid out by a kangany – the planters were fully aware that these middlemen kept a disproportionate share to themselves – remained insufficient to buy food, clothes and other necessities, the upshot being that the workers were never able to pay off their debts and were trapped into permanent bondage to the estates. Another factor which contributed to the creation of a permanent work force was the planters' fear of nationalist and communist ideologies. They reasoned that complete isolation of the work force would prevent undesirable sentiments and ideas from infiltrating the estates.

The planters' policy of establishing a permanent labour force was allied to three other strategies. The first tactic was that the planters tried to keep the workers as illiterate as possible. As a planter explained to the Unemployment Enquiry Commission, 'the less "educated" a man was, the better character he had'.[184] As a consequence, most field labourers, including the children,

[183] RCL 1931 Evidence Vol. 7, part 2: 207; Mackay 1984: Tape 2 Side 2.
[184] Unemployment Enquiry Committee 1928: 9.

were not given much education. Only the children of the planters and of some of the staff members obtained a proper education.[185] The literacy rate of the High Range division for the years 1901, 1921 and 1931, especially shocking with regard to females and Hindus, is shown in Table 5.12. By comparison, 29 per cent of the total population of Travancore was literate in 1931.[186]

TABLE 5.12

Literacy in the High Range Division in 1901, 1921 and 1931[187]

Year	*Total*		*Sex: Female*		*Religion: Hindu*	
	Lit.	*Illit.*	*Lit.*	*Illit.*	*Lit.*	*Illit.*
1901	2,470 (11%)	19,119 (89%)	340 (4%)	8,511 (96%)	1,164 (7%)	14,862 (93%)
1921	7,363 (13%)	49,558 (87%)	1,217 (5%)	23,921 (95%)	3,718 (9%)	37,726 (91%)
1931	12,369 (12%)	92,855 (88%)	1,735 (4%)	43,916 (96%)	5,456 (7%)	74,229 (93%)

The second ploy by which the plantation managers attempted to keep their workers in check was to hinder labourers obtaining information from 'outside' sources. This also explains why some children – presumably of Indian staff – who visited a meeting organized around Mahatma Gandhi were punished severely. They were expelled from the Munnar High School, an institution controlled by James Finlay and Company.[188] Thirdly, the planters tried to prevent 'undesirable' persons from entering the estates, which is why no Malayali Muslims from the Malabar District in the Madras Presidency, often indicated as Moplahs, were recruited. Ever since the so-called Moplah Rebellion (1921) they had been feared by the British planters. The Muslims' struggle, which was predominantly directed against the wealthy landlords and their political allies, the British bureaucrats, made

185 RCL 1931 Evidence Vol. 7, part 2: 203; Soutar 1976: 24.
186 *Census* 1931, part 1: 283.
187 *Census* 1901, part 3: 256–7; *Census* 1921, part 2: 26; *Census* 1931, part 2: 126.
188 Proceedings TSCSC 9 December 1936.

possible undertones of anti-British feelings crystal clear to the planters.[189]

The planters' new policy of building up a stable labour force, complemented by strategies to ban unwelcome information, sentiments and ideologies, explains why both the nationalist and the labour movements were virtually unknown in the plantation districts in the late colonial period. With regard to nationalism, Mackay noted that the Quit India Movement did not affect the planters in any way. 'We heard little of it', Tewson remembered. Even in 1947, the year of India's Independence, there was 'no anti-European stuff'. Closely tying in with the analysis presented above, Mackay explained the lack of nationalism in the estate regions thus: 'Our isolation was our strongest asset'.[190] The emergence of labour unions in the estate regions during the last years of the British Raj also did not cause many ripples in the European planting community.[191]

Nevertheless, as we will observe in the next chapter, the European planters were indeed confronted with 'nationalism' and 'socialism' in later years. Following Indian Independence, they, as well as their indigenous counterparts, functioned within an increasingly leftist-oriented Indian region, a 'socialist' unit within a 'nation(alist)' state.

189 Clough 1984: Tape 1 Side 2; Mackay 1984: C17. The fear of Malayali Muslims also delayed the construction of the western outlet road of the Kannan Devan Hills. The planters, many of whom served the Southern Provinces Mounted Rifles during the Moplah Rebellion, opposed it (see also page 108, note 192).
190 Mackay 1984: Side 2 Tape 2; Tewson 1984, part 3: 1.
191 Soutar 1976: 24; Clough 1984: Tape 1 Side 1; Mackay 1984: Tape 2 Side 2.

Six

The Successful Planters' Lobby in a Leftist 'Problem' State: *c.* 1947–1963[1]

After all, our government is not so autonomous. Constitutionally: we are under the directions of the central government. Politically: plantation interests, Christian community interests, all these operate. (. . .) Planters pressure, Christian lobby, these special interests also operate together to get a new government

E.M.S. Namboodiripad on the question why plantation crops such as coffee, tea and rubber were excluded from the land ceiling in the Kerala Agrarian Relations Bill[2]

6.1 Syrian Christian Prominence

6.1.1 The Birth of a 'Problem' State

15 August 1947, the date of Indian Independence, is naturally regarded as an important turning-point in South Asian history. The transfer of authority ended the British Raj and inaugurated the newly-formed nations of India and Pakistan. However, 15 August 1947 also symbolizes changes which were less direct and less visible. Princely states were gradually abrogated and new Indian states were created, mostly founded on a linguistic basis. In the case of south-west India, the princely states of Travancore and Cochin were amalgamated into the Indian state of Travancore-Cochin (1949), which was again replaced by the

[1] A fascinating long-term development, the dominant political powers – the colonial state and the Indian government – attached different labels to south-west India in different periods of time. Travancore was called a 'Model State' in the second half of the nineteenth century, whereas Kerala was labelled a 'Problem State' in the post-1947 period. As we will see, these labels are closely related to the attitude and policy of the various south-west Indian governments towards plantation development. Madhava Rao's Travancore favoured (British) planters while Namboodiripad's leftist Kerala openly questioned (Indian and European) plantation development.

[2] Interview E.M.S. Namboodiripad.

Kerala state in 1956. Internally, the handover of power stimulated (a) the abolition of the monarchy; (b) the substitution of the powerful Dewan by a 'responsible' Chief Minister; and (c) the replacement of the bicameral body by the Travancore Representative Body (1948) and the Legislative Assemblies of Travancore (1948–9), Travancore-Cochin (1949–56) and Kerala (since 1956).

In the early post-Independence era, some of the movements started in the late colonial period sprang into prominence, fundamentally affecting the nature of plantation development in south-west India. At the outset, Syrian Christian agro-entrepreneurs took over the leading role from their European counterparts, particularly in their joint struggle against the leftist parties in general, and the communist Namboodiripad government (1957–9) in particular.

There has already been some reference to the fact that the importance of the British plantation segments was on a irrevocable downward slide. Just as the rise of the British plantations had been closely related to the establishment of a colonial regime in the nineteenth century, the decline of the foreign planters was inextricably bound up with the emergence of an increasingly hostile political environment in the twentieth century.

Slowly but surely, labour unions continued on their way to becoming powerful organizations in the plantation regions, demanding a rapid improvement in the working and living conditions on the estates. Taking into account the wider context of the late 1940s and the 1950s and the quality of the plantation workers' lives, there is no need to question why the labour unions gained considerable strength. Falling on willing ears, the message 'Proletarians Unite!' met with a wide response on the estates of south-west India.

In these years, various global, national and regional events bolstered the idea that, following the overthrow of the colonialists, the capitalists would also be defeated. This was reflected in Western European countries, including Great Britain, where social-democratic parties came to power. Even more forcibly, the communist-led Soviet Union became a military superpower and presented itself as an alternative to capitalist states and societies. And last but not least, China functioned as a source of inspiration to those who were searching for fundamental change. Unlike the Russian revolution, the Chinese path to socialism was often considered a more valuable experience because of its genuine 'agrarian' and 'Asian' nature.

At the all-India level, many politicians used Gandhian, Marxist, Leninist and/or Maoist rhetoric. The Indian constitution, inaugurated in January 1950, is to a very large extent based on socialist principles. Although the Right of Assembly and the Right to Form Associations or Unions were fundamental in this regard, the Right Against Exploitation illustrates most compellingly the spirit with which the law was laid down.[3]

To an even greater extent than these various events and developments, the enactment of the Indian Plantation Labour Act (November 1951) gave reasons to believe that the mid-twentieth century would mean a fundamental break with the past. The Plantation Labour Act was born of both nationalist and leftist ideologies. More explicitly, the legislature gave the impression that it wished to convert the plantation industry, once a major centre of 'white' colonialist exploitation, into an ideal 'Indian' working place in which the interests of labourers, employers and officials were equally protected. To achieve this goal, the workers especially needed legal protection, an assertion which had also been put forward by the International Labour Office. The preamble states that the law was meant 'to provide for the welfare of labour, and to regulate the conditions of work in plantations'. The most salient points contained in this law, casting a swift eye over the component sections, dealt with the appointment, functions and powers of inspectors (Chapter 2), health and welfare (Chapters 3 and 4), hours of work (Chapter 5), paid leave (Chapter 6) and penalties in case of obstruction of an inspector (Chapter 7). Some additional, specific details, *including the date of actual implementation*, were to be decided at a later stage by the individual state governments concerned.[4]

In south-west India itself as well, there were many hopeful signs for the plantation workers. The Communist Party of India (CPI), in conjunction with other political parties like the Kerala

[3] Respectively article 19 and articles 23 and 24, in: Pylee 1984: Ch. 1, 9, 10.
[4] Plantation Labour Act 1951.

Socialist Party (KSP), the *Praja* Socialist Party (PSP) and the Revolutionary Socialist Party (RSP), gradually managed to build up a very influential leftist movement in the coastal regions.

The growing influence of the leftist movement was to a very large extent due to the perseverance of its leaders. Despite the official ban on communist activities in the early post-colonial phase (until 1952), they were unswerving in their political work in the villages, giving lectures and establishing reading rooms.

Propitious constitutional changes also played key roles in the rise of the leftist movement. In February 1948 Travancore was the first Indian state to hold an election on the basis of universal suffrage. For the progressive forces, the amalgamation of Travancore and Cochin into the state of Travancore–Cochin also signified a step forward. Cochin, like Travancore, had witnessed the growth of a strong labour movement. The formation of the state of Kerala – the state of Travancore–Cochin merged with the Malayalam-speaking regions of the Madras State in the north and lost the Tamil-speaking regions in the south – strengthened the leftist movement even more. The first obstacle removed was the problem of separate Malayali and Tamil electorates. The disappearance of Tamil political parties after 1956 illustrates that language could no longer form a barrier to class organization at state-level. The second source of strength was that the region which became part of Kerala, the erstwhile Malabar district of the Madras Presidency, was a communist stronghold, a circumstance which was closely related to the region's past. In the previous chapter there was a passing reference to the fact that in Malabar district, in contrast to Travancore especially and to a lesser extent Cochin, the opposition between the influential landlords and their tenants had been exacerbated in the colonial period.[5]

The Congress Party held continuous power in south-west India until 16 March 1954. Then, a period which had been greatly feared by the planters, the planting organizations and the Indian National Congress was ushered in: the leftist political parties, including the Communist Party, came to power. Chief Minister Pattom Thanu Pillai (Praja Socialist Party) headed the

[5] Nossiter 1982: Ch. 4 and 5; cf. Proceedings TCLA 14 March 1950; Alexander 1989: 382.

Kerala government in the period 16 March 1954 to 14 February 1955. The installation of E.M.S. Namboodiripad as Chief Minister of Kerala on 5 April 1957 formed a much greater threat. This was a unique event: he was to lead the first democratically-elected communist government in the world.

Though sympathizers of leftist ideologies considered the transfer of power a clear step forward on 'the parliamentary road to socialism', the planters feared the future course of events. To quote the President of the Mundakayam Planters Association in a definition of the problem and a search for solutions: 'During the year the Communist Government have come into power in Kerala, and we are all awaiting the future trend of events'. 'It is too early as yet to express any opinions', the President continued, 'but from statements made by the various Ministers, it would appear that the Planting Industry is in for a very thin time, with their threats of increased taxes, labour wages, etc. and it behoves us all to follow the situation very carefully, and to voice our protest, where necessary, through the proper channels'.[6]

The leaders of the Indian National Congress, especially at the all-India level, including Prime Minister Jawaharlal Nehru, were greatly worried as well. For one thing, the communist rise to power confirmed their idea that Kerala had been correctly labelled a, if not, *the* 'Problem State' of India.[7] One saving grace was that Congress leaders realized that specific issues, particularly those related to ownership rights (of landed properties), had to be approved by the – Congress-controlled – central administration. The table below attempts to provide a concise overview of the political situation in south-west India. Taking over from the elected state administrations, the central authorities twice governed Kerala affairs directly: in the periods 23 March 1956–5 April 1957 and 1 August 1959–22 February 1960.

This brief introduction to early post-colonial south-west India is also intended to function as a guideline for this chapter. Following a short discussion about the development of rubber and tea, the growing influence of Syrian Christian planters and the declining prominence of European agro-entrepreneurs will

[6] Minutes MPA: 1957.
[7] Cf. Miller 1988: 596.

Table 6.1
Governments in Travancore-Cochin and Kerala, 1949-64[8]

Period	*Chief Minister (Political Party)*
1 July 1949-26 Jan. 1950	T.K. Narayana Pillai (INC)
26 Jan. 1950-3 March 1951	T.K. Narayana Pillai (INC)
3 March 1951-6 Sept. 1951	C. Kesavan (INC)
6 Sept. 1951-12 March 1952	C. Kesavan (INC)
12 March 1952-16 March 1954	A.J. John (INC)
16 March 1954-14 Feb. 1955	Pattom Thanu Pillai (PSP)
14 Feb. 1955-23 March 1956	P. Govinda Menon (INC)
23 March 1956-5 April 1957	President's Rule
5 April 1957-31 July 1959	E.M.S. Namboodiripad (CPI)*
1 Aug. 1959-22 Feb. 1960	President's Rule
22 Feb. 1960-26 Sept. 1962	Pattom Thanu Pillai (PSP)**
26 Sept. 1962-10 Sept. 1964	R. Sankar (INC)***

* The government was supported by some independent members within the Kerala Legislative Council

** In that period Pattom Thanu Pillai headed a INC-PSP ministry

*** R. Sankar headed a INC-PSP coalition government in that period

be discussed. Attention will then shift to the demands of the labour unions, the planters' anti-union strategies and the government's role in these conflicts. The last section will concentrate on three major issues in south-west Indian politics in the late 1940s and the 1950s: nationalization of plantations, land reforms, and the 1958 Kannan Devan Hills strike. Here, the hypothesis is adduced that the leftist movement, though powerful at Kerala-level, was unable to defeat the planters' lobby, which was particularly forceful at the all-India level.[9]

[8] Nossiter 1982: 379-80.

[9] Realizing that my focus and findings are somewhat unusual, I will reflect briefly on the historiography regarding the rise of communism in south-west India. Most scholars focus on the influence of the labour unions and the leftist political parties (see for example: Nossiter 1982; Alexander 1989; Kurup 1989; Lukose 1991). Although I acknowledge the importance of these scholarly efforts, I feel that insufficient attention has been paid to the tactics and

6.1.2 Rubber: Liquid Gold

Seeking a foothold in the midst of this rapidly changing political landscape, the planters did their best to achieve some stability, including a secure, remunerative price level. In the case of rubber their objective was reached by the intervention of the central government. In the foregoing chapter it was mentioned that prices of this commodity had been regulated since 1942. The Allied Forces were in desperate need of rubber, so the colonial ruler enacted the Rubber Control and Production Order and installed the Indian Rubber Production and Marketing Board. Afraid that the end of the Second World War and the transfer of power would throw a spanner in the opportune climate of fixed minimum prices and other incentives offered for the cultivation of rubber, the growers requested the government of India to continue the policy. Their petition did not go unheard. In 1947 the Indian government passed the Rubber (Production and Marketing) Act and appointed the Indian Rubber Board. Although this particular legislation was amended in 1954 and 1960, shortening the names of the law and the organization into the Rubber Act and the Rubber Board respectively, the purpose of the legislation remained unchanged throughout the 1940s and 1950s: namely, providing assistance to the rubber cultivators. Two important incentives to the extension of rubber cultivation and the enhancement of productivity were that prices were fixed at a remunerative level and planting expertise was more widely disseminated. The Rubber Research Institute of India, established in 1955, played a key role in the latter. The rubber cultivators were not the only beneficiaries; the Indian government profited from the policy as well. Far outweighing all other considerations, the plan was instrumental in its effort to achieve the goal of national self-sufficiency in production. A target which was never fully accomplished: the rapidly growing internal demand outstripped internal supply since 1948.[10]

---

strategies of the communists' opponents. My second argument is closely related to the first. Although some researchers argue that the rise of the leftist movement was extremely successful in Kerala (see for example: Franke and Chasin 1992: Ch. 4), my findings – i.e. those regarding the estate sector in south-west India until the early 1960s – point in a different direction.

[10] Plantation Enquiry Commission – Rubber 1956: 4; Haridasan and Nair

In sharp contrast to the situation in the rubber market, international developments brought a large influence to bear on the value of tea. The International Tea Agreement, initiated in 1933, was not repealed during the early post-colonial era. Its mechanism, adjusting supply to world demand, did not alter, but its overall significance lost ground. The most important factor in this was the aftermath of the Second World War. The British government carried on its wartime policy of direct purchases from the growers and rationing in Great Britain until the early 1950s. The London tea auctions were reopened only in 1951. The development of tea cultivation in other parts of the world also affected the scheme. Opting out of the control, in 1947 the cultivators and governments of Kenya, Uganda and Tanganyika ceased to regulate production and exports in accordance with the International Tea Agreement. Though gradually representing a smaller share in world production, South and Southeast Asian countries remained the main forces behind the agreement. Settlements under the scheme carried on till 1955 when the agreement was rescinded because of the development of unbridgeable differences in opinion among the producers and administrations of Ceylon, India, Indonesia, and Pakistan. The matter has remained in abeyance since then. Away from the government level, combination among British trading companies contributed to the maintenance of a profitable price level for the south-west Indian estates. As mentioned earlier, most of these so-called coastal firms were directly involved in the production of plantation crops.[11]

Table 6.2 shows that the various strategies to regulate the prices of rubber and tea affected both crops differently. The development of rubber was enormous during the late 1940s and the 1950s. Leaving the category 'additional rubber holdings' out of the picture and enumerating the figures regarding central Travancore, Mundakayam and south Kerala (originally named south Travancore), the acreage under rubber amounted to

1980: 9–10; Tharian George, Haridasan and Sreekumar 1988: M-160; Thomas and Mackil, in: The Association of Planters of Kerala 1989: vii–viii; Haridasan 1992: 50.

[11] Plantation Enquiry Commission – Tea 1956: Ch. 11; Harler 1956: 248–50; Griffiths 1967: Ch. 17 and 36; Tharian George 1984: 39–40; Tharian George and Tharakan 1986: 225–7; see also page 124, note 34.

TABLE 6.2
Acreages Under Rubber and Tea in Kerala in 1960[12]

Region	*Rubber*		*Tea*	
Central Travancore and Mundakayam	123,539	(43%)	37,705	(38%)
Cochin	19,765	(7%)	2,404	(2%)
Kannan Devan Hills	-	(0%)	33,308	(34%)
Malabar	38,565	(14%)	1,402	(1%)
Malabar Wynaad	12,241	(4%)	13,449	(14%)
South Kerala	52,212	(18%)	10,943	(11%)
Additional Rubber Holdings	42,127	(15%)	-	(0%)
Total	*288,449*	*(100%)*	*99,211*	*(100%)*

175,751 acres in the erstwhile region of Travancore in 1960. By way of comparison only 110,012 acres were planted with rubber trees in that same area in 1948 (consult Table 5.1).[13] By contrast, the development of tea remained slow. In the former tea districts of Travancore – i.e. central Travancore, Kannan Devan Hills, Mundakayam and south Kerala – the acreage under tea only increased from 78,700 in 1948 to 81,956 in 1960 (compare Table 5.1 and Table 6.2).

The reorganizations of the political entities meant that the successors of Travancore, respectively Travancore-Cochin and Kerala, became the most important plantation regions in south India. This is illustrated by the figures provided in Table 6.3. In 1960 about 94 per cent of the south Indian acreage under rubber was situated in Kerala. The state's position with regard to tea was less prominent. Approximately 55 per cent of south Indian tea bushes were planted on Kerala territory in 1960.

[12] *UPASI Planting Directory* 1960: 390. In 1985-6, the land use pattern of the three main crops and tea was thus in Kerala: total geographical area: 3,885,497 ha. (100%); net area sown: 2,190,985 ha. (56.4%); coconut: 683,821 ha. (17.6%); rice: 663,803 ha. (17.1%); rubber: 337,700 ha. (8.7%); tea: 34,760 ha. (0.9%) (*Kerala at a Glance* 1988: 3–4).

[13] As will be discussed at length in Section 6.3.2, the favourable price-level was not the only factor contributing to the increase in rubber cultivation. Fearing the land reforms the communists tried to implement in the state, many agriculturalists converted their holdings into rubber lands.

TABLE 6.3

Acreages Under Rubber and Tea in South India in 1960[14]

Region	*Rubber*		*Tea*	
Kerala	288,449	(94%)	99,211	(55%)
Madras	12,631	(4%)	76,776	(43%)
Mysore	4,517	(1%)	4,519	(3%)
Total	*305,597*	*(100%)*	*180,506*	*(100%)*

The significant position of south-west India becomes even more apparent when the region's ranking within the whole of India is considered. Almost all Indian rubber originated from Travancore-Cochin in the mid-1950s. The same region also accounted to 11 per cent of the total acreage under tea in India in 1953–4. The figures are provided in Table 6.4.

TABLE 6.4

Acreages Under Rubber and Tea in India on 31 December 1955 and 1953–54 Respectively[15]

Region	*Rubber*		*Tea*	
Travancore-Cochin	167,229	(81%)	83,282	(11%)
Other Regions	40,011	(19%)	703,035	(89%)
Total	*207,240*	*(100%)*	*786,317*	*(100%)*

6.1.3 *Syrian Christian Power*

The group which profited most from the upward development of plantation crops, particularly rubber, was the Syrian Christian community. Its key position was reflected in the UPASI enumeration of 1960. More than half of the sizable, indigenously-owned, private rubber estates in Kerala belonged to Syrian Christians. Moreover, 32 per cent of the large, privately-owned, Indian tea plantations in the state were the property of members of this particular community. In fact, the prominence of the Syrian

[14] *UPASI Planting Directory* 1960: 390.

[15] Plantation Enquiry Commission – Rubber 1956: 5; Plantation Enquiry Commission – Tea 1956: 17–18.

Christians in terms of ownership of land cultivated with either rubber or tea was much more impressive. The major share of the category 'Christian' was also most probably Syrian Christian. Moving to figures referring to both smaller estates (between ten and a hundred acres) and smallholdings (less than ten acres), the significance of Syrian Christian proprietors was just as important. Syrian Christians were also prominent in the segment of company-owned rubber and tea plantations, the shares in the enterprise often being distributed among relatives.[16]

TABLE 6.5
Community Background of Indigenous Proprietors of Large, Privately-Owned Rubber and Tea Estates in Kerala in 1960 (total area in acres and number of estates)[17]

Community	*Rubber*				*Tea*			
	Area		*Estates*		*Area*		*Estates*	
Syrian	15,430	(51%)	70	(50%)	2,161	(32%)	9	(31%)
Christian	7,032	(23%)	32	(23%)	2,497	(37%)	12	(41%)
Brahmin	388	(1%)	3	(2%)	699	(10%)	2	(7%)
Nayar	336	(1%)	3	(2%)	–		–	
Muslim	3,376	(11%)	13	(9%)	573	(9%)	2	(7%)
O,U,U	3,465	(12%)	19	(14%)	758	(11%)	4	(14%)
Total	*30,027*	*(100%)*	*140*	*(100%)*	*6,688*	*(100%)*	*29*	*(100%)*

1. Large estates indicate the undertakings with more than 100 acres (40, 47 hectares) under cultivation (cf. Section 5.2.1)
2. Syrian = Syrian Christian; a 'Christian' can either be a Syrian Christian or newly-converted Christian, most of whom originated from castes considered lower in the existing hierarchy; O,U,U = Others, Uncertain, Unknown

In solving the question of why more Syrian Christians than members of other communities began or expanded the cultivation of plantation crops, especially rubber, the factors which explained the opening up of estates by Syrian Christians since

[16] *UPASI Planting Directory* 1960: 214–387; interview: Tharian George; and see: Ramakrishna Pillai 1974: 5.
[17] *UPASI Planting Directory* 1960: 214–387. See also Appendix 8.

the early 1900s were still largely valid during the early post-Independence period. Syrian Christians were very well-placed to be able to acquire the requisite know-how. Those who were already engaged in cultivating rubber, mainly Syrian Christians, realized its profitability, which sometimes spurred them on to buy up new properties but more often to convert holdings cultivated with other crops (particularly coconuts), to say nothing of waste lands, into rubber areas. As far as those members who were not yet engaged in planting rubber but thinking about doing so were concerned, Syrian Christians were in a particularly advantageous position: usually some relative or acquaintance, known through the agriculturalist's wider social networks within the Syrian Christian community, was cultivating rubber and was willing to provide information about the planting, processing and marketing aspects of rubber.

Syrian Christians also had the upper hand regarding access to land and capital. Most of the lands suitable for rubber cultivation were already owned by Syrian Christians, especially in the 'core' region of Kottayam district, encompassing the areas around Kanjiripalli, Mundakayam and Palai. Because of the prominence of Syrian Christian capital in planting, trade, newspapers and banking – perhaps south-west India's four most lucrative economic sectors – Syrian Christian investors were the best placed to be able to start or expand rubber cultivation.[18]

Within the Syrian Christian planting segment, however, a process which had already started in the late colonial period solidified during the first few years of Independence: the formation of a Syrian Christian planting *elite*. It is undeniable that among the Syrian Christian agriculturalists a group with great economic strength, close social ties and enormous political power had emerged. Perhaps it might be even better to characterize the newly-established elite as an *aristocracy*. Status was inherited: most 'members' were the sons and grandsons of famous planting pioneers, including the male offspring of the above-mentioned P.J. John (Aiymanam), K.C. Mammen Mappillai (Kandathil), P.T. Thomas (Palampadom), Dominic Thommen (Karipparambil) and Chacko Abraham (Kallivayalil). Other

[18] For any references regarding the prominent position of Syrian Christians in the plantation sector, consult Section 5.2.2.

'associates' included the male members of the Kallarakkal, Karimpanal, Kollamkulam, Koltukapally, Kuruvinakunnel, Ooppoottil and Pottamkolam families.[19] It is with good reason that 'members' and 'associates' are placed in inverted commas in the above text, as the planting aristocracy was not a formal organization: membership as well as a person's place within the elite was informally determined on the basis of criteria such as economic power, social status and political influence.

Focusing on the group's economic significance first: it held the lion's share in both the Syrian Christian and overall Indian planting segments. There were two reasons for this. In terms of quantity, the new planting elite controlled a major part of the planting industry, sometimes through individual ownership, otherwise through shareholding. Secondly, it took the lead in terms of quality: because of its strong financial position as well as its access to the latest information regarding planting techniques and materials, including the High Yielding Varieties, the group produced relatively superior rubber and tea, putting the smallholders in particular in the shade.

Their economic power was consolidated by various social strategies. Most, if not all, sons and grandsons found employment in the family enterprises started by their respective fathers and grandfathers, usually beginning as managers, later taking over full responsibility as directors.[20] This ensured that all estates remained entirely under family control. Networks were built up by marrying off daughters and granddaughters to the sons and grandsons of other influential families, who in many cases were themselves engaged in planting.[21] This strategy established, or more often strengthened, the social ties between the Syrian Christian planting aristocrats. Indeed, many of the

[19] Chacko Kallivayalil 1969: 2; interviews: Tharian George; Hannibal Dominic; Michael Tharakan; Michael A. Kallivayalil. Intriguingly, all these families were Syrian Catholics with the exception of the Aiymanams (Marthomites), Kandathils, Pampadoms and Oopoottils (all Jacobites). Religious identity influenced marriage patterns: Syrian Catholics married amongst each other, whereas weddings among Marthomites and Jacobites were fairly common. Notwithstanding the various denominations, the planting families identified with common interests.

[20] Interview: Abraham Eapen and Abraham Kurien.

[21] Interviews: Michael Tharakan; Hannibal Dominic.

above-mentioned families were connected to each other through marriage, enabling 'insiders' to acquire the necessary information or capital in order to manage their enterprises more successfully, and preventing 'outsiders' from entering their field of interest.

The formation of the Syrian Christian planting elite affected the planters' organizations. Significantly, prominent Syrian Christians gradually replaced the British as the most influential representatives of the south-west Indian plantation industry. The first Indian to be installed as President of the Mundakayam Planters' Association was Jose A. Kallivayalil, who took office in 1954.[22] Of even greater importance, the first Indian to assume the function of chairman of the Association of Planters of Kerala, the successor to the Combined Travancore Planters' Association and the Association of Planters of Travancore, was also a Syrian Christian, namely P. Kurian John, the son of the indigenous rubber pioneer P.J. John.[23] It goes without saying that all other Indian chairman in the 1950s were also Syrian Christians. Besides P. Kurian John, chairman in the year 1951–2, M.M. Varghese and M.K. Kuriakose presided over the Association of Planters of Kerala in the years 1952–3 and 1953–4 respectively 1959–60.[24]

The Syrian Christian planting community attempted to influence both the state and central governments through the Congress Party, supporting Congressmen in all sorts of ways, but above all financially.[25] As seen in the previous chapter, Syrian Christians were closely associated with the Travancore State Congress in the late colonial period. At that juncture, prominent leaders within the Indian National Congress, and more particularly their policy of noninterference in the internal affairs of the princely states, prevented any close co-operation between the Travancore State Congress and the INC, probably

[22] Minutes MPA.

[23] See also Section 5.2.2.

[24] Thomas and Mackil, in: The Association of Planters of Kerala 1989: xviii. In later years, the appointment of British chairmen became even more sporadic. Of those Indians who presided over the Association of Planters of Kerala after 1960, most were Syrian Christians (Thomas and Mackil, in: The Association of Planters of Kerala 1989: xviii).

[25] Interviews: E.M.S. Namboodiripad; Michael A. Kallivayalil; Michael Tharakan.

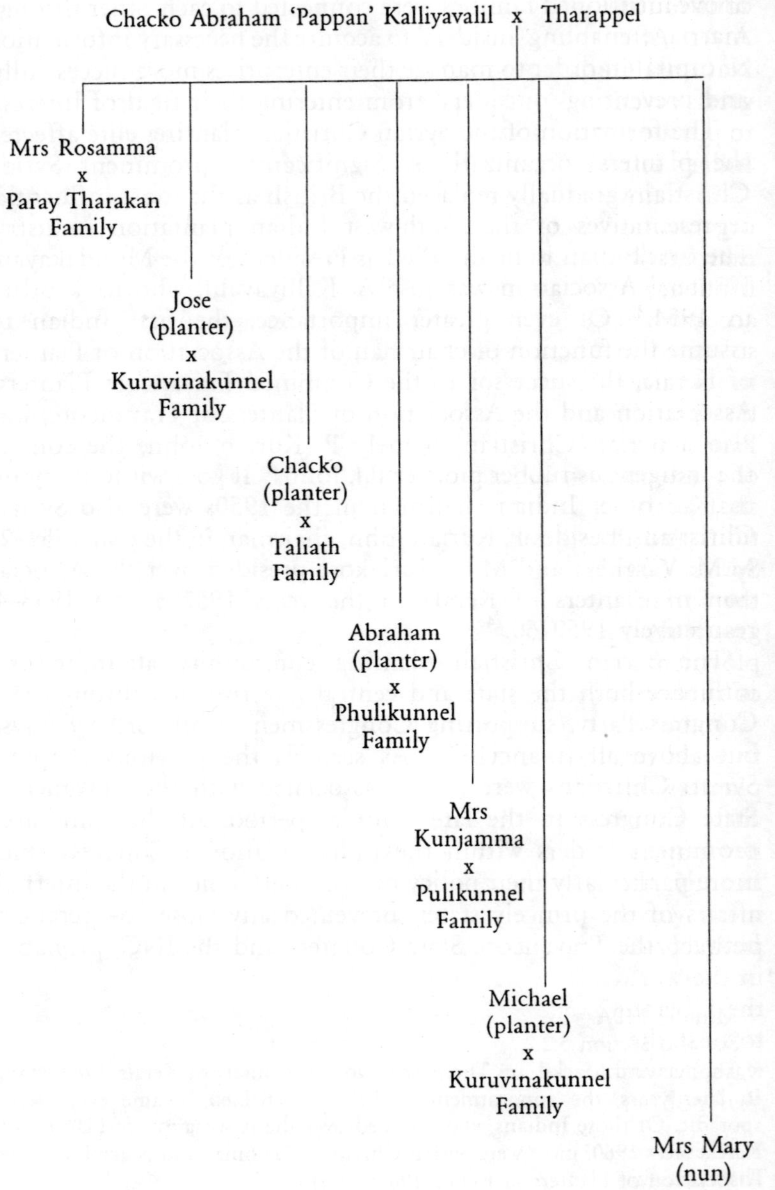

Figure 6.1 The Kallivayalil Family

contributing to the perpetual dewanship of C.P. Ramaswami Aiyer. After Independence, the national leaders of the Indian National Congress linked up with the Travancore State Congress, investing the latter organization with many more powers to pursue its transmuted main objective: the defeat of south-west India's leftist movement.

The close relations between Syrian Christians, the planting organizations, and the Indian National Congress requires closer examination. There was much more to the growing political influence of Syrian Christian planters than a reflection of their accumulating wealth and status. Political participation consolidated and improved their socio-economic position, which in its turn augmented their authority. Pertinently, Syrian Christian participation in the planting associations and the Congress Party was mutually beneficial. It expanded the bargaining power of these political organizations, especially for negotiations with the Kerala and central administrations. The planting organizations and the Indian National Congress, in both of which the Syrian Christian planters played significant roles, represented the Syrian Christian plantation segment. In other words, this group represented a major source of government income, employment and – most importantly – were the greatest producers of rubber in India.

There were other ways by which the planting aristocracy tried to wield influence in political decisions and developments. The planters' class gave generous financial support to Christian, anti-communist institutions, including churches, schools and newspapers (like the *Malayala Manorama* and *Deepika)*. Significantly, the planting elite was involved in the Moral Rearmament Movement (from the mid-1950s), an American-based and probably CIA-financed crusade against the rise of communism in the world which was entrenched in Kerala as well, as also in the Liberation Movement (1959), which successfully attempted to oust the Communist Namboodiripad Ministry. These were ways by which the Syrian Christian planting elite tried to make *their* political agenda the generally-accepted viewpoint.[26]

[26] Interviews: E.M.S. Namboodiripad; Michael A. Kallivayalil; Michael Tharakan. For further particulars concerning the political influence of the Syrian Christian planting elite, see also Jose A. Kallivayalil's life story in Section 7.2.2.

6.1.4 *European Withdrawal*

In the years following Independence, the presence of British planting companies in India was vociferously contested in the various South Asian political arenas. Even worse than having to deal with the often outspokenly unfriendly atmosphere, at an official level the British entrepreneurs were confronted with government policies which affected their companies adversely. Table 6.6 illustrates the ebbing significance of the British plantation segments. In Kerala the area owned by sterling companies under rubber and tea amounted in 1960 to 7 and 64 per cent respectively. By way of comparison, in Travancore the same figures amounted to 15 and 69 per cent respectively in 1948 (consult Table 5.4).

TABLE 6.6
The Relative Strength of the Foreign Plantation Segments in Kerala in 1960 (acreages)[27]

Ownership	*Rubber*		*Tea*	
Sterling Companies	19,888	(7%)	63,059	(64%)
Others	268,561	(93%)	36,152	(36%)
Total	*288,449*	*(100%)*	*99,211*	*(100%)*

Indicating that the times had changed for the British planters, agencies of both the state and central governments became increasingly adept in the acquisition of information about the foreign companies' financial backgrounds. Spurred on by the initially weak position of the tax officials and the planters' reluctance to give access to their companies' bookkeeping, the Kerala government became a shareholder in some major foreign plantation firms, thereby shrewdly attempting to get a better understanding of the companies' financial situation. The state government took shares in such firms as the Kannan Devan Hills Produce Company and the Anglo-American Direct Tea Trading Company.[28] Disappointingly, the state officials were unable to make public the particulars thus obtained. The tax

[27] *UPASI Planting Directory* 1960: 415–17.
[28] Proceedings TLA 16 February 1949; Proceedings TCLA 14 March 1950.

laws did not permit the disclosure of such information.[29] Only in the late 1950s did tax officials and the public became 'fully informed about the foreign companies' balance sheets, profit and loss accounts and other financial details.[30]

The minutes of the Kannan Devan Planters' Association also indicate that the British planters were greatly worried about the new taxation system to be introduced by the central government. Following his explanation that 60 per cent of taxable income was subject to the Kerala agricultural income tax, the general manager of the Kannan Devan Hills Produce Company, J.B. Soutar, stated that 'on 40 per cent of the taxable income, the rate of Central Government taxation on sterling companies has been increased from 61.5 to 63 per cent'. Soutar continued, 'this rate contrasts sharply with the 45 per cent rate of Central Government income tax proposed to be levied on Indian companies'.[31] Other sources confirm that the sterling companies had to submit to a higher rate of taxation than the rupee undertakings.[32]

The European planters' right to land was often questioned in the legislative assembly. In this regard the situation in the Kannan Devan Hills became the cynosure of all eyes. Most assembly members were furious that the nineteenth century agreements between the Travancore government and James Finlay and Company, the mother company of the Kannan Devan Hills Produce Company and the Anglo-American Direct Tea Trading Company, were still largely intact. Some were satisfied with the explanation usually trotted out by the state officials claiming that the government was not in a position to recover ownership of the area. As the agreement stated: 'land is to be held in perpetuity as heritable or transferable property'. Unimpressed, others proposed nationalizing the estates situated in the district.[33]

Labour union activists also focused their attention on the

[29] Proceedings TCLA 8 November 1952; 16 April 1955.
[30] Minutes of the KDPA 23 May 1959.
[31] Minutes of the KDPA 23 May 1959.
[32] Plantation Enquiry Commission – Rubber 1956: 124; Plantation Enquiry Commission – Tea 1956: Ch. 15.
[33] Proceeding TLA 9 August 1948; proceedings TCLA 20 August 1949; TCLA 5 April 1955; 16 April 1955; compare Sections 4.1.3 and 6.3.1.

foreign estates, which was not surprising. Concentrating on the European plantations afforded many distinct advantages. Such a policy gave certain ideological benefits. Though the labourers' working and living conditions on the indigenous estates were just as appalling as on the foreign plantations, it was mainly the British enterprises which were identified as culpable by the Malayali public. Building on a tradition rooted in the late colonial period,[34] the construction of that particular image was carried on by the leftist activists and like-minded authors in the post-colonial period. Examples of polemics abound. In his book *Kerala: Past and Present* the influential communist A.K. Gopalan spoke about the European estates as 'centres of inhuman slave labour and white exploitation', while making no mention whatsoever of similar conditions prevailing on the indigenously-owned plantations.[35] The other reasons for these authors' setting their sights on the British plantation segments were merely practical. On an average, the European estates were larger than their indigenous counterparts and thus represented the major centres of working-class concentration. Of even greater weight, on the British-owned estates contradictions in class coincided with differences in race. Vertical solidarity was therefore checked.[36] Lastly, the labour movement assured itself of the approval of the indigenous planters, though only in particular contexts and always tacitly.[37] In Section 6.2 the rise of labour unions will be discussed at length.

Notwithstanding these drawbacks, the European enterprises remained highly profitable in the late 1950s and the 1960s. This is reflected in the level of dividends paid on ordinary shares. The Kannan Devan Hills Produce Company, for example, only once distributed a dividend less than 10 per cent. In all other years in the period 1947–60, the dividend paid on ordinary

[34] See Section 5.3.1.

[35] Gopalan 1959: 46.

[36] To counteract the 'foreignness' of their estates, the board of directors of the British companies gradually replaced the European employees by Indian personnel. Although the posts of assistant manager and manager became 'Indianized', the functions of group manager and general manager mostly remained the domain of Europeans in the 1950s and 1960s (Griffiths 1967: 237–8; Soutar 1976: 24; Mackay 1984: 118; Mackay 1984: Tape 2 Side 2).

[37] Cf. Ramachandran Nair 1973: 113.

shares amounted to 10 per cent or more. No less than four times did the shareholders receive a dividend of more than 20 per cent: 25.5 per cent in 1947; 23 per cent in 1950 and 1951; and 27.5 per cent in 1954.[38] As evidence that the cultivation of both rubber and tea remained profitable ventures for European entrepreneurs, the financial background of Malayalam Plantations is provided in Table 6.7. Unlike the Kannan Devan Hills Produce Company, Malayalam Plantations produced both rubber and tea. As shown by the figures, in all years 10 per cent or more was paid on ordinary shares in the period 1947–8 to 1959–60. In the years 1954–5, 1955–6 and 1956–7 the dividend paid reached a height of 25 per cent annually.

TABLE 6.7

The Financial Situation of Malayalam Plantations During the Period 1947/48–1959/60 (in British pounds)[39]

Year	*Rubber*		*Tea*		*Profit*	*Income Tax*	*Dividend*
	Average Sale Price	*Average Cost Price*	*Average Sale Price*	*Average Cost Price*			
1947/48	1/0.96	8.25	2/0.24	1/4.66	560,674	353,456	15
1948/49	1/2.18	10.25	2/1.76	1/8.32	370,898	234,488	15
1949/50	1/3.67	10.75	2/3.62	1/8.60	608,299	357,488	15
1950/51	1/4.98	10.88	2/6.75	1/10.98	708,012	390,056	15
1951/52	1/10.84	11.75	2/5.75	1/9.81	994,658	573,218	10
1952/53	1/10.57	1/0.25	2/1.63	1/10.86	538,723	248,513	10
1953/54	1/11.75	1/0.38	2/8.63	1/8.64	1,112,938	685,445	20
1954/55	2/0.57	1/1	3/3.13	1/11.37	1,406,497	683,064	25
1955/56	2/2.88	1/2.75	2/9.63	2/0.45	1,060,799	582,340	25
1956/57	2/3.56	1/3	2/11.82	2/0.29	1,101,243	685,091	25
1957/58	2/5.27	1/4.25	2/5.59	1/11.78	827,060	496,200	20
1958/59	2/3.13	1/4.88	2/5.57	1/11.39	960,031	555,300	18
1959/60	2/3.43	1/5.13	2/8	2/1.44	856,638	455,978	18

[38] Various *Tea Share Manuals*, in: IOL&R: MSS Eur F 174.

[39] Figures obtained from various *Tea Share Manuals*, in: IOL&R: MSS Eur F 174.

The situation described above raises an important question: how did the British entrepreneurs manage to keep their heads well above water financially, despite operating in a politically disadvantageous context? There is no denying that the various price regulation strategies were of great importance in this regard. As we will observe later, the planters' reluctance to implement legislation regarding the labour and living conditions of the workers played a key role as well. Granted that they were significant, these factors still prove insufficient to explain the above issue: foreign as well as indigenous planting interests, often in combined efforts, adopted these policies, but only the former was confronted with a hostile political environment. Therefore, a third factor should be introduced. Most British companies rapidly reduced the scale of their investments. As the Plantation Enquiry Commission reported, this was particularly marked in the case of tea estates. The commission emphasized that the British planters, unlike their indigenous counterparts, had no intention of replanting old, uneconomical tea bushes by new and higher yielding plants. As the planters themselves acknowledged, the economic life of a tea bush receiving reasonable care was about sixty years. At the time of the investigation, most areas under tea, including those opened up during the 1880s, had never been replanted. The British planters' refusal to replant fitted in with their overall strategy: the gradual repatriation of capital culminating in the sale of their estates. In fact, that process was accompanied by another development. At the time of disinvestment in south-west Indian estates, many British entrepreneurs turned their attention to other parts of the world, of which colonial Africa became the most important region for investment.[40]

[40] Plantation Enquiry Commission – Rubber 1956: Ch. 4; Plantation Enquiry Commission – Tea 1956: Ch. 5; Griffiths 1967: 238–9; Jeffery 1982: 246–7; Tharian George 1984. To provide some additional information to those particularly interested in the Kannan Devan Hills, James Finlay and Company gradually sold all its properties to Tata, a large north Indian conglomerate, in the 1970s and 1980s.

6.2 Towards Tripartite Talks

6.2.1 Proletarians Unite!

As described in more detail in the previous chapter, the rise of labour unions in the plantation districts was relatively late. The members of the so-called Rege Committee, who studied the conditions of plantation labour in India in the mid-1940s, were also struck by this particular feature. With regard to south India they reported: 'In striking contrast to the powerful organization of the employers [the UPASI], there is no organization of any kind worth the name among the large mass of plantation workers'. The next part refers to south-west India in particular. 'But one estate, viz., the Koney Estate in South Travancore, mentions in its reply the existence of a trade union with very small membership among the workers.' 'The reply also states', the Rege Committee continues, 'that in the event of any dispute the union officials are expected to co-operate with the Manager!' 'It is, therefore, unnecessary to comment on the strength or the representative character of this trade union or its power of bargaining *vis-à-vis* the Manager', the Committee concluded.[41]

In the years after this the development of labour unions representing estate workers was limited. As the Travancore Administration Report of 1947–8 observes: 'Plantation labour has been slow and halting in trade union activity. There were ten registered trade unions for this class of labour at the end of the year.'[42] Later on again, the officials were still continuing to emphasize the lack of organization among the workers, implicitly arguing that the trade unions of estate workers were weak compared to those in other sectors. For example, the Administration Report of Travancore-Cochin for the year 1949–50 stated that 'including the trade unions registered in the previous years, there were twenty trade unions of plantation labour at the end of the period', and it went on to say that not one of them was 'sufficiently well-organised'.[43]

Only in 1953–4 was the growing strength of *all* labour unions acknowledged. No longer considering the plantation labourers

[41] Rege 1950: 183.
[42] TAR 1123: 91.
[43] TCAR 1125: 76; cf. TCAR 1124: 82.

as an exception, the Administration Report of that year states: 'The labour classes increasingly reorganized the advantage of organizing themselves into trade unions'.[44] Though having gained strength, a phenomenon especially observable since the early 1950s,[45] the position of the trade unions representing estate workers remained fragile throughout the 1950s.[46]

The trade unions' complaints and demands concentrated on the low wages, the harsh working and living conditions, and the exploitative kangany system. A memorandum submitted by the Palappilly Rubber Estate Workers' Congress to the estate managers of Cochin Malabar Estates Limited, Kerala Calicut Estates Limited and Malayalam Plantations Limited illustrates this nicely. Referring to the level of the wages, it noted that 'the rate of wages obtained here is too low and we hope nowhere in India it is as low as here'. The document claimed that the wage rate had not changed in a decade, despite the fact that the cost of living had risen several times. The memorandum continues: 'In spite of this the wages of the labour remains the same here and neither are they given a better Basic Wage nor a sufficient Dearness Allowance'. The actual payment of wages presented another problem, as the workers had to be satisfied with what they got and the estate accounts could not be questioned. Nor were working and living conditions acceptable to the labour union activists.[47] The last wrong against which the union took

44 TCAR 1953/54: 203.

45 Proceedings TCLA 17 March 1951; 8 November 1952; 25 March 1953; 20 April 1955; TCAR 1953/54: 203; TCAR 1954/55: 173; KAR 1956/57: 418; Clough 1984: Tape 1 Side 1.

46 See especially the last section of this chapter.

47 As the paper pointed out: 'When accidents occur, labour are not compensated and when compensated not sufficient at all.'; 'The type of lines built for the resident labour are not at all satisfactory. There is no proper ventilation. Walls are not plastered and white washed in some places and hence are invested with flies, bugs and ants, etc.'; 'The wells are not nearer to the lines. There is not sufficient water in them at times. Neither the water is clear and drinkable, nor the wells well kept. When the managers and some staff get clear pipe water supply, the labour is forced to drink the unfiltered and unclean water from the worst type of wells.'; 'For name sake there are dispensaries in the estates. (. . .) They are not even getting good mattresses to lie down.'; 'Owing to the inability for affording good and nourishing food and for the lack of good drinking water and sufficient and proper medical aid, the labourer's standard of health is very poor in the estates.'

up cudgels was the system of recruiting labour through the contractors. It demanded direct employment and direct payment by the managements.[48] To make it clear that we are not dealing with an insignificant association, it should be noted that the Palappilly Rubber Estate Workers' Congress represented 1700 members in 1952.[49]

6.2.2 *Anti-Union Strategies*

In comparison to the managers of other labour-intensive industries, the planters were in a disadvantageous position to counteract the rise of trade unionism. Unlike many of their colleagues, the plantation managers were unable to introduce fundamental changes to transform the production process: the tapping of rubber and the plucking of tea could not be mechanized. Therefore, three other options remained open to those planters who wanted to adopt strategies to avoid trade union demands. Of these, geographical relocation of the enterprise was only feasible for the large British companies, privileged in terms of capital, trading networks, and political influence in British colonies. The second alternative, co-optation of trade unions, had been applied only cautiously since the early 1950s, a matter which will be examined in more detail in the next section. The third option, quite baldly repression of labour union activism, was considered by both indigenous and foreign planters alike to be the most effective instrument to fight the rise of the labour movement throughout the 1940s and 1950s.

Several times in earlier chapters there have been allusions to the fact that the planters tried to keep the estates as

[48] The document argued that the kangany system caused countless hardships to the workers. 'It is the labour contractors who are benefitted. They get from 12 to 16 per cent commission on the total earnings of the labour. (. . .) The labour contractors enjoy at the cost of poor labour. (. . .) However, experienced be the labour and whatever be his service in the estate there is no security of permanency to his work. His lot is entirely left with the contractor. He can be sent off at any time at the sweet will of the employer and contractor. Labour who are victims to this sort of treatment are rotting here as they are penniless to reach their villages and homes' (proceedings TCLA 27 March 1951; and cf. proceedings TLA 9 August 1948; proceedings TCLA 6 November 1950; 18 September 1951; 27 September 1951).

[49] Proceedings TCLA 8 November 1952.

isolated as possible from the outside, increasingly 'hostile' world. In line with the anti-communist strategies they had been adopting since the mid-1930s, the managers did their utmost to prevent leftist activists from setting foot on their properties. As a representative of the planters declared in the early 1950s: 'the management could not close their eyes to unauthorized entry into estates by outsiders, holding meetings without permission and shouting of slogans near factories during working hours'.[50]

The planters did all in their power to forestall the labour union organizers from meeting a 'revolutionary' work force. Conscient of the weak position of migrant workers, the managers favoured the recruitment of non-local labour. To take one example, on the plantations under the aegis of the Palappilly Rubber Estate Workers' Congress most resident labourers were not natives of Cochin, where the estates were situated. The bulk of these (semi-)permanent workers came from the neighbouring regions of Travancore and Malabar. However, the great majority of the casual, non-residential labourers did come from Cochin, living mostly six to ten miles away from the estate.[51] One highly enlightening point is that the planters preferred the recruitment of non-local labour, in the sense of non-Malayali workers. Well aware that the majority of the leftist activists were to be found among the people in the Malayalam-speaking, coastal regions of south-west India, the planters continued their policy of

[50] Quoted in: Ramachandran Nair 1973: 239. In the period of political transition, the Rege Committee observed: 'Any tendency at combination among the workers is jealously watched by the Superintendents who view with disfavour the contacting of their workers by outsiders. In the large majority of estates outsiders, other than friends and relative of labours, have to take previous permission of the Superintendents before they can enter the precincts of the estates, especially if such outsiders come with the purpose of organizing the workers in Trade Unions. There have been instances when such outsiders were severely man-handled at the instigation of the Superintendents. (. . .) If the workers in plantations are left to their own devices in the matter of organization, not for another fifty years can we see a trade union coming up. Perhaps it is this confidence that is at the root of the planters insisting on the preclusion of outsiders' (Rege 1950: 184).

[51] Proceedings TCLA 27 March 1951.

recruiting Tamil workers.[52] This huge immigration into Travancore–Cochin/Kerala is reflected in the official figures. In 1955 the States Reorganization Commission reported that 72 per cent of the population of Devicolam and Peermade were Tamils.[53] Turning to the census, it reveals that 43,361 males and 37,569 females migrated from the Madras State to Kottayam district in 1961. The authors say that this large number of immigrants was primarily, if not solely, due to the plantations in the region.[54]

The importation of Tamils by the planters was a thorn in the sides of both the labour movement and the state bureaucracy. Occasionally, Malayali trade union activists, through persuasion, violence or some other tactic, succeeded in expelling individual Tamil workers from the estates,[55] but in public debates leftist activists studiously avoided this 'sons of the soil' strategy as it conflicted with both their class-based communist ideology and the interests of thousands of Malayalis working outside southwest India.[56] The attitude of the state bureaucracy towards the Tamil migrant workers was likewise ambiguous. Afraid that areas inhabited predominantly by Tamils, like the Kannan Devan Hills, would be excluded from the future state of Kerala, bureaucrats discriminated against Tamil labourers in such matters as political franchise, right of association and assignment of lands. Publicly, naturally, state officials denied such accusations when these were made by various Legislative Assembly members.[57]

The planters showed a strong preference for the recruitment of illiterate, low-caste family labour. The idea behind the strategy was that this category of workers was likely to be most alienated from the labour movement. Illiteracy prevented, or at least discouraged, the spread of communist ideas.[58] Cold blooded

[52] Proceedings TCLA 8 March 1950; 21–23 November 1955; interviews: V. Haridasan; Abraham Kurien; P. Madhavan; K.T. Thomas and T.J. Joseph; Marthanda Varma; K.U. Warier.

[53] Proceedings TCLA 21 November 1955.

[54] *Census* 1961, vol. 7, part 1A: 301.

[55] Interview: Abraham Kurien.

[56] Interview: K.U. Warier.

[57] Proceedings TCLA 16 August 1949; 8 March 1950; 12 July 1954; 21–23 November 1955; Nossiter 1982: 119–20.

[58] Interviews: Suresh; Alex Manuel; M.C. Mathew.

calculation made planters well aware that the workers' ranking within the caste system brought low expectations regarding the quality of life and the ability to climb the social ladder.[59] The preference for the recruitment of low-caste labourers is reflected in the sources. About 30,000 (!) *harijans* – Gandhi's term for members of the lowest castes – worked on the plantations situated in the taluk of Devicolam, Kannan Devan Hills.[60] The employment of families was also calculated to put a spoke in the rise of unionism. Women and children were not supposed to take part in political activities, or were supposed to do so only to a lesser extent.[61]

Realizing that their stratagems of keeping out unwelcome activists and welcoming illiterate, low-caste Tamil families were still insufficient to counteract the rising tide of labour unions in the planting districts, the planters supplemented these policies with other strategies. One such tactic was outright victimization: the ill-treatment of trade union activists. In examining the planters' retaliatory measures two major developments stand out. The first to suffer from these policies were those staff personnel who sympathized with the Estate Staff Union of South India (ESUSI) in the late 1940s and the early 1950s. There is ample evidence to show that employers like Tropical Plantations (Kottayam), Chinnar Estate (Peermade), Eldorado Group (Mundakayam) and A.V. George and Company (Kottayam) dismissed K.M. Abraham, T.O. Chacko, K.T. Mathai, and P.O. Philip respectively, because of their active participation in the ESUSI.[62] Another widely applied planters' ploy was to transfer 'politicized' staff members either to another function or to another estate.[63] Later on, the planters shifted their attention to a lower level on

[59] Interview: V.J. Joseph and V.J. Shaju.

[60] Proceedings TCLA 23 January 1950. Pascale Chaput, an anthropologist who analysed the working of a Kerala rubber estate in the period 1987–90, argued that the 'traditional' caste hierarchy was reflected on the 'modern' plantations: the management consisted mostly of high-caste Malayalis whereas most workers were of low-caste origin, many of whom had Tamil roots (Chaput 1992). It seems, therefore, that the situation described in this and earlier chapters has continued up to the present.

[61] Cf. Department of Statistics 1959: 3; Haridasan 1967: 4.

[62] Proceedings TCLA 26 August 1949.

[63] Cf. Proceedings TCLA 3 November 1950.

the plantation hierarchy: the workers engaged in trade union activities. Here discharges and transfers were commonly used anti-labour strategies as well.[64]

As the planters' focus shifted from staff to workers, a second development occurred. The nature of the disciplinary actions changed. Violence became an increasingly significant weapon in the struggle against the labour movement. In May 1954 the managing proprietor of Palkulam Estate assaulted three plantation workers, most probably all members of the Estate Workers' Union, Nagercoil.[65] With malice aforethought, in February 1955 the managers of Merchiston Estate hired in thugs to stone those present at a meeting organized by the Travancore-Cochin Estate Workers' Union.[66] At the same time, there are no indications that the heavy-handed treatment meted out to labourers by policemen and kanganies ended. The reverse seems to have been true. In the Travancore-Cochin Legislative Council the representatives of leftist political parties accused the police of frequent and unnecessary use of violence against the plantation workers.[67] The leftist activists also publicized the fact that the violent encounters between field supervisors and workers were sometimes fatal, for which a fight on Cheruvalli Estate provides ample evidence.[68]

Perhaps accepting that the labour movement was there to stay, the planters did their best to oppose the formation of *one single* labour movement. To ensure that their room for manoeuvre remained optimal, the estate managers encouraged the division of the labour movement into various segments, each with its

[64] Cf. Proceedings TCLA 9 August 1948; 27 March 1951; *Kerala Gazette* 6 October 1959; Mathew 1984: 1678.

[65] Proceedings TCLA 28 July 1954.

[66] Proceedings TCLA 3 February 1955.

[67] For example, a motion of adjournment put forward by Sivarama Bharati stated: 'On the 15th instant [March 1955] on the pretext of a criminal case, Nelliampathy Plantation Workers' Union Secretary Mr Xaviour and some of the Union workers were arrested and taken to the lock-up of Nemmara. Thereafter they were brutally assaulted in the lock-up. Following this the police are terrorising the working population on the Estate including the womenfolk and children. These excesses of the police have made peaceful life impossible in the Estate area and the situation is grave' (proceedings TCLA 25 March 1955; and cf. proceedings TCLA 8 November 1952).

[68] Proceedings TCLA 26 February 1953.

own particular interests, a stratagem which met with considerable success. There is no doubt that the planters' recruitment of both outside, (semi-)permanent labour and of local, casual workers created many problems for those labour organizers who fought for unity within the labour movement. The enormous difference in legal status between each category of workers proved to be a major obstacle. Organizing combined protests became frightfully complicated as most permanent workers enjoyed at least some rights, whereas the majority of the casual labourers toiled without any form of legal protection whatsoever.[69]

As the planters had hoped, the different ethnic identities of the workmen kept the labour movement divided. Treated as 'foreigners', most Tamil workers felt alienated from their Malayali counterparts. Naturally resenting the anti-Tamil strategies of Malayali activists and officials, most Tamil labourers eschewed the Malayali-led political organizations. In the Kannan Devan Hills, for example, the Tamil workmen solidly supported the Travancore Tamil Nad Congress, a party struggling to exclude the predominantly Tamil-inhabited districts from the future state of Kerala. The Tamil labourers also operated their own unions, like the Travancore Tamilnad Estate Workers Union (Thittuvila).[70]

The division of the estate population into different task categories, represented by various – often conflicting – social groups[71] and organizations, was another tactic which enlarged the planters field of political manoeuvrability. As referred to above, the staff was represented by the ESUSI. One step lower in the plantation hierarchy, kanganies formed organizations

[69] Cf. Proceedings TCLA 27 March 1951; Ramachandran Nair 1973: 48; Ravi Raman 1986: 961.

[70] Proceedings TCLA 16 August 1949; 8 March 1950; 8 November 1952; 12 July 1954; 21–23 November 1955.

[71] As on the European-managed estates in colonial times, the mother tongue and community background of the employees often coincided with their position in the task-based division of the labour force. Formulating this another way, upper and middle-caste Malayalis especially were recruited for staff functions whereas most field workers were low-caste Tamils. To reinforce their superior positions on the estates, the Indian managers, like their British counterparts, spoke English with each other. This policy has still been continued up till today (interviews: AITUC workers Gudurale Estate; P. Madhavan; M.C. Mathew; George Tharian).

such as the Estate Kanganies' Association (Munnar), the Estate Kanganies' Union of Travancore (Mundakayam), the Kerala Estate Labour Contractors' Association (Thenmalai) and the South Travancore Estate Kanganies' Association (Kulasekharam). The tea factory workers (All Travancore Estate Factory Workers' Union, Punalur) and rubber-tappers (Tappers Union, Kottayam) were also represented by separate trade unions.[72]

In short, the creation of a work-force which was divided legally, ethnically and by task, allowed room to play the different interest groups off against each other. Differences in caste background and in political preferences offered the planters even more opportunities for forestalling the creation of unity among the workers.[73] Seen in the context of the emerging political landscape of independent India, most trade unions were affiliated either with the All India Trade Union Congress (AITUC) or with the Indian National Trade Union Congress (INTUC). The influence of the United Trade Union Congress (UTUC) was relatively insignificant. The AITUC, INTUC and UTUC maintained close links with the Communist Party of India, the Indian National Congress and the Revolutionary Socialist Party respectively. In the last section of this book there is a discussion of the fact that the INC-controlled unions were started by the planters themselves.[74]

Although they held some reckoning with the trade union movement, this does not automatically imply that the planters negotiated with the trade unions either continuously and/or openly. The contrary was in fact true. With might and main, the plantation managers hammered at the invalidity of most labour unions. Leaving aside their often openly displayed aversion to socialist ideologies, the planters' arguments are worth looking into. Pretending that historically trade union activity emerged in and remained confined to the industrial sector, they argued that labour unions were out of place on the estates.

[72] Proceedings TCLA 14 March 1950; 17 March 1951; 8 November 1952.

[73] Mathew 1984: 1678; Chaput 1992; interviews: Abraham Kurien; Suresh.

[74] See Section 7.2.2. At the same time, the reader should realize that the planters' 'control' over INC-directed unions was not peculiar to Kerala. According to Das Gupta, the Congress leaders in Jalpaiguri (Bengal) did little to initiate any struggle in the district because of their close social and familial links with the Jalpaiguri-based Bengali planters (Das Gupta 1992a: 113).

Plantation production should be considered 'agriculture' and not 'industry'.[75] They also stated that they themselves, aided by individual, non-organized labourers, were perfectly capable of defending the workers' interests. Suiting actions to words, the planters appointed labour officers, whose official task was to look after the welfare of the workers, and founded so-called Works Committees, in which non-radical representatives of the labourers could discuss their problems with the management.[76] Initially, the planters' refusal to accept the validity of the trade unions occurred in a favourable political context. As indicated above, until 1952 the communist party and its associated organizations were declared unlawful. Even after the ban on communist activities was lifted, however, the planters refused to recognize most trade unions, an attitude reflected in the text submitted by the Palappilly Rubber Estate Workers' Congress. Demand No. 11 stated: 'That this Labour Union known as Palappilly Rubber Estate Workers' Congress be recognised'.[77] As we will observe in Section 6.3.3, the planters were still refusing to recognize many labour unions in the late 1950s.

6.2.3 Conflicts Governed

The steadily growing strength of the labour movement on the one side and the anti-trade union strategies of the planters on the other culminated in a considerable increase in conflicts on the estates. In the late 1940s the *inception* of the open clashes between labour and management can be observed. As the government report of 1946–7 mentions, 'during the year, estate labour also [in the sense of: apart from workmen in other sectors] began to agitate for higher wages and better conditions of employment'.[78] Two years later, the administration noted an increase in the confrontations in the plantation sector. 'Trade disputes were comparatively numerous in this industry during the year. Some of them culminated in direct action like *satyagraha, hartal* and strikes.'[79] In the early and mid-1950s the number of conflicts

[75] Cf. Ramachandran Nair 1973: 102–3.

[76] TAR 1122: 95; TAR 1123: 91; Ramachandran Nair 1973: 57.

[77] Proceedings TCLA 27 March 1951.

[78] TAR 1122: 95.

[79] TCAR 1124: 82. The successful protest acts of the freedom movement were

rose sharply. 'Numerous industrial disputes arose in plantations during the period and some developed into strikes', the official document states.[80] The rise in clashes is also reflected in the table given below. The number of industrial disputes in the plantation industry jumped from 570 in 1950–1 to 2,227 in 1956–7. To give an indication of the growing participation of workers in strikes, the number involved in work-stoppages amounted to 71,772 in 1955–6. With regard to man-days invested in work-stoppages, 1958–9 was a climax. During that year – i.e. during Namboodiripad's second year as Chief Minister – the planters lost 700,189 man-days. Of the number of work-stoppages, the great majority were strikes, and in some years these protests were supplemented by a few lock-outs: the closing of a place of employment, like the rubber/tea factories, the staff quarters and/or the whole estate, either by the employers or the employees, to step up the pressure to agree to terms.[81]

TABLE 6.8

Industrial Disputes in the South-west Indian Plantation Industry, 1950/51–1958/59[82]

Year	*Number of Industrial Disputes*	*Number of Work-Stoppages*	*Number of Workers Involved in Work-Stoppages*	*Man-days Invested in Work-Stoppages*
1050/51	570	12	6,210	38,993
1951/52	682	38	18,212	177,589
1952/53	974	42	12,428	50,201
1953/54	1,404	30	69,683	360,301
1954/55	1,905	64	52,972	246,847
1955/56	2,040	58	71,772	122,741
1956/57	2,227	38	41,799	43,423
1957/58	1,777	70	13,120	50,237
1958/59	1,940	83	69,844	700,189

adapted by the labour organizers. Though satyagraha refers to many forms of non-violent resistance and hartal stands for the act of closing shops or suspending work, in this particular context satyagraha, hartal and strikes probably all hint at the same phenomenon: work-stoppages.

80 KAR 1956–7: 412.

81 Cf. Proceedings TCLA 15 July 1954; Mackay 1984: I21; Clough 1984: Tape 2 Side 2; Mackay 1984: Tape 2 Side 2.

82 Figures of the Labour Department, in: Ramachandran Nair 1973: 407.

State involvement in these labour-management confrontations assumed many guises. At the grass roots, policemen often played a crucial role in preserving 'law and order', applying violence in particular cases, as when police forces intervened on the Pasumala Estate following an attack on a European manager by five hundred workers. A *lathi* charge and firing brought the situation under control.[83]

At the negotiating level, Labour Department representatives were increasingly involved in the settlement of the disputes. At estate, district and state levels they guided meetings in which the representatives of the workers and those of the employers met. Figures demonstrate the significance of this in governing conflicts. The Labour Department dealt with the greater number of disputes: 501 in 1950–1, 621 in 1951–2, 878 in 1952–3, 1026 in 1953–4, 1513 in 1954–5, 1928 in 1955–6, 1739 in 1956–7, 1627 in 1957–8 and 1658 in 1958–9. In most cases, settlement was reached either by conciliation – i.e. 'assisted bargaining' – or state arbitration – i.e. the arbitrator's decision being binding.[84]

In the late 1940s and the 1950s, the tendency for labour activists to set the agenda for the discussions was highly typical. This was certainly the case during the so-called Tripartite Conferences, gatherings at which the representatives of labour organizations, the state government, and planting associations met. During three industry-wide meetings, which took place in the early 1950s, the recognition of trade unions, the payment and level of the wages, and the abolition of the kangany system were central themes. The trade union demands won the installation of the Minimum Wage Committee for Plantations and the Minimum Wage Advisory Committee for Plantations as well. Appointed by the government of Travancore-Cochin in August 1952 and November 1953 respectively, the task of the commissioners was to enquire into the possibilities of revising and fixing the financial reimbursements for estate work. The Plantation Labour Committee had more permanent character. Established subsequent to and at the instance of the third Tripartite Conference, the committee met monthly and, apart from the more

83 Ramachandran Nair 1973: 123.

84 Proceedings TLA 9 August 1948; proceedings TCLA 18 September 1951; 5 November 1952; 15 July 1954; 19 and 20 April 1955; 8 March 1956; Ramachandran Nair 1973: 407.

pressing above-mentioned issues, discussed topics like cattle-grazing schemes and the carrying of leaf and latex over long distances. Besides being a clear step forward in improving the workers' lot, the consultations produced some serious dilemmas for the trade union leaders as well. One major example was the eternal dilemma: what should be decided if higher wages implied less jobs on the plantations?[85]

In sharp contrast to their public image, the state officials who hosted the industry-wide joint consultations were neither just observers nor were they neutral. They were there to protect their state's interests. Realizing that the plantation industry employed the largest number of labourers in the state,[86] the bureaucrats were well aware that the adoption of an indulgent attitude towards the plantation labour activists was of fundamental importance in preserving law and order. With their feet firmly on the ground, the officials acknowledged that giving in to the trade unions' demands either too completely or too quickly would harm the interests of the state as well. It would do so directly, in the sense that lower profitability or an eventual loss of the plantation industry would considerably affect the state's financial situation. No one could deny that the plantations provided a major source of income for the treasury. In a more indirect sense, the bureaucrats feared that too powerful a trade union movement would discourage entrepreneurs from investing in their territory. Finding a solution for the lagging behind of industrialization, especially in view of the huge unemployment figures in the state, was high on the list of priorities of both politicians and officials.[87]

Although they had never at any point favoured public discussion regarding labour conditions on the estates, the planters tried to take maximum advantage of the tripartite sessions and shrewdly tried to exploit the dilemmas of both the labour movement and the state government. Paradoxical as it may at

[85] Proceedings TCLA 4 April 1950; 31 March 1952; 8 November 1952; Minimum Wage Committee for Plantations 1953; Minimum Wages Advisory Committee for Plantations 1954; Ramachandran Nair 1973: 237–42.

[86] TCAR 1950–1: 80; TCAR 1951–2: 77; KAR 1956–7: 412; and cf. TCAR 1124: 82; TCAR 1125: 76.

[87] Cf. Proceedings TLA, TCLA and KLA; various TAR, TCAR and KAR.

first sound, the planters used the industry-wide meetings to attenuate the effects of the trade unions. Pragmatically, the ongoing discussions at state-level were used to reduce the hostility and anxiety of the labourers on the individual estates. The planters drummed into the members of the trade unions that their problems were being dealt with at a higher level. In other words, the industry-wide joint consultations offered an opportunity to lessen the tensions and conflicts in their original settings: the plantations.

The tripartite meetings themselves were also used to slow down the tempo of the process of change proposed by the trade unions and their members. Manipulating the interests of both labour leaders and state officials, during the sessions the planters continually harped on the awkward financial situations of the plantations, implying that higher costs would lead to labour cutbacks on the estates and would lower the revenues from the industry.

Calculatedly delaying final decisions about important issues, the representatives of the planters sometimes left the meetings under protest, frequently adjourned to consult their planter colleagues, encouraged the appointment of various sub-committees and subsequently disagreed about the findings of the commissioners. There are numerous examples of these procrastinating tactics, one being the putting into effect of the Indian Plantation Labour Act. Passed in 1951, and brought into force in 1954 by the all-India legislature, the specific details regarding the working and living conditions of plantation labourers for south-west India were regulated only in 1959. Until then the employers in particular cavilled and refused to agree on the particular contents of the Kerala Plantation Labour Rules.[88]

Not content with slowing down the negotiation process, the planters were also reluctant to implement the decisions which were finally taken. Most regulations, including the Kerala Plantation Labour Rules, were implemented neither immediately nor in full. One often-used strategy was to divide large estates into various smaller holdings, thereby circumventing and/or reducing

[88] Proceedings TCLA 8 November 1952; 6 March 1953; 22 March 1954; 15 July 1954; 25 January 1955; 4 February 1955; 19 March 1955; *The Kerala Plantation Labour Rules* 1959; Haridasan 1967: 28.

the effects of the labour legislation. Speaking in very general terms, the larger the estate the more responsibilities which accrued to the management, especially those to do with the provision of housing and medical facilities.[89] The planters' policy is also reflected in the official figures. For instance, in 1960 140 rubber and twenty-nine tea undertakings were classified as covering more than a hundred acres in the state of Kerala. To make a telling comparison in time and space, 140 rubber and 137 tea enterprises covered more than a hundred acres in Travancore – smaller in territory than Kerala! – in 1948 (see also Table 5.3). In other words, despite an expansion in the cultivation of both crops (see also Section 6.1.1), the number of estates with more than a hundred acres under cultivation remained stable in the case of rubber, but experienced a rapid drop in the case of tea.

TABLE 6.9

Size and Number of Rubber and Tea Undertakings in Kerala in 1960[90]

Size of the Undertaking	*Rubber*		*Tea*	
100 Acres and Less	1,117	(89%)	240	(89%)
More Than 100 Acres	140	(11%)	29	(11%)
Total	1,257	(100%)	269	(100%)

The upshot of all this shilly-shallying and legal cavilling was that improvements in the conditions of estate workers proved slow and circumscribed.[91] Both adult partners, often aided by their children, still had to continue to work for their sustenance, an example which is valid for the rubber plantations as well as the tea estates.[92] Even then their income was inadequate to

[89] *Plantation Labour Act* 1951; *The Kerala Plantation Labour Rules* 1959.

[90] *UPASI Planting Directory* 1960: 214–387.

[91] Kerala was certainly not unique in this respect. Guha writes that despite the passing of the Indian Plantation Labour Act 'basic problems were hardly solved' in Assam (Guha 1977: 332).

[92] *The Report on the Nutritional Diet Survey of Workers in Tea Plantations*, based on a survey conducted throughout Kerala in 1958, emphasized the arduous living condition of these families. 'In every household both the grown-up male and female go out for work. The working hours is normally from 7 a.m. to 5 p.m. with an interval of one hour in the afternoon for mid-day meals.

maintain a proper standard of living. As the *Report on the Nutritional Diet of Workers in Tea Plantations* showed: 'The calorie requirement of an adult worker in plantations thus worked out comes to 3,294, and that consumed at present is 2,900'.[93] The story was the same on the rubber estates. A survey conducted under the guidance of the Rubber Board in Kottayam district in 1963 revealed that 'the quality of food consumed by the rubber plantation worker is poor. With the assumption of the requirement of energy items of 2,800 calories per day per adult consumption unit the survey reveals that the under nourishment is 9.6 per cent (. . .)'.[94]

It is an understatement to say that the housing conditions on the tea and rubber estates were far from ideal. Quoting again from these studies and starting with the situation on the tea plantations: 'The workers live in quarters provided by the estate owners. These are found to be over-crowded and unhealthy. It is not rare to see that one or two families are accommodated in a single room (. . .).' The report concluded that housing conditions were far below the general standard.[95] On rubber estates, 'it is seen that 17 families live in houses having living space less than 200 sq.ft. and 74 families in houses having living spaces between 200–500 sq.ft. (. . .) The labourers are generally

---

Hence they are early risers by habit and get up by about four in the morning. Since the male and female workers have to go out for work, both breakfast and mid-day meals will be prepared in the morning itself. Thus these households cook their meals twice a day, the second being in the evening after returning from the day's work' (Department of Statistics 1959: 2).

[93] Department of Statistics 1959: 13. The report also stated: 'The coefficient of elasticity of expenditure on food with respect to total expenditure is found to be greater than one, thereby indicating that an increase in the total expenditure will not bring any decrease in the proportion expended on food. It can, therefore, be concluded that the workers in general have not reached a level of satisfaction in regard to food, in the sense that any additional income they get will be utilized to purchase more food items. (. . .) This is further confirmed by the coefficient of elasticity in clothing with respect to total expenditure which works out to be less than one. Such a type of expenditure pattern is nothing uncommon in a society of workers as this, in which both man and woman have to do hard manual labour for not less than eight hours a day' (Department of Statistics 1959: 15).

[94] Haridasan 1967: 12.

[95] Department of Statistics 1959: 2.

accommodated in one room tenements with common latrine, bathroom and well'.[96]

Another indication of the slow pace of improvement of the condition of the estate workers was their level of education. In this particular respect, the plantation workers remained the most backward citizens of south-west India. Again leaving the argument to statistics, in the Peermade taluk, inhabited predominantly by estate labourers, the literacy rate was up to 25 per cent below the average for Kerala in 1961. The same figure was even worse for Devicolam taluk, the major administrative unit in the Kannan Devan Hills. Here the literacy rate was between 25 per cent and 50 per cent below state average.[97]

Burdened as they were by these other handicaps, it was the prolonged existence of the kangany system which brought the greatest disappointment to the plantation workers. The delaying tactics employed by the planters prevented its immediate abolition. The first victory of the planters was the passing of a resolution to 'defer final decision in the matter until there is an opportunity of consulting the general membership on both sides' (First Tripartite Conference, 1950). One year later, they supported the abolition of the kanganies on the condition that a commission was appointed to give more detailed advice on the matter (Second Tripartite Conference, 1951). After having agreed that the kanganies should be replaced by a new cadre of labour supervisors (Third Tripartite Conference, 1953), the planters encouraged the introduction of a fourth party into the political arena: the kanganies, represented by their trade unions[98] and leaders. Subsequently, in 1957 the government appointed a committee to study the rehabilitation of the kanganies – restoring them in the position of salaried field supervisors – and the payment of compensation to those kanganies who were not

[96] Haridasan 1967: 8–9.

[97] *Census* 1961, vol. 7, part 1A: 370. Scholars interested in the labour conditions on rubber and tea undertakings in later years should consult: Labour Bureau 1964; National Commission on Labour 1968; Unny and Jacob 1972; Jacob 1974; Ramakrishna Pillai 1974; Krishnan Kutty and Haridasan 1976; Ravi Raman 1986.

[98] For the names and demands of the various unions, representing the *c.* 15,000 kanganies in Kerala, see: *Kerala Gazette* No. 41 dated 20 October 1959: Report of the Committee for Enquiry into the Kangany System.

willing to continue estate work. The recommendations were slightly modified by the Kerala government in 1959 but the planters still refused to implement the regulations. Only in September 1962, i.e. 25 years after the first plantation workers' unions appeared, 15 years after Indian independence and 4 years later than in the neighbouring state of Madras, was labour recruitment by kanganies legally abolished.

Therefore, the situation in which labourers, after accepting advances, were trapped in a debt spiral to the recruiters and, because of the low wages and the dishonest payments, were unable to pay off their debts, lumbered on at least until the early 1960s. To try to understand the meaning of bonded labour, 'slavery' in the workers' own terms, information provided by some elderly labourers in the Kannan Devan Hills is most illuminating.

Speaking about the late 1940s and the 1950s, the workers emphasized that at the time of recruitment they were told untrue stories about plantation life by the kanganies. Upon arrival on the estate, the level of the wages turned out to be much lower than had been promised. On pay day, which occurred with lapses of great intervals, the majority of workers obtained a *black* receipt, meaning that they still owed money to the kangany. Often the recruiters-cum-field supervisors disguised their holding back of payment and manipulation of the accounts by informing particular workers that they had borrowed more than they had actually earned.

The plantation work itself was much more arduous than they had expected. The kanganies used their position as creditors to force the workers to accept long working weeks. Working days lasting from 5 a.m. to 10 p.m. were no exception. To reinforce their position as superiors, the kanganies also found other ways to dominate the work force. Under the guise of teaching them how to show respect, the workers were not allowed to protect their feet nor were they given permission to cover their heads from the burning sun.

Though often regarded as the only way out of their intolerable circumstances, running away was hardly an option. In the Kannan Devan Hills, for example, the area was surrounded by checkpoints. Only those who were able to show a *white* receipt were allowed to pass through. Those workers who tried to escape and were caught were often subjected to grievous bodily harm. As one labourer recalled, the runaway labourers were beaten, kicked and doused with agricultural chemicals by the kangany and his helpers in front of the other members of the working gang.[99]

In the Peermade/Vandiperiyar district, various elderly workers told the same story about that period: human suffering stemming largely from the violence of kanganies, indebtedness, illiteracy and an (initial) lack of workers' organizations. They were also able to add certain new information. Many workers, male and female alike, seeing no way out of the cruel labour regime, committed suicide by taking poison or hanging themselves in the labour lines or from trees. The informants stressed that these were not isolated incidents, sporadic cases from which the evils of plantation life could be exculpated. The opposite was the case. These suicides ended the lives of a great number of workers and were directly related to the working and living conditions on the estates. This evidence leads me to the conclusion that the shift from a temporary to a more permanent labour force certainly did not lead to any extenuating or softening of the labour regime. There are good grounds for assuming it most probably led to an intensification of labour control.[100]

6.3 Kerala, 'The Problem State', Controlled

6.3.1 The Impossibility of Nationalization

The planters' fear of a leftist-oriented Kerala, of having to deal with and being situated in a 'problem state', was aggravated by the propaganda of the Communist Party. Especially since the early 1950s, the party leaders had proclaimed they would nationalize the plantations if the Communist Party were to win the Kerala elections and subsequently head the state government. It would have been very strange had the planters'

[99] Proceedings TCLA 29 March 1955; *Kerala Gazette* No. 41 dated 20 October 1959; Labour Bureau 1964: 15; Ramachandran Nair 1973: 240, 245–7; Jacob 1974: 9; interviews: AITUC workers Gudurale estate; Alex Manuel; P. Madhavan; K.T. Thomas and T.J. Joseph; with regard to the continued indebtedness of the plantation labourers after 1960 see for example: Haridasan 1967: 8.

[100] Interview: estate workers Pasupara Estate; compare also Section 5.3.4.

associations not protested against these ideas. Although the communist rhetoric was directed against foreign-owned estates in particular, both European and Indian entrepreneurs feared that they would be victimized. Different nationalities notwithstanding, all were representatives of the 'bourgeois class'.

The planters' disgust at communist plans was shared by the members of the Congress Party, especially those in the central government. Prime Minister Nehru, for example, had this to say about the proposals of the Kerala communists: 'I am not going to tamper with the honour and good name of India just to gain a few crores of rupees'.[101] Besides protecting the interests of the indigenous entrepreneurial classes, the Congress Party feared that the nationalization of the plantation industry would halt the influx of foreign capital and technology, both regarded as necessary conditions for the economic development of India.[102]

A debate carried on in the Lok Sabha in late 1956 and early 1957 represented the zenith of the controversy between the Communist Party and the Congress Party regarding the nationalization of plantations.

On 14 December 1956 the influential Kerala communist A.K. Gopalan moved the following resolution: 'This House is of opinion that the predominant position of British Capital in the Tea Industry has proved detrimental to national interests and that the Tea Industry should be nationalized forthwith'.

He then adduced several arguments to support his proposal. The first grievance was that the British companies refused to replant the tea bushes. Drawing his information from the findings of the Plantation Enquiry Commission, Gopalan argued that 'the Britishers are not replanting here, because they want to develop this industry in East Africa. If the [Indian] production is less', he elucidated, 'then there will be more profit [in Africa]. So, in order to have more profit, they want to curtail production here'. Moreover, the monopoly position of the British trading firms was a thorn in Gopalan's side. 'The monopoly has been extended as far as production, warehousing, brooking, buying, internal distribution and exports

[101] *The Hindu* 2 February 1957, quoted in: Lieten 1979: 31.
[102] Cf. Chaudhury 1984.

are concerned.' He was also concerned about the capital drain. 'According to the Reserve Bank data', Gopalan informed the parliament, 'the dividends, profits and interests payable abroad on foreign capital in the plantations amounted to as much as Rs 10.4 crores in 1948–9, that is more than 33 per cent of the total remittances.' Lastly, he focused on the social costs of plantation development. To quote one of his main concerns: 'the wages in tea plantations are generally lower than the wages in other industries in spite of the adverse working conditions and the higher costs of living in the plantation areas'. 'Under the Plantation Labour Act', Gopalan finally argued, 'it was said that about 380,000 families would require a house each (. . .). (. . .) These houses which have to be built have not been built. (. . .) None of the prescribed rules for the provision of medical facilities, drinking water, education facilities etc., have been properly implemented'. He wound up his plea thus. 'So, it is very essential that in the interests of the economy of the country, in the interests of the working people and also in the interests of solving the problem of unemployment (. . .) Government should see that this industry (. . .) is nationalized'.[103]

The resolution was taken up again three months later, on 22 March 1957. In that session, A.M. Thomas, a Kerala Congress Party member, strongly criticized Gopalan's proposal. He was particularly concerned about the implications of the plan, arguing that when an agricultural industry is brought under state control the production is likely to suffer. Echoing the special Gandhi-Nehru rhetoric, Thomas continued: 'But the very basis of socialism, as we all know, is increased production and the other matter is with regard to distribution and meting out social justice'. Opposing Gopalan's statement regarding the capital drain, Thomas informed his audience that any nationalization of the tea industry would swallow up crores of rupees because the compulsory acquisition of foreign interests should be compensated fairly and equitably. This had already been announced as a government statement of policy.

Thomas' argument met with approval from the Minister for Commerce and Industries, M. Desai. This influential representative of the central – Congress – government also emphasized

[103] Proceedings LS dated 14 December 1956.

the negative consequences of nationalization. Starting from the point that 'it is not possible to nationalize merely the foreign-owned estates, that it will be necessary to nationalize all the tea estates, whether they are Indian or non-Indian, whether they are big or small', Desai expected that nationalization would harm the efficiency of the enterprises, which, in its turn, would affect the working and living conditions of the workers.[104] Desai's next argument was of even greater importance. 'If we had any surplus capital and if we wanted to utilize it in this manner, it would have been, to some extent, useful. But today, we are asking for foreign capital to be invested in this country on certain conditions because it is necessary for raising our industrial production and for having many industries which we have not got at present and which we badly need.' As Desai himself summed up: 'On the one hand, we invite foreign capital to come and settle here on fair terms, and on the other hand we wish to liquidate or buy over'. 'I do not see how we can be benefited by this sort of a contradictory process', he observed. Desai therefore concluded his speech thus: 'I am quite sure that our interests are served better by continuing the estates as they are and by controlling them as we are doing at present. (. . .) Nationalization, instead of being useful or beneficial to our country, would be quite suicidal as the facts are (. . .).'

Gopalan's resolution was subsequently put to vote. With a Congress majority in the Lok Sabha the outcome was foregone: the resolution was rejected.[105] Although the communist activists did not relinquish their vision of nationalized estates, the issue never again assumed the same degree of gravity.[106]

The debate regarding the nationalization of plantations was of great significance. Not only did it mean a 'victory' for the planters' associations/Congress Party and a 'defeat' for the leftist movement, the struggle reflected much more. The controversy and its outcome revealed the political context of plantation

[104] In Desai's words: 'When the transfer takes place, it will require also a large number of experts and other personnel, which we cannot immediately produce. That will immediately bring down the production and will also worsen the condition of these estates and the living conditions of the labour employed in these estates.'

[105] Proceedings LS dated 22 March 1957.

[106] Cf. *The Planters' Chronicle* 1 May 1958: 6; Griffiths 1967: 231.

development in the post-colonial era. The labour unions, supported by the Communist Party, were hamstrung by a powerful planters' lobby. Although the leftist movement became very influential in Kerala, and headed the state government in certain years, it never became *the* dominant force in south-west India. The weak political position of 'Trivandrum' *vis-à-vis* 'Delhi' lay behind this failure. The planters and Congress Party in Kerala were backed by the central authorities, dominated by the Indian National Congress.

This also explains why the communist leaders never discussed their ideas of nationalizing the plantations in any great detail: they were fully conscient that the Indian political constellation made such a step impossible. As Namboodiripad recollected: 'What will happen after nationalization, that we had not put together thoughts in the concrete because that was not on the agenda, that was not in any way coming up. But we did not think of dividing the plantations. (. . .) Our vague idea was state-owned. (. . .) It was not practically on the agenda because it was not going to be accepted by the central government'.[107] Having no intention of overthrowing India's political framework undemocratically, the communist leaders accepted their inability to transfer the ownership of the estates.

This is not to say that the communists abandoned their dreams altogether. As Namboodiripad stated shortly after his party's election victory in Kerala in 1957: 'Our proposal with regard to the taking over of foreign plantations is so reasonable and so essential for the development of the economy in Kerala that we have every hope that we can, through consultations, convince the Centre of the same'.[108] As Namboodiripad explained later about the communists' motives in continuing their struggle for nationalization, a goal beyond their reach: 'even though the [central] government may not accept it, we wanted to press for it'.[109] Clearly, for ideological and practical reasons – votes in particular – the communist leaders held on to their ideas.[110]

[107] Interview: E.M.S. Namboodiripad.
[108] *The Hindu* 3 March 1957, quoted in: Lieten 1979: 31.
[109] Interview: E.M.S. Namboodiripad.
[110] Cf. Interview: K.U. Warier.

6.3.2 *The Land Reforms Movement Defeated*

The controversy regarding land reforms echoed the debate with regard to the nationalization of plantations. The planters' lobby, powerful at the all-India level, was able to defeat the leftist movement in Kerala.

The first serious attempt to introduce land reforms in south-west India was made by the first non-Congress government in the area. Five years after a Land Policy Committee of the Travancore-Cochin state had recommended a ceiling on future holdings, the Praja Socialist ministry introduced the Restriction on Possession and Ownership of Land Bill in August 1954. They proposed imposing ceilings on existing holdings and on future land acquisitions. Only plots larger than 25 acres which were 'used for growing' coffee, tea, rubber and cardamom were to be exempted from the provisions of the bill.

The proposed legislation raised a storm of protest. Some wondered why plantations were given special treatment. As K.C. Krishna Iyer, a resident of Peermade taluk, informed his government: 'The exemption on estates should not be granted as it is against the principles and objects of the bills. Such plantations that require a factory to manufacture the produce before it is brought to the market may be exempted. In the case of rubber, coffee and cardamom, factories are not needed (. . .).' 'So in the ceiling on possession section', Krishna Iyer concluded, 'cardamom, coffee and rubber should be included'.[111]

Though the repercussions for the plantation industry – if the proposals were accepted – would be limited, certainly when compared with other agricultural sectors like paddy and coconut, the planters hastened to voice their protests against the bill. An individual agriculturalist named A.K. Krishnan complained that 'land used for growing' should be enlarged to include 'land intended to be used'. His argument ran: 'there are many rubber planters in the State (. . .) who have lands fit for planting rubber (. . .) but who could not plant because of the uncertainty in the rubber market and lack of facilities in obtaining good seed material'. 'If these areas are now dispossessed', he persisted, 'the State can never hope to expand areas under these valuable crops

[111] OBLR 1954: 254.

as the future possessors would neither have the resources nor the courage to plant up new areas'. 'Dispossessing', he concluded, 'is something like killing the goose that laid the golden eggs'.[112]

Carrying more political weight were the memoranda submitted by the Quilon District Planters' Association and the Association of Planters of Travancore. The planters' organizations played upon the importance of plantation development, particularly in the field of rubber, not just for the south-west Indian but also for the national economy. The latter association claimed that 'Travancore–Cochin affords immense scope for the further expansion of this key industry'. 'But the Bill, which proposes to parcel out all unplanted area and to prevent the acquisition of extensive lands in the future, would operate as a clear and effective prohibition of any expansion in this direction.' They ended by warning of the dire consequences for the country in the fields of commerce and strategy.[113]

The bill never became law as the Praja Socialist ministry had to resign in February 1955. Political insiders insist that the Congress Party withdrew its support to the minority government of Pattom Thanu Pillai mainly on the issue of land reform.[114]

The second effort to introduce land reforms was launched by the Communist Party.[115] Seven months after the communist ministry of Namboodiripad had been installed, the Kerala Agrarian Relations Bill was introduced in the State Legislative Assembly. In accordance with their more general policy of seeking to abolish 'landlordism', the main purpose of the bill was to fix ceilings on land ownership and to distribute the land thus generated among small peasants and landless labourers.

Nevertheless, with regard to plantation development the Kerala

[112] OBLR 1954: 126.

[113] OBLR 1954: 287.

[114] Earlier, Pattom Thanu Pillai's position had weakened because of his strained relations with Dr Lohia, the PSP leader at the all-India level (Thomas 1985: 87).

[115] In the first week the communists assumed office, the Namboodiripad government issued the Stay of Eviction Ordinance, protecting tenants and hutment dwellers until the full agrarian programme had been legalized and implemented (Nossiter 1988: 69).

Agrarian Relations Bill of December 1957 had even less impact than the Restriction on Possession and Ownership of Land Bill of August 1954. Instead of excluding 'plots larger than 25 acres which were used for growing coffee, tea, rubber and cardamom', all such holdings were exempted from the provisions of the bill. Defining plantations as 'any land used immediately before 18th December 1957, principally for growing tea, coffee, rubber or cardamom or such other kind of special crop as may be specified by the Government by notification in the Gazette',[116] the original reference with regard to area was omitted.

Again, protests were raised on all sides. Many communists were not satisfied with the bill. They stressed that all cultivators, irrespective of the crops they produced, should be treated equally. The debates during the formulation of the bill are very illuminating. Some prominent communists saw no reason whatsoever to exclude plantations, and rubber estates in particular, from the provisions of the legislation, but other communist leaders including Namboodiripad himself understood that land reforms were only possible in the non-plantation sector. They realized that the planters, the Christian community, and the central authorities would never accept land reforms which affected the plantation industry fundamentally.[117]

Paradoxically, the paddy growers' lobby used the same arguments as the discontented communists. They were not willing to accept the special treatment extended to coffee, tea, rubber and cardamom cultivators either. As the Ottapalam Small Landholders' Association complained: 'This is discrimination quite unwarrantable'. The organization could not understand why plantations were completely excluded from the ceiling limit. 'If they are meant to improve our commercial products and thus save foreign exchange, *then the rice fields of Kerala are as much needed – if not more for the substance of our agricultural economy.* If large-scale production and the need for large capital is the reason then in paddy cultivation also the same rule applies.' Concluding: 'Whatever reasons apply to plantations should apply to paddy lands also. What is sauce for the goose

116 Article 61 (ii) The Kerala Agrarian Relations Bill 1957.

117 Interview: E.M.S. Namboodiripad. See also the quotation at the beginning of this chapter.

is sauce for the gander'.[118] Nevertheless, the statement above reveals quite clearly that the goals of the disappointed communists and the protesting paddy cultivators were diametrically opposed. While the former wished to include plantations in the land reforms, the latter struggled to exclude paddy fields from the provisions of the bill. Nor were paddy growers a voice crying in the wilderness. Coconut largeholders also had their objections to the bill.

Notwithstanding the unique status accorded to 'plantations' in the Kerala Agrarian Relations Bill, the planters found many points on which they disagreed with the proposed legislation. Though by no means the planters' only protest letter, the memorandum submitted by the Rubber Board elucidates the concerns of the plantation owners admirably. In the memorandists' view, the definition of the concept 'plantation' presented the first stumbling block. For example, the bill did not allow for the establishment of any new large estates. Therefore, the Rubber Board suggested that the definition of 'plantation' should also include lands reserved *bona fide* for the cultivation of rubber and areas used for the preparation of the product for the market. The memorandists also strenuously opposed the plan that any person already holding a plantation was not permitted to own or possess any other land for any other purposes than plantation production.[119]

A memorandum submitted by the Quilon District Planters' Association adduced the same arguments against the Kerala Agrarian Relations Bill. 'It becomes impossible for an enterprising man to plant a hundred acres of rubber tomorrow. He can neither buy land nor possess any land beyond the ceiling area.' The Association saw the bill as economic suicide because it put a brake on all future planting.[120] The subsequent statement, apart from presenting an argument against the proposed legislation, also reveals the mentality which the planters fostered towards their subordinates on the estates. Though most labourers were housed on the plantations, the owners themselves lived elsewhere because of unhealthy conditions. 'It is commonly believed that Rubber plantations, unlike other plantations, are not quite

[118] Emphasis in the original text, Opinions KARB 1957: 14.
[119] *The Planters' Chronicle* 15 May 1958: 236.
[120] Opinions KARB 1957: 24.

healthy spots to live in. Naturally he has to reside elsewhere.' This was used as an argument to support the assertion that any planter, whether he be a planter of rubber, tea, coffee or cardamom, should be able to own land elsewhere, 'besides his plantations'.[121]

The struggle continued in the Kerala Legislative Assembly. In this arena the Congress Party presented itself as the wholehearted champion of the planting industry. The Congress members demanded the total and unconditional exemption of plantation holdings from the operation of the ceiling clause in the bill. Their spokesman and leader of the opposition, P.T. Chacko, also stated that the Kerala legislature had no power to sanction legislations limiting the expansion of tea, coffee and rubber plantations. He claimed that the Tea, Coffee and Rubber Acts of the Indian government brought these issues under the jurisdiction of the central authorities, not of the state government. Though the protests were partially successful – the expansion of rubber holdings was allowed to up to 20 per cent of the planted acreage in such cases where there were contiguous reserve lands – the communists managed to get the bill passed in June 1959.

Whether the passing of the Kerala Agrarian Relations Bill would prove an unqualified success for the Communist Party remained to be seen. While leftist leaders praised the bill as a 'unique achievement of the Communist Party', it soon became obvious that the passing of the bill had ushered in the fall of the communist ministry. To clarify this statement, soon after the bill was passed by the state legislature all the political parties in opposition joined forces to end communist rule in Kerala. They organized a so-called Liberation Struggle *('Vimochana Samaram')*, a state-wide agitation which made it impossible for the Kerala government to function properly. Besides the major victims of the land reforms, the jenmies, Namboodiripad assessed the forces behind the mass upsurge as: 'Planters pressure, Christian lobby, these special interests also operate together to get a new government'.[122]

After Prime Minister Nehru made an on-the-spot enquiry, discovering in the course of it that Kerala was divided into two

121 Opinions KARB 1957: 23.
122 Interview: E.M.S. Namboodiripad.

separate worlds, communist and non-communist, he decided to dismiss Namboodiripad's ministry, to proclaim President's Rule, and to prepare the state for new elections.

The elections changed Kerala's political landscape and finally provided Pattom Thanu Pillai with a second opportunity to serve as Chief Minister. Although it was headed by a prominent member of the Praja Socialist Party, the Indian National Congress dominated the new government. The transfer of power from the Communist Party to the Congress Party is best illustrated by the land reforms which were passed by both the state and central legislature: the Kerala Agrarian Relations Act of 1960.[123] The legislation reflected the outcome of the power struggle and, where it touched upon plantations, this was clearly won by the coalition of planters and Congress Party.

Without one single exception, the planters' protests against the definition of 'plantations' mentioned in the Kerala Agrarian Relations Bill of 1957 were all taken into account. This was of crucial importance because of the exclusion of these lands from the ceiling provisions. The definition of the concept 'plantations' contained in the Kerala Agrarian Relations Act of 1960 was:

> any land used by a person principally for the cultivation of tea, coffee, rubber or cardamom or such other kind of special crops as may be specified by the Government by notification in the Gazette, and includes:
>
> a) land used by such person for any purpose ancillary to the cultivation of such crops or for the preparation of the same for the market
>
> b) land contiguous to, or in the vicinity of, or within the boundaries of, the area cultivated with such crops, not exceeding twenty per cent of the area so cultivated and reserved by such person and fit for the expansion of such cultivation;
>
> c) agricultural lands interspersed within the boundaries of the area cultivated by such person with such crops, not exceeding such extent as may be determined by the Land Board as necessary for the protection and efficient management of such cultivation.[124]

[123] The Kerala Agrarian Relations Bill of June 1959 was sent to Delhi for central approval, though was returned to Trivandrum without this having been granted, embellished with critical comments and in a time in which the Congress-PSP ministry was in power (Nossiter 1988: 69).
[124] Clause 2 (39) The Kerala Agrarian Relations Act 1960, Act IV of 1961.

Abstracting a few of these points to underscore the significance of the alterations, several matters are conspicuous. One obvious victory was that the new definition did not mention any date: plantations could be established after 18 December 1957. Significantly, the first sub-clause stated that lands used for factories, quarters for workmen, office buildings, hospitals and schools were also included in the term 'plantation'. The next sub-clause created legal space for *all* plantations to expand. In a former definition only rubber holdings had been granted that right. The last sub-clause determined that islands of coconut, pepper, and other cultivation within holdings planted with coffee, tea, rubber and cardamom were also exempted. Thus, the planters initial fear of 'alien enclaves' within the boundaries of their estates was removed.

And these were not the only ways in which the Kerala Agrarian Relations Act of 1960 strengthened the position of the planting community. Apart from particular categories of peasants, the chief beneficiaries of the Kerala Agrarian Relations Bill of 1957 had been the *kudikidappukarans.* In judicial terms, the kudikidappukaran.

> means a person who has no homestead or land of his own to erect a homestead and
>
> i) who has been permitted with or without an obligation to pay rent by an owner of land to have the use and occupation of a portion of the land in his possession for the purpose of erecting a homestead, or
>
> ii) who is in occupation of a hut constructed by a person owning and possessing the land in which the hut is situated and who has been permitted by such person to occupy the hut with or without the obligation to pay rent,
>
> but otherwise has no interest in land.[125]

To rephrase that definition in concepts more familiar in the social sciences, kudikidappukarans were landless labourers who were given permission by the landowners to reside on the land but who also functioned as a source of cheap labour to these proprietors. As stated above, these hut dwellers profited from the Kerala Agrarian Relations Bill of 1957: the landowner's right

[125] Article 2 (15) The Kerala Agrarian Relations Bill 1957.

to evict the kudikidappukarans was restricted and, secondly, these hut dwellers obtained fixity of tenure.

In the Kerala Agrarian Relations Act of 1960, however, plantations were exempted from these provisions regarding kudikidappukarans. It was no accident that this particular amendment was moved by a member of the coalition government, the Minister for Law and Revenue, K. Chandrasekharan. Members of the opposition objected to the proposed change in the bill. Putting forward their point of view, P. Balachandra Menon explained: 'There are about 118,000 plantation workers in our State. (. . .) only about half the number of workers are getting any housing. (. . .) about half the huts belong to the workers themselves'. Drawing attention to both the historical and socio-economic context of plantation development, Chandrasekharan continued: 'Workers go to the plantation areas in large numbers along with their families and settle permanently in those places. The employers do not provide housing to them, but they allot to them certain sites where they should live. They have been living there for generations, and when they have constructed houses and have been living there for generations on the plots given by the employers, they become kudikidappukarans'. As Chandrasekharan concluded: 'it will be rather unfortunate if lakhs and lakhs of our workers are to be sent out of their homes, because of the very unfortunate amendment'. In addition, 'this would result in the victimization of the workers, because the employer will come and say: "here is a hut which is mine: you will have to go out" '.[126] Despite such rhetoric, the opposition's arguments were fruitless. The amendment was accepted and included in the Kerala Agrarian Relations Act of 1960.

Though passed by the state and central legislature, the Kerala Agrarian Relations Act of 1960 could not be implemented. The Supreme Court found that the Kerala Act conflicted with Indian constitutional rights regarding property, equality and freedom. This occurred at a time that similar fundamental questions were being raised by the Indian parliament. The final result was that the defunct Kerala Agrarian Relations Act of 1960 was replaced by the Kerala Land Reforms Act of 1963. Without going any

[126] Proceedings KLA 5 October 1960; cf. proceedings KLA 26 September 1960.

deeper into the details, it can be said the new legislation again represented an improvement in conditions for the planting community. While retaining their earlier privileges, the planters also succeeded in dilating upon the definition of 'plantation'. The concept now also included plots cultivated with cocoa and cinnamon.[127]

To sum up the points discussed in this section, the land reforms did not impede plantation development in any way. All battles with regard to the Restriction on Possession and Ownership of Land Bill 1954, the Kerala Agrarian Relations Bill 1957, and the Kerala Agrarian Relations Act 1960 were won by the planters, culminating in the Kerala Land Reforms Act 1963. In fact, the latter acts not only *protected* the planting community, the legislation actually *stimulated* plantation development.[128] To escape certain detrimental provisions, many farmers, who had initially cultivated non-plantation crops, shifted their attention to the production of plantation crops, particularly rubber.[129]

6.3.3 *The Labour Movement Curtailed*

Besides the threats of nationalization of plantations and land reforms, during the Namboodiripad government the planters had to contend with a growth in trade union activism. That are already intimations of this in Table 6.8. This information shows that a new record was set in 1958–9, when more than 700,000 man-days were invested in work-stoppages.

The strike in the Kannan Devan Hills in October 1958, which became national news following the violent death of two communist activists, symbolized the culminating point. The reasons for focusing on this particular event are both abundant and well-founded. The clash illustrates both the strategies and dilemmas of the major interest groups involved and, moreover, sheds

[127] Clause 2 (44) The Kerala Land Reforms Act 1963, Act I of 1964.

[128] Despite the land legislation, and because of the fragmentation of estates following the inheritance system of the Syrian Christian (see Section 5.2.2) and the Plantation Labour Act (see Section 6.2.3), the plantation sector did not grow, at least not in a strictly formal sense (Chacko Kallivayalil 1969: 2; interviews: Hannibal Dominic; Michael A. Kallivayalil).

[129] *The Planters' Chronicle* 1958–60; minutes KDPA 30 April 1959 and 15 July 1959; Oommen 1971; Nossiter 1982: 149–53; Javernesa Begom 1983; Raj and Tharakan 1983; Tharian George, Haridasan and Sreekumar 1988; m161.

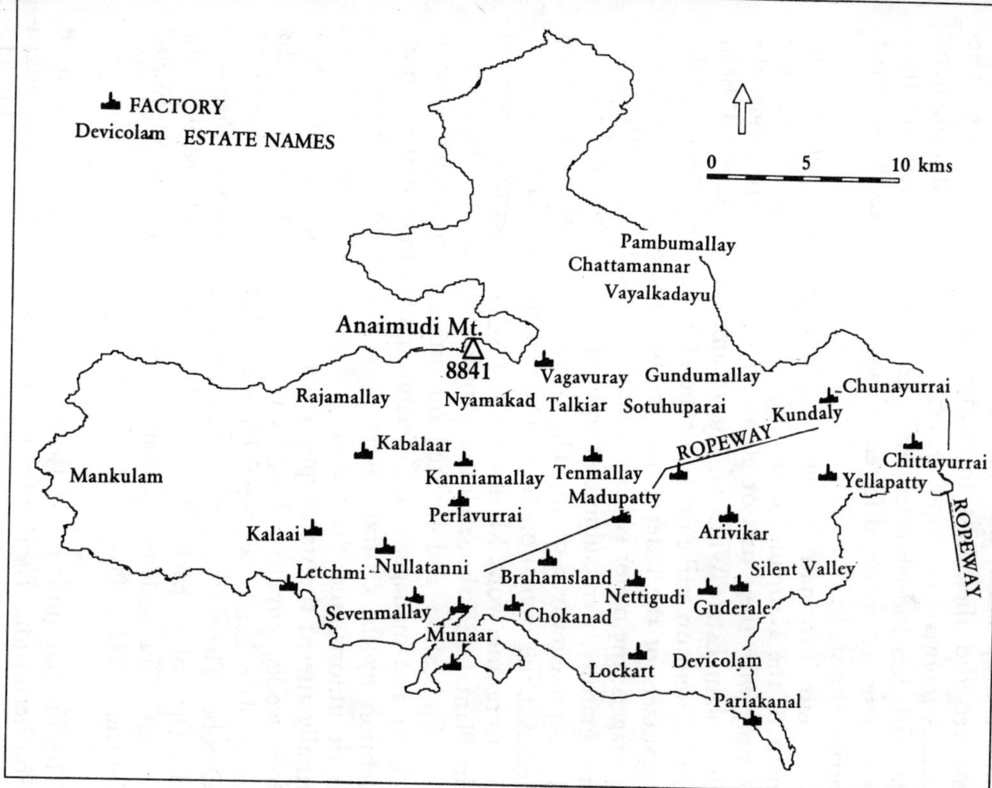

Map 6.1 The Kannan Devan Hills District in the Early Post-Independence Period

a lot of light on the wider framework in which plantation development in south-west India occurred. In studying the events related to the October 1958 Kannan Devan Hills' strike, one such strategy stands out immediately. All the interest groups involved did their best to manipulate public opinion. The various reports give ample evidence that not only did planters, state officials and labourers interpret the events differently; they also presented totally different views on what exactly occurred during the strike.[130]

Certain information, however, is undisputed. As the management of the Kannan Devan Hills Produce Company did not comply with their lists of 30 and 7 demands respectively, the Devicolam Estate Workers' Union and the Plantation Labour Union went on strike on 4 October. On that very same day the organizers of the latter union reached an agreement with the managers: the major issues, including those dealing with bonus and gratuity, were submitted to adjudication. Still dissatisfied, the representatives of the Devicolam Estate Workers' Union declined the invitation to join the plan and told the workers not to resume work. Meanwhile, the Kerala government rejected the Plantation Labour Union's application to refer the dispute to adjudication and insisted on direct negotiations between labour and management, i.e. without state intervention. Intervention by the central government, forcing the planters, the trade unions, and the Kerala government to come to terms, finally altered the status quo. After a work-stoppage of almost four weeks, punctuated by many violent confrontations, the Devicolam Estate Workers' Union called off the strike on 30 October. The political backgrounds of each union were different: the Devicolam Estate Workers' Union was affiliated with the AITUC/CPI while the Plantation Labour Union was supported by the INTUC/INC.

The official bulletin of the United Planters' Association of Southern India, *The Planters' Chronicle*, expressed very outspoken opinions about the events in the Kannan Devan Hills. The journal averred that the supporters of the Devicolam Estate

130 With regard to the conflicting interpretations of and stories about violence on nineteenth century Sumatran plantation, see Stoler's highly-interesting article *In Cold Blood* (1992).

Workers' Union used strategies such as 'bullying, blackmail and terrorism' both against workers who were willing to go to work and against the management. The former especially were 'obstructed, molested or intimidated' by the communist activists. The bulletin rubbed in the idea that the leftist government was largely responsible for the continuation of the 'unfortunate strike'. The journal blamed the Namboodiripad administration for not accepting the application made by the management and the Plantation Labour Union to refer the dispute to adjudication. The ministry's refusal was taken to be unequivocal evidence that the state government placed party politics above state interests. The communist government supported the Devicolam Estate Workers' Union in its struggles against both the planters' organizations and rival unions. Nor did *The Planters' Chronicle* leave room for any doubt with regard to the role of the management in October 1958. The planters did not put a foot wrong during the strike. 'The managements of course refused to negotiate under conditions of violence and intimidation', the journal stressed.[131] In addition to *The Planters' Chronicle*, the planters published their views in a booklet entitled *The High Range Strike*. This particular publication, which was issued by the Kannan Devan Planters' Association shortly after the events, propagated a similar opinion, but one which was meant to reach a wider audience.[132]

A major Indian newspaper, *The Hindu*, presented an analogous version of the events. The reports tended to focus on the violent methods employed by the communist trade union activists and the resulting 'insecurity' and 'lawlessness' in Kerala. Most articles highlighted the opinions of prominent members of the Congress Party. A very good example in this regard is a text published on 15 October. To quote from *The Hindu*: 'Mr K.A. Damodara Menon, KPCC (Kerala Pradesh Congress Committee) President, who went to Munnar to study the situation there, told me to-day that there was a state of tension in the Munnar Estates area as a result of the tactics adopted by the Communist-led labour unions who were now leading the strike'. The journalist continued that Damodara Menon

[131] *The Planters' Chronicle* 1 November 1958: 487–8; 15 November 1958: 511–12.
[132] KDPA 1958, in: IOL&R: MSS Eur F 174.

'said the workers of the INTUC-led unions, who formed the majority of plantation workers, were willing to go back to work and a large proportion of them had done so, but owing to the violent tactics of the Communist Party who were intimidating and assaulting the workmen, they refused to go to work'.[133] The Congress member stated that the Kerala administration only exacerbated the problems in the Kannan Devan Hills. No single government agency fulfilled its tasks correctly: the police did not provide proper protection to those who were willing to work; the government itself encouraged further instability; and the Labour Department was not trying hard enough to find a solution to keep the escalating situation under control.[134] Showing an evident bias, *The Hindu* published no critical statements about the plantation managers. The newspaper also suggested that in the fatal confrontation, in which two supporters of the strike died, the police acted in self-defence. As the headlines stated: '*Sequel* to attack by striking workers: 2 killed & 4 injured', and subsequent information explained: 'the police *had* to open fire'.[135]

This particular image of the event, which stigmatized the Devicolam Estate Workers' Union and the Namboodiripad ministry as the troublemakers, was reinforced by the publication of particular 'letters to the editor' in *The Hindu* as well as in other widely-read journals. In *The Indian Express*, for example, the role of the communist administration in this particular strike was described as 'an incitement to class war in all its naked brutality and crudeness'.[136] In another letter to the editor, in this case published in *The Hindu*, the communist groups were reviled

[133] *The Hindu* 15 October 1958: 5.

[134] The article in *The Hindu* concluded: 'the Police posted there (. . .) had not been able to prevent the acts of violence perpetrated by Communist-led workers. Mr Damodara Menon expressed the view that Kerala Government were neglecting their duty in preserving the law and order in the Estates and in effecting a speedy settlement of the strike, though the Management had conceded some of the demands of the workers and had agreed to refer the other issues for adjudication. He appealed to the Labour Department of Kerala to see that the strike was not continued' (*The Hindu* 15 October 1958: 5).

[135] Italics mine, *The Hindu* 21 October 1958: 1.

[136] *The Indian Express* 23 October 1958, quoted in *The Planters' Chronicle* 1 November 1958: 489.

as 'violent mobs led by outside elements' which used 'terror tactics' like 'knifing and stoning'.[137]

The close collaboration between the planting community and the leading newspapers suggested in these examples was also reflected in the captions and the contents of the articles dealing with the strike. Examples of some additional headlines like 'Loyal Workers Assaulted' and 'Continuance of Strike; Adverse Effect on Output'[138] also betray this tendency. Other sources confirm that major newspapers were instrumental in the spread of pro-planter and anti-communist information.[139] In this regard, the minutes of the Association of Planters of Kerala provide a good insight into the relations between planters and journalists. In his welcoming speech addressed to those present at the annual meeting of the Association of Planters of Kerala, the chairman, referring to this particular strike, stated: 'I must also record our sincere appreciation of the generous co-operation of the news agencies and the press. We particularly appreciated the impartial reporting and editorial comments of the respective organs of the press'.[140]

A diametrically opposite point of view was presented by S. Govinda Menon, a member of the Board of Revenue who was officially requested to investigate the incidents in question. Appointed by the state government on 30 October, he submitted his conclusions regarding the conduct of the trade union activists, the policemen, and the planters to the Kerala Legislative Assembly early in 1959. Govinda Menon's findings, summarized in The Report on the Enquiry into Matters Arising from the Labour Strike in Munnar, contrasted sharply with those presented in *The Planters' Chronicle* and *The Hindu*.

Govinda Menon paid much more attention to the communist activists' allegations that they had been intimidated, beaten and tortured by the police. Though the report provided no definite answer to the question whether or not these cruelties had actually been inflicted, the text was unique in the sense

137 *The Hindu* 24 October 1958, quoted in *The Planters' Chronicle* 1 November 1958: 489–90.

138 *The Hindu* respectively 18 October 1958: 5 and 19 October 1958: 5.

139 Cf. *New Age* 26 October 1958: 4.

140 Minutes of the APK, in: *The Planters' Chronicle* June 1959: 268; cf. *New Age*.

that the communist activists were not seen only as aggressors. In fact the reverse was true. The workers were portrayed primarily as victims. To quote from the report: 'plantation labour (. . .) has to look to the management for its sustenance'. 'Hence in the event of a plantation strike', Govinda Menon continued, 'the labour with its limited power for "holding on" may develop a sense of frustration sooner than we expect and this might end in violence leading to a deterioration of law and order'.[141]

Govinda Menon believed that the policemen had given adequate protection to the managers. He suggested that the planters might be spreading false information in order to bring the communist administration into discredit. Referring to this he stated: 'a letter of appreciation from the Manager, Thalayar Estate, of the good work done by the police and (. . .) a letter of appreciation from the management of the K.D.H.P. Co. (Kannan Devan Hills Produce Company) of the good work done by the police, leave no room for doubt about the adequacy of police protection given to the management'.[142] Giving credit were credit was due, Govinda Menon *praised* the policemen for their efforts to protect labourers willing to work from intimidation, threats of violence and attacks; 'police protection was always available for the willing workers and their [the management's] assertion to the contrary is without bonafides'.[143]

The conclusion of his report was probably most shocking to the planters. 'A reorientation of the outlook of the managerial

[141] RLSM 1959: 14.

[142] RLSM 1959: 9. Govinda Menon subsequently stated: 'In the light of the foregoing facts, the question naturally arises why the Kannan Devan Planters' Association, of which Mr J.B. Soutar is the Chairman, chose to complain of inadequacy of police protection in their booklet entitled *The High Range Strike*, a publication which has received the widest publicity. (. . .) Why he receded from the stand originally taken is not clear. Two things are possible. Either Mr J.B. Soutar must have been inspired into writing this booklet for reasons known to himself or he must be trying to get himself embroiled in local politics. It looks as though I am confronted with Mr J.B. Soutar the Politician and not with J.B. Soutar the General Manager of the K.D.H.P. Co. (. . .) Plantation Companies, especially sterling companies, would do well to get themselves away from local politics so long as profit making is their motive' (RLSM 1959: 16).

[143] RLSM 1959: 13.

staff is necessary for the sake of ensuring peace. The English personnel of the several tea estates have not yet reconciled themselves to the changed political conditions in India.'[144] In glaring contradistinction to the planters' own ideas on the subject, Govinda Menon allotted particular blame to the management for the hostilities of October 1958.

New Age, the communist newspaper, presented yet another picture of the events. In an article entitled 'United Action in Kerala Plantations', K.U. Warier praised the members of the AITUC and blamed the leaders of the INTUC.

He reported that all central trade union organizations in Kerala had been preparing for a general strike in the plantations during the third and fourth weeks of October. The Plantation Labour Union jumped the gun and served a strike notice which was to mature on 4 October. 'In these circumstances, leaders of AITUC unions correctly sensed the need for a united general strike in all plantations in the state as otherwise, partial and sporadic strikes would result in frittering away the energy of the workers and giving an upper hand to the employers.'

Consequently, Mrs Rosamma Punnose, the President of the Devicolam Estate Workers' Union and a member of the Kerala Legislative Assembly, suggested that the representatives of the Plantation Labour Union join hands and wait till 13 October before launching the strike. The reply elicited was a resounding negative. Warier disapproved of the INTUC decision: 'the INTUC leaders were adamant about beginning the strike on the 4th itself when there was no chance of a general strike anywhere else in the plantations in Kerala'. It was therefore 'the hasty action of the INTUC and its refusal to see the gains of a united industry-wide strike that led to an isolated and premature action in the biggest and most powerful plantations of Kerala on October 4'.

'As unity was the most important factor for the success of any working class action', Warier continued, 'the AITUC union also joined the strike on the 4th'. As a direct consequence, 36,000 workers went on strike on that day. Simultaneously, attempts were got under way to end the strike. Warier carried on his

[144] RLSM 1959: 15.

accusations against the INTUC leadership unabated: 'Meanwhile, a most heinous and disgraceful attempt to sabotage the strike was being made by the INTUC leaders behind the backs of the own workers as well as the AITUC leaders who were leading the struggle along with them'. Warier wrote that 'these gentlemen were closeted with Mr Soutar, the acting General Manager of the Kannan Devan [Hills Produce Company], and they came out only after they had surrendered all along the line and betrayed the struggle on the very day it had started'.

Warier claimed that this caused the INTUC leaders nothing but trouble. Incensed that the 'settlement' contained not even one concession or even an assurance from the management, the INTUC followers were so infuriated by the decision of their leadership that they converged upon the INTUC office and smashed the furniture and beat up any leaders they found there. Warier seized this opportunity to underline the non-violent behaviour of the AITUC workers, stating: 'it was the intervention of AITUC workers and the police that saved these "leaders" from further trouble at the hands of their embittered followers'.

Like the large-scale violence by communist trade union activists, the existence of two separate, hostile camps – AITUC and INTUC – was another myth, particularly at the lowest levels of these organizations. As Warier explained: 'There is a continuous flow of workers from nearby estates into the town [Munnar] carrying the Tricolour and the Red Flag[145] and banners proclaiming unity. Public meetings and demonstrations are being held in which workers belonging to the AITUC and INTUC jointly participate and pledge to continue the struggle unitedly. "Workers' Unity *Zindabad*" is the most popular and common slogan today'.

Apart from examining the behaviour of the trade union leaders and their followers, Warier also discusses the strategies of both the Kerala administration and the planters. The role of the Namboodiripad government was judged favourable. The insight of the Kerala Labour Minister, T.V. Thomas, was singled out for special praise. He arrived at Munnar on 13 October to direct the negotiations personally, with every intention of bringing about an amicable settlement. In Warier's view, it was meaningless

145 The symbols of the INTUC/INC and the AITUC/CPI respectively.

indeed to talk of adjudication while *the majority of workers* continued the strike and opposed adjudication. 'Simply because a handful of INTUC leaders have asked for adjudication, it does not become that labour will accept it if it is imposed.'

The planters were lambasted for misbehaviour. 'The cry of the planters that "loyal" workers are attacked by "Communist-led workers" and that the "law and order situation" in the High Ranges is in danger is absolutely false and intended only for the ears of the central government.' 'The anti-Communist newspapers in the State are already speculating on "Central intervention" in the plantations to establish "law and order"', Warier concludes. There is no room for doubt that Warier regarded the INTUC leadership, the anti-communist writers and the central authorities as the henchmen of the planters and the opponents of the communist activists.[146]

The additional information which was provided by AITUC activists who were eyewitnesses to the police firing on Gudurale Estate on 20 October is highly enlightening. As indicated above, the confrontation proved fatal for two communists: a nineteen year old woman named Pappammal and a man called Hussain Rowther, aged fifty-five. Like Warier in *New Age*, the interviewees emphasized that all members of the Devicolam Estate Workers' Unions did not resort to violence during the strike. They were also agitating in a peaceful manner on 20 October. The information provided next was unique. The informants stated that the planters' refusal to recognize the validity of their union led to the strike. Of even greater importance is the assertion of the eyewitnesses, who were also shot at on that particular occasion, that they were convinced that the policemen were not Malayalis. They claim that the police force had been imported from outside Kerala and operated under the control of the central government. The eyewitnesses stated that the central police force encouraged the members of the rival INTUC Plantation Labour Union to throw stones at them, machinating to provoke a clash which they – the central policemen – would inevitably be sucked into and which would enable them to intervene brutally and break the morale of the communist union. The eyewitnesses

[146] *New Age* 26 October 1958: 3–4.

believed that the management, unable to break the strike with the help of the Kerala police – who took an independent stance – was aided by the policemen under central control. Those interviewed are convinced that the police were not on their side (as the management and *The Hindu* suggested), nor were they independent (as Govinda Menon, the government official, indicated). The police, and particularly the section which was controlled by the centre, was opposed to the communist union and was helping the planters. Accordingly, the fatal shootings were not acts of self-defence, but the outcome of the planters' anti-communist strategies.[147]

Some major conclusions should be drawn from the events during the strike in the Kannan Devan Hills in October 1958. The planters shaped the dominant view through their channels of communication. They were the most successful in promoting their image of the strike. Inevitably, the planters' point of view, presented as 'objective knowledge' or 'the truth', became widely accepted and was hardly questioned throughout India.[148] Compared to the reports in *The Hindu* and other national newspapers, the official report of Govinda Menon and the publications of Warier and other communist authors had only a very limited impact.

The contradictory historical sources unequivocally indicate the strategies and dilemmas of the major interest groups involved. Reflecting themes already discussed in the first section of this

[147] Interview: AITUC workers Gudurale Estate. The information provided by the AITUC workers was later confirmed by a close relative of a plantation manager who was employed by the Kannan Devan Hills Produce Company during October 1958. According to the informant, a high-ranked police officer employed by the central authorities stayed at the bungalow of the particular manager in order to direct the confrontations against the militant workers. The informant preferred to remain anonymous.

[148] One interesting side-light on the affair is that during my stay in the Kannan Devan Hills, the general manager of Tata Tea, the successor of the British plantation company, gave me their romanticized history of the district but refused to allow me to study their historical material, afraid that I would 'write against the company' (interview: general manager Tata Tea). Thus, like in 1958, the management was creating its own history and was afraid that *that* story – a glorification of the past as well as of the present – would be openly questioned.

chapter, the events in the Kannan Devan Hills in late 1958 illustrated the trade unions' focus on European-owned estates, the planters' anti-communist policies and the state government's role as negotiator. At the same time, the particular events throw the dilemmas of the interest groups concerned into high relief. For the strike committees, everything hinged on whether or not labourers willing to work should be halted, eventually by the use of force. Certainly, the violent creation of a united labour front would improve the bargaining position of the particular trade union *vis-à-vis* the employers. Viewed negatively, such tactics would encourage further tensions within the labour force and would lead to bad publicity. The planting community juggled with the question of whether or not to negotiate with trade unions. Ignoring the labour activists' demands could lead to prolonged strikes, but this had to be weighed against the fact that negotiations would probably bring higher wages and other costly measures. Taxing as these problems were, the Kerala administration was perhaps confronted with the greatest dilemma of all. The Namboodiripad government was unwilling to undermine the position of the communist trade union movement in the plantation regions, but it could not ignore the reality that the administration profited financially from a flourishing plantation sector and was responsible for preserving law and order in the state.

Before mentioning a third consequence, the outcome of the strike should be analysed a little. The unpublished sources, and the archival material of the planters' associations in particular,[149] indicate that the aid of the Indian government was of crucial importance to the planting community in its struggle to resist the demands of the labour movement. To mention one example, the President of the United Planters' Association of Southern India, as requested by the Kannan Devan Hills Produce Company, telegraphed the following message to Prime Minister Nehru and four other important Cabinet Ministers[150] on 27 October 1958:

149 Minutes MPA 1957; minutes KDPA 23 May 1959; correspondence APK, UPASI and SIA 1958–60, in: IOL&R: MSS Eur F 174.

150 The Union Minister for Commerce and Industries, the Union Minister for Labour, the Union Minister for Finance and the Union Minister for Home Affairs.

Kerala plantations labour situation completely out of control/stop/mob rule & widespread violence reported/stop/many tea estates in high ranges abandoned/stop/pray your good offices immediately to stop violence & restoration normal conditions[151]

It was no coincidence that Union Home Minister Pant discussed the matter with Chief Minister Namboodiripad and Kerala Labour Minister Thomas the next day. A compromise was reached which satisfied both the Indian and Kerala governments: the central authorities withdrew their insistence on adjudication, while the state administration agreed to try to end the strike. The compromise broke the deadlock: the leaders of the Devicolam Estate Workers' Union called off the strike on 29 October and advised their members to resume work the following day.[152] Thus after the alarmist telegram the strike was over within four days!

The planters' victory was also underlined by the final outcome of the conflict. To the disappointment of thousands of plantation workers, the bonus issue was only decided at an industry-wide, tripartite meeting on 25 January 1960. Already chagrined by the time taken to reach the decision, the contents of the agreement did not meet the expectations of the labourers either. Instead of the trade unions' initial demand of a bonus of 25 per cent for the year 1957, it was finally decided that the workers would receive a bonus of only 4 to 7.5 per cent, depending on the type and quantity of the crops produced on their estates.[153] In comparison, the shareholders of the Kannan Devan Hills Produce Company received a 10 per cent dividend on ordinary shares in the year 1957.[154]

The planters' success is also reflected in Table 6.10. Compared to earlier years, 1958–9 in particular (see Table 6.8), the number of strikes and the amount of man-days invested in the work-stoppages were negligible in 1959–60 and 1960–1. The Kannan Devan Hills strike in October 1958 taught the labour movement an important lesson: even when south-west India was governed by a communist ministry, work-stoppages, the ultimate weapon

[151] Correspondence SIA, in: IOL&R: MSS Eur F 174.

[152] *The Hindu* 29 October 1958: 1 and 30 October 1958: 4.

[153] *The Planters' Chronicle* 1960: 79–80.

[154] *Tea Share Manuals*, in: IOL&R: MSS Eur F 174.

of the trade union leaders, proved inadequate to break the planters' power. No wonder both labour union leaders and workers lost their enthusiasm for going on strike in the period after the dismissal of the Namboodiripad government.

TABLE 6.10

Industrial Disputes in the South-west Indian Plantation Industry, 1959/60–1960/61[155]

Year	*Number of Industrial Disputes*	*Number of Work-Stoppages*	*Number of Workers Involved in Work-Stoppages*	*Man-days Invested in Work-Stoppages*
1959–60	1,657	12	2,389	10,239
1960–61	1,028	16	4,590	12,072

The third consequence, therefore, reflects certain earlier conclusions which were drawn with regard to the issues of nationalization of plantations and land reforms. The plantation managers in the Kannan Devan Hills, assisted by the various planters' organizations (the Kannan Devan Planters' Association, the Association of Planters of Kerala and the United Planters' Association of Southern India), successfully mobilized the press, the Congress Party and the central authorities to win the battle against the communist Devicolam Estate Workers' Union and the Namboodiripad ministry. Indeed, the Kannan Devan Hills strike, like the issues of nationalizing plantations or implementing land reforms symbolizes the success of the planters' lobby in a leftist 'Problem State'.

[155] Figures of the Labour Department, in: Ramachandran Nair 1973: 407.

Seven

Conclusion

7.1 Nothing Ever Alters – Change is Eternal

7.1.1 Continuities

In the introduction to this book, estates and plantations were defined as large-scale agricultural enterprises, mostly specialized in one particular crop meant for distant markets, and employing large numbers of wage labourers who worked under close supervision in a strictly hierarchical organization. This definition was especially useful in pinpointing the focus of the study: the planters' enduring struggle for land, labour, and government assistance.

This definition simultaneously brings out another strand of continuity in the trajectory of south-west India. Here, plantation development from the early colonial period was preceded by a proto-plantation system in the eighteenth century: the production of pepper for non-local markets by a substantial number of small cultivators functioning in a strictly authoritarian, hierarchical set-up. In other respects too there are similarities: (a) between the original pepper cultivators and the later coffee, tea and rubber workers; and (b) between the earlier Varma monarchs and the later British and Syrian Christian planters. Taking production as an example, both the pepper producers and the estate workers were bonded through advances to fulfil certain tasks during a particular period of time. Conversely, both the Varma monarchy and the planting elites tried to use the administrative, judicial and military agencies of the state to influence labour conditions and maximize export production.

Arriving at the point at which I have to draw conclusions from the study, I do not now feel fully at ease with the definition of plantations presented above. Certainly, it functioned as a useful guideline throughout the book but now that this study had deepened our knowledge and expanded our horizons in the field of plantation development, we are in a better position to appreciate its shortcomings, particularly when applied to south-west India. My exact predicament is that the scope of the statement is insufficient to reflect the tactics, strategies and policies of the planters, other agriculturalists, workers and government. Taking this into due consideration, I would like to introduce four different meanings of the word 'plantation', each suited to a particular interest group.

Let us start with the planters. The owners and managers saw the estates as commercial enterprises, sources of wealth and prestige of which the establishment and profitability were based on various non-market factors and strategies, including community background, labour bondage, and government assistance.

Indeed, group identity, unfree labour and state aid were of crucial importance to both the British and the Syrian Christian planters. Of these factors, perhaps the issue of community background requires the most pressing elaboration. For both experienced planters and novice agro-entrepreneurs *being* European, particularly in the nineteenth century, or *being* Syrian Christian, especially in the twentieth century, gave access to planting expertise, financial resources, agricultural lands and political power. This is exemplified by the first generation of British coffee and tea planters who formed the planting associations, obtained loans from European banks and were well-acquainted with the newly-introduced rules in south-west India regarding private property and sale of lands. In the power game, the European planters were backed by the British Resident, the puissant representative of the colonial state, should difficulties arise with non-European subjects, whether workers, peasants or ministers. Twentieth century entrepreneurial Syrian Christians also were in a privileged position regarding access to information (thanks to the *Malayala Manorama* and planting experience on British estates), capital (by means of chitties, later transformed into Western-style banks) and land (geographically and especially socially: the breakdown of the Nayar taravad system). To promote and defend their interests in rubber, the Syrian Christian planters formed a very strong lobby in various legislative bodies (expressed in the protests against the Brooke Bond scheme), political parties (the Congress Party) and planting organizations (the Kanjirapalli Planters' Association, the Travancore Rubber

Planters' Association, the Mundakayam Planters' Association, the Association of Planters of Kerala, and the Rubber Board).

Apart from being instrumental in the emergence and development of British and Syrian Christian plantation segments, community was of significance in two other closely related respects. Group identity functioned as a weapon in the struggle to minimize the number of competitors. In many cases, Europeans and Syrian Christians consciously excluded non-community members from practical knowledge, landed property and political influence. And outsiders were unable to become insiders: membership was determined by birth. Psychologically, being British or Syrian Christian gave yet another advantage: feelings of superiority. This seems to have been almost inevitable since the Union Jack was seen on top of the Rock of Gibraltar, along the Suez Canal and in Colombo, the route most British planters traversed on their journey to Travancore. Admittedly, Syrian Christians could not claim a world empire but they developed feelings of religious superiority, over the Hindus and Muslims, whom they considered pagans and infields, as well as over the low-caste and European Christians. Whereas the Syrian Christian tradition had been initiated by the apostle Thomas, Christianity influenced northern Europe and low-caste Indians only much later and less 'directly'. These ideas of superiority aided the planters in important ways. They bolstered the managers' confidence in justifying their actions and imbued them with a kind of 'natural' authority over the labourers.

Another constant factor was the planters' efforts to create an unfree labour force. The planting community, European and Indian alike, plainly tried to restrict the labourers' choice of employer and to influence labour conditions. On this issue, the Criminal Breach of Contract Act (1865–1935) usually attracts attention: and no wonder, for it enabled planters, policemen and magistrates to track down, arrest and sentence absconders. Its enforcement intensified the labour regimes on the estates. Wielding the threat of trial, conviction and imprisonment, the planters used the law to constrain the labourers to pay off their debts and to meet their other obligations, including serving the period of contract. Yet throughout its existence, the Criminal Breach of Contract Act only provided a temporary legal context encircling the nucleus of unfree labour: the kangany system.

Planters advanced money to labour recruiters (kanganies), who, in turn, gave advances to workers.

The kangany system limited the freedom of the labour force in three ways. Most pertinently, it encouraged debt slavery. The kanganies owed money to the planters: the advances offered by the planters were used by the kanganies to recruit workers. In their turn the field workers became indebted to the kanganies: the advances immediately being used, in most cases, to pay off existing debts, for basic needs (food, clothing, and so forth) and for the travel expenses to the estates. Secondly, the kangany system encouraged the spread of false information about plantation life. Interested in luring as many workers as possible to the estate, the recruiters told rosy stories about the plantations, withholding information which would enable the labourers to make a more deliberated choice. Finally, it was an incentive to various other kinds of abuse, including the use of violence.

Having put this in a nutshell, I want to elucidate on it in more detail. The kanganies, who initially functioned as labour recruiters on the plains and subsequently worked as overseers on the estates, derived the bulk of their incomes from commissions on their gangs' earnings. This encouraged the kanganies not only to keep the labourers indebted (through falsifying contracts, accounts, and so forth) but also encouraged the overseers to abuse their position, detaining the workers beyond the period of contract, forcing all workers, including the sick, to fulfil their daily tasks and work long hours, and punishing unwilling workers severely. The worst illustration of this is the beating, kicking, and covering with agricultural chemicals of runaway labourers by kanganies in the Kannan Devan Hills in the post-1947 period. These cruelties could reach such heights that they even led to a great number of plantation workers committing suicide, of which there is evidence from the Peermade/Vandiperiyar area. The kangany system was formally abolished only in September 1962.

The planters also had other strategies to restrict the labourers' room for manoeuvre. One such tool was the policy of creating three different groups, all clearly distinct in terms of responsibilities, geographical and social background and lifestyles. This effectively prevented low-caste field workers from climbing the plantation hierarchy. The highest they could hope to rise was the position of kangany. The estate managers, both British and Syrian Christian, preferred to attract well-educated, Malayali, Christian, high-caste employees for the staff functions: namely Syrian Christians and, only in the initial phase, Christian Shanars. At the other end of the scale, low-caste, Hindu, Tamil, illiterate workers were automatically assigned to the category of field labourers. It was a division designed to forestall the planters' nightmare: a united labour front. Their disparate identities and interests ensured that the staff did not side with the field workers against the third group, the managers. Since the mid-1930s in particular, the carefully selected field workers were kept out of reach of undesirable sentiments and ideas. A state within a state was created, complete isolation being the ultimate goal. Indeed, most outsiders were unwelcome (particularly the anti-British Moplahs and the leftist activists) and it was hoped the work force would exist in a state of suspended animation, with its mental status fossilized (explaining the change from a basically temporary to a more permanent labour force, the severe punishment inflicted on children visiting Mahatma Gandhi, and the victimization of workers sympathizing with the labour unions).

Another continuous thread is the planters' attitude to the state. They constantly tried to influence government policies, taking an active role in the south-west Indian political framework. This was logical since the Travancore state, and later the Travancore-Cochin and Kerala administrations, exercised an enormous influence on plantation development. The government regulated land. This is perhaps best illustrated by the Rules Regarding Grant of Land for Cultivation of Coffee (the Memorandum of 8 March 1892), the Brooke Bond scheme (1926–7) and the various land reforms (1950s and early 1960s). In addition, the state administration decided on important issues relating to labour, including the above-mentioned Criminal Breach of Contract Act and the maintenance of law and order. Therefore, the planters constantly discussed matters with leading political figures, formed various planting organizations and were represented in the different legislative bodies.

This immediately raises another point. In their efforts to control the south-west Indian political arena, the planters' relations with the dominant administration were of crucial importance. The nineteenth century British planters were adept of mobilizing the British Resident, the representative of the powerful Madras government, for their own purposes. Some favourable results: the enactments of the Memorandum of 8 March 1862 and the Criminal Breach of Contract Act (1865). Twentieth century events indicate that the planters' position within the wider political context remained one of crucial importance. There is sound evidence that the decline of the British planting community was closely related to its slackening grip on the dominant government. Two factors were significant in this regard. The more direct of these was that colonial officials, including the British Agent, were not so ready to satisfy the British entrepreneurs' demands, fearing negative publicity at home (leading to the abolition of the Criminal Breach of Contract Act in 1935) or social and political unrest in India (contributing to the non-realization of the Brooke Bond scheme). Less directly, the colonial state itself lost dominance, brought about by the gathering strength of the nationalist movement. Simultaneously, the rise of the Syrian Christian planters was intimately connected with the fall of the colonial state and its ultimate replacement by the independent state of India. During the early post-Independence years, the close relations between Syrian Christians, the Congress Party, and the central Indian government counteracted the strength of the labour movement. Being on good terms with the central authorities, the planters were able to withstand various attacks, including the cry for nationalization and land reforms.

Besides influencing the formulation of government regulations, the planters manipulated the execution of state policies as well. This was already at work during the nineteenth century. Then, the planters joined hands at public land auctions, bought large tracts of land for non-productive purposes (especially to restrict competition and out of speculative mania), fraudulently appropriated government lands, paid taxes too late, only partly, or not at all, and recklessly evaded the rules which protected the natural environment. The planters' manipulation of government regulations continued into the twentieth century, culminating in the (legal) division of estates into smaller-sized holdings. This measure was meant to circumvent and/or reduce the often costly implementation of the Kerala Plantation Labour Rules

(1959), the regulation which put the Plantation Labour Act (1951) into operation.

Having considered the planters, let us now focus on the labourers' motives and strategies. Here, the search for government assistance was not a constant factor. In addition, it should always be borne in mind that the nature of the historical information available precludes drawing conclusions regarding the exact role of communal ties and individual networks in the labourers' actions. We do know, however, that for the workers an estate represented a potential or actual instrument to save or at least improve the quality of their lives, depending on the worker's initial situation, his/her ability to deceive the labour recruiter or planter, and the labourers' struggle for respect.

Certainly, at the time of their recruitment a small group of labourers was completely misled by the kanganies and regarded the plantations as acceptable opportunities to earn fine wages. Did the attractive advance offered not indicate that the planter was able and willing to pay a good wage? A larger number of workers gave the kanganies and planters the benefit of the doubt. Putting ourselves in their position, we should not be surprised that they were *inclined* to believe the recruiters' glowing stories of plantation life. Most of these poor, unemployed, low-caste workers would have been hard put to imagine circumstances more difficult than those in which they already found themselves. The majority of workers, however, were shrewd enough to realize the one-sidedness of the information and were acquainted with the hardships of plantation life, either through relatives/friends or from personal experience. Still, for many of them estate work offered a way out, a temporary solution, and often a last resort in periods of personal problems, unemployment, and/or food shortages.

The labourers were quite capable of exploiting the deficiencies in the planters' system. To mention a widely-known example, workers and kanganies accepted advances and then did not meet their contractual obligations. Though running the risk of severe punishment, the labourers, sometimes in collusion with kanganies, returned home earlier than initially agreed or deserted and left for another estate, particularly if a neighbouring planter (mostly secretly) offered more attractive wages. Many workers including ex-slaves, *free* to choose their employer, volunteered for *unfree* contract labour, *freeing* themselves from these bonds whenever they thought this was to their advantage. The workers' meagre incomes were also supplemented by thefts of coffee, tea and rubber.

In fact, the workers not only used and circumvented the planters' rules, leaving the estate system itself largely unchallenged, they also interfered with the foundations of the plantation regime. To be more precise: the labourers struggled to regain, maintain and enlarge their self-respect and prestige. This endeavour was noticeable throughout the period studied.

Often the first step was for large numbers of plantation labourers to embrace Christianity after entering estate work. Indeed, this religion was instrumental in their fight for dignity. In this respect, the missionaries' stories about the life of Jesus undoubtedly attracted them most. They derived comfort from His Resurrection: human suffering, even if death followed, was neither meaningless nor final. They discovered an unexpected friend. Jesus had mercy on all who were despised, rejected and humiliated, giving them hope for a better life. The workers gathered new strength from His message that the last will be the first, providing them with the confidence that they were also welcome, worthy human beings and creatures of God.

In another bid for self-respect, from the early twentieth century labourers began to call in the help of policemen and magistrates. Again in 1905, for example, an employee named Koshi accused the prominent coffee and tea planter F.G. Richardson of tying him up. Also in 1905, the kangany Moses informed Travancore state officials that a European tea planter called Lee had assaulted a kangany named Jacob with hands, horsewhip and boots, eventually causing Jacob's death. In late 1912 or early 1913 the rubber planter Hamond was arrested on the word of a kangany. And in 1913, the plantation worker Mylen Chothan informed the Peermade magistrate that his twelve year old son had been kidnapped, eventually resulting in the so-called Slavery Case in the Travancore Highlands. All these instances hint at the same feature: if the planters and/or the kanganies went too far in their humiliation of their inferiors, they could not do so with impunity. By reporting the behaviour of the planters and kanganies to the police (as well as by other, non-official means, as shown by the stabbing of the planters Seabrooke and Gillespie in 1930 and 1933 respectively), the 'weak' clarified *their* boundaries, *their* limits of endurance. In other words, the lower sections also shaped the line between inferior and superior, the rules between labourers, kanganies, and planters. This was certainly not unique to one specific region: Richardson, Lee and Hamond worked in the Peermade/Vandiperiyar district, the Kannan Devan Hills, and the Mundakayam area respectively.

From the late 1930s onwards, the workers needed no longer to rely just on the missionaries' moral values and the magistrates' concepts of justice to resist and transform plantation life. From this period, the labourers' struggle for respect was given added encouragement by the rise of a leftist movement in the state. The first trade union activists who (often secretly) visited the plantation regions offered the workers great advantages. Apart from particular forms of organization (trade unions, political parties) and apposite methods of improving labour and living conditions (including joint petitions and well-planned strikes), the leftist activists imported certain values and sentiments. It was especially this last which aided the labourers in their fight for a more dignified life. Socialism and other revolutionary ideologies placed proletarians in the vanguard, justified the workers' struggle and promised a worthy, bright future. According to some of my informants, this was undoubtedly the greatest single accomplishment of the leftist movement in the plantation regions: the labourers' increased self-assurance.

Weighing up the workers' use of Christianity, the judicial agencies of the Travancore state, and the leftist movement, one characteristic immediately catches the eye: the labourers' fight for self-respect and dignity depended heavily upon 'outside' individuals, movements, and institutions. This again explains the planters' ambiguous, and often negative, attitude towards the missionary activities in the estate regions, their anger about particular magistrates, including P. Raman Tampi, the Sessions Judge who was in charge of the Slavery Case in the Travancore Highlands, and the managers' fear of labour union activists. Indeed, to prevent the labourers from gaining moral strength it was the planters' aim *to keep out* unwanted sentiments and ideas, while *keeping in* an obedient labour force, bereft of any ethical or ideological arms to fight the planters' dominance. For the labourers it was the other way round: they tried *to bring in* self-assuring ideologies *to bring out* their human dignity.

The group of peasants, landlords and other non-planting agriculturalists was far more heterogeneous than the categories of planters and workers. Their modes of production and community background varied greatly. This disparate body of cultivators included high-caste paddy producers, Christian coffee and rubber peasants, and tribal shifting cultivators. Though hardly a homogeneous category, these cultivators shared the fact that their ability to profit from plantation development was closely connected to their religious characteristics, communal identity and individual networks. On the downside, their inability to guard against estate development was due to a lack of political influence.

This latter issue requires some elaboration. The non-planting cultivators tried to resist the negative implications of plantation development by enlisting government support. This endeavour forms a recurrent theme in the study, observable for the first time during the 1850s. At that time the high-caste, paddy-producing landholders opposed the missionaries-cum-future planters' plan to free the slaves, fearing social unrest and a labour shortage. Some decades later, in 1876 and again in 1903, C.J. Eapen and Abraham Pillai respectively requested the government to protect their properties against the encroachments of the British planters. Of greater value in our search for the other cultivators' strategies (Eapen and Pillai were possibly planters), in 1896 hillmen protested against the acquisition of 400 acres of land situated in the Kannan Devan Hills by John Muir, the prominent director of James Finlay and Company. They visited Trivandrum, the capital of Travancore, to emphasize their claim that that particular piece of land was legally theirs. In the mid-1930s, rubber peasants regularly lobbied to influence the implementation of the International Rubber Regulation Agreement, requesting the government for access to the profitable world market and permission to open up new areas. After Indian independence, various non-planting agriculturalists, sometimes acting as the mouthpiece for particular organizations (such as the Ottopalam Small Landholders' Association), openly criticized the various south-west

Indian ministries for exempting plantations from the land reforms. It says something that none of these political battles were won by the planters' opponents: slavery was formally abolished in 1855; surviving evidence suggests that the requests of C.J. Eapen and Abraham Pillai as well those of as the hillmen met with refusal; alterations to the implementation of the International Rubber Regulation Agreement favoured the indigenous rubber *planters;* and the Restriction on Possession and Ownership of Land Bill (1954), the Kerala Agrarian Relations Bill (1957), the Kerala Agrarian Relations Act (1960) and the Kerala Land Reforms Act (1963) increasingly supported plantation development. It is even more telling that these political battles were lost by the planters' adversaries because they lacked support from the dominant state: the colonial administration and, later, the central Indian government.

Though resistance to the planters' unfavourable strategies was often unsuccessful or had only a limited effect, the efforts of the non-planting agriculturalists to copy the planters' agricultural practices met with more regular success. Three factors were of crucial importance in this: religious identity; communal characteristics; and individual qualities. Religion is the factor which explains why large numbers of Shanars and Syrian Christians especially became involved in the cultivation of coffee and rubber respectively. Examining the case of the former group we discover that in mid-nineteenth century south Travancore the low-caste, poor Shanars in particular responded positively to the proselytism of the missionaries. No wonder: spiritually the preachers offered the despised Hindu Shanars an attractive religious alternative and at a more practical level taught them the commercially valuable English language. Moreover, the missionaries, of whom many were also planters (e.g. John Cox and Samuel Mateer), provided the Shanars with employment as field workers, kanganies, clerks and managers. Having obtained the requisite planting expertise and realizing the profitability of the venture, many Shanars opened up new coffee lands. Religious identity also played a crucial role in the emergence of a Syrian Christian rubber segment who were largely responsible for the opening up of large areas in and outside the Travancore region in the first part of the twentieth century. Appointed as staff and superintendents on European estates, Syrian Christians obtained knowledge about planting, processing and marketing and launched into rubber cultivation themselves.

Community provides the clue to why most Shanars were unable to follow the example set by P.D. Devasahayam, fellow-Shanar and successful coffee planter, and became coffee peasants, tilling a relatively small plot with family labour. The Shanars had learned how to plant, tend and harvest the coffee bushes but lacked sufficient prestige and financial means to open up and run an estate. In sharp contrast, many Syrian Christians were successful in following in the footsteps of their own model, the Syrian Christian rubber-planting pioneer P.J. John (1905). Undeniably, in the case of the Syrian Christians communal background was a much more stimulating factor, particularly because of the prestige and comparative wealth of this group.

Lastly, social networks are the key to elucidating why, in contrast to other Shanars, P.D. Devasahayam became a wealthy and respected coffee planter, and why individual Syrian Christians, following the example of P.J. John, became successful rubber planters. Having close relations with First Prince Vishakham Tirunal and Dewan Madhava Rao, P.D. Devasahayam was in an exceptional position. As his financial means and status improved, Devasahayam freed himself from the restrictions of his community. Networks, however, were not only important in stepping *across* social boundaries. In the case of Syrian Christian entrepreneurs, contacts *within* the community were of crucial importance. This is beautifully illustrated by P.J. John: he maintained close relations with Rev. C.P. Philipose Kassessa, leader of the Evangelical Association of the Mar Thoma Church, and K.C. Mammen Mappillai, a prominent member of a wealthy Syrian Christian banking family.

The government, our fourth focal point, regarded the estates as useful instruments for increasing its power, in so far as plantations would not challenge the state's credibility, authority and validity. Remarkably, the individual convictions and social backgrounds of the leading political figures played only minor roles in the formulation of government policies.

The Travancore, Travancore-Cochin and Kerala governments all encouraged the rise and further growth of estates for their own benefit. Logically so, as plantation development enhanced the administration's ability to rule, for instance by building up the state revenues. From the tax collector's point of view, plantations needed to maintain elaborate accounting systems and were therefore less elusive than other agrarian enterprises, small family plots in particular. The state also benefitted from the planters' opening up of estates in 'peripheral' zones. This strengthened the state's grip on these areas. The planters were joined in the plantation regions by state officials like policemen, fiscal specialists and magistrates. Just as significantly, plantation development induced the government to create new functions (e.g. the Superintendent and Magistrate Cardamom Hills and the Conservator of Forests) and to establish or reorganize various public institutions, including the Survey Department and the Public Works Department. The plantation areas themselves became political units, namely state districts, and (thus) the basis for gathering all kinds of information, to which the various Census reports bear testimony. Estates were certainly instrumental in the colonization of hitherto marginal, peripheral zones. Moreover, the ministers were happy to use the plantations to justify their responsibility. To be more precise, the government showed both external powerholders as well as internal pressure groups its concern for its subjects' welfare by pointing out its policies regarding plantations: had not the government encouraged the introduction of profitable crops, modern production techniques and full employment by stimulating estate development?

All the while, the government's authority was inexorably threatened by plantation development. As has been discussed in more detail above, planters evaded taxes, encroached on government land and so on and so forth. Other challenges to the government's ability to preserve law and order included the instances of extreme violence between field workers, kanganies and superintendents. Peasants, frustrated by a lack of government protection, sometimes resorted to force as well: for example, in the late 1910s or early 1920s a local cultivator took revenge on his neighbour, the encroaching rubber planter Brander, and killed him. Most of these cases did not present the government with much of a problem. Often knowledge of these particular 'contra-governmental events' did not spread beyond a limited circle of people and therefore did not interfere too much with the government's authority. Moreover, in most of these cases the government fought back 'silently': violators of the law were corrected, sometimes punished subtly, without arousing much or any public attention.

The power of the state was more seriously threatened when the government's attitude was questioned in a wider, more significant and/or more public political context. In most of these cases, the ministers were forced to reconsider their original standpoint. Three of these conflicts illustrate this splendidly. In the first example, the planters J.D. Munro, H.G. Turner and A.W. Turner challenged the Travancore government's claim that the north-eastern areas were state property. And they practised what they advocated or *wanted* to believe: on 11 July 1877 they acquired the vast region of 215 square miles from a local zamindar, the so-called Pooniat Chief. Bowing to the intervention of the powerful British Resident, the Travancore government finally gave in to the planters' wishes and recognized the validity of this particular agreement. The second example dates from the first half of the twentieth century when newspaper journalists, those of the *Malayala Manorama* in particular, and members of the various legislative bodies challenged the government about the question of citizenship. They argued, often implicitly, that the state had renounced its validity because the administration favoured 'foreigners' over 'its people'. And they met with considerable success: after the late 1920s preference was given to subjects of the Travancore state with regard to the acquisition of lands. Equally importantly, in the implementation of the International Rubber Regulation Agreement the Travancorean planters' interests were also taken into account. Again, the government had to change its initial course, afraid of losing authority. The third example dates from after 15 August 1947, the date of Indian independence. Then the planters, in close co-operation with the Congress Party and the central Indian government, accused the leftist ministries of promoting 'lawlessness' and 'insecurity'. As the strike in the Kannan Devan Hills in 1958 shows, these accusations could change government policy. Fearing central government intervention, including President's Rule, the Namboodiripad ministry changed its initial 'independent' stance to one of actively attempting to end the strike.

It is significant that the role of the leading political figures was determined much more by 'strictly governmental interests' (characterized by the dilemma discussed above: plantations as both a stimulus and threat for the state) than by personal beliefs, communal issues or party politics. For instance, personally fearing the total destruction of the natural balance in the state because of the planters' ruthless methods of deforestation, Dewan A. Seshiah Sastri (1872–7) *could not* fundamentally change the government policy regarding estates. The British Dewan M.E. Watts (1925–9), trying to help British planting interests (and possibly local cultivators) through the Brooke Bond scheme, *could not* prevent government policy from changing radically and favouring indigenous applicants for land above their foreign competitors. Dewan C.P. Ramaswami Aiyer (1936–47), well-known for his anti-Christian/pro-Hindu attitude, *could not* stop the rise of Syrian Christian planters. Finally, Chief Minister E.M.S. Namboodiripad (1957–9), a convinced communist with a Brahmin background, *could not* adhere to his original position regarding the Kannan Devan Hills strike of 1958. There was a sound logic behind all these examples. These ministers were fully aware that their government's authority would be undermined, and that they would ultimately lose power, if they pursued their personal, communal and/or party interests too far. Seshiah Sastri was afraid of repercussions on the colonial administration; Watts and Ramaswami Aiyer foresaw public unrest; and Namboodiripad feared central Indian government intervention. Nor were their fears unfounded: in 1907 Dewan Gopalachariar was dismissed from office because of his unwillingness to comply with the planters' requests (he was, for example, not prepared to maintain roads illegally constructed through state properties and 'solely benefitting the planters') and in 1959 the passing of the Kerala Agrarian Relations Bill, 'a unique communist achievement', ushered in the fall of the Namboodiripad government. In both cases, the ministers' policies, though congruent with their personal convictions, conflicted with the policies of the dominant authorities and finally led to their removal from office.

7.1.2 Changes

The above discussion focuses on continuity. However, the development of plantations was also characterized by fundamental alterations. In short, the transformations were related to three closely connected processes: the widening and deepening of the political arena; the amalgamation of 'old' and 'new' production systems; and the successive emancipation of the British planters, the Syrian Christian agro-entrepreneurs and the low-caste plantation workers.

To come to grips with the various phases through which plantation development passed, it is essential to have an understanding of the period prior to the establishment of estates, the pre-plantations years. Eighteenth century south-west India witnessed the rise of the Varma monarchy. The Rajas Marthanda Varma (1729–58) and Rama Varma (1758–98) transformed the tiny, weak political entity of Venad into an expanding and increasingly powerful state, which became widely-known as Travancore. Gradually embracing all the territory from Cape Comorin in the south up to Cochin in the north, the Travancore rulers brought commercial and agricultural resources under stringent state control and often ruthlessly subordinated defeated competing powerholders, thereby radically changing the complex, fluid and loose political structures which had characterized early eighteenth century south-west India. It is very significant that the rise of the Travancore monarchy was closely related to its increasing military power which in turn was linked to its control over the production of and trade in pepper.

The Varma monarchs undoubtedly profited most from the situation in which pepper, a very valuable Asian product for various European trading companies, was exchangeable for weaponry, the most useful European article for south-west Indian rulers attempting to defend and augment their power. Fully cognizant of the importance of pepper for attaining his political objectives, Marthanda Varma declared the pepper trade a state monopoly in 1743. He introduced the Commercial Department, which organized the buying, warehousing and selling of the product, and determined that pepper cultivators were only allowed to offer their produce to authorized traders. Besides transforming the position of the hitherto independent merchant class, the state monopoly also affected the situation of the pepper producers. Their already difficult position became even more troubled. Now, the smallholders were unable to play the traders off against each other and found themselves up against closely-supervised government agents who were backed by the powerful judicial and military agencies of the Travancore state. In other words, the pepper cultivators' ability to raise the price, i.e. advance, for pepper was considerably cut back during the mid- and late eighteenth century.

The most formidable agent in the decline of Varma power was the British East India Company. The British gradually achieved the exclusion of other European trading companies from participation in the pepper trade along the Malabar coast, thereby breaking the powerful position of the Travancore state in one blow. In contrast to earlier times, the Commercial Department was unable to sell its products to the highest bidder and hence was less able to manage the conditions of trade. Inexorably, the British became increasingly involved in south Indian political affairs. Rama Varma, unable to defeat the expansionistic Mysorean armies led by Tippu Sultan, saw no other solution than to ask the East India Company for assistance. The alliance, anchored in the treaties of 1788 and 1795, meant a loss of military sovereignty and aggravated the already parlous financial situation of the Travancore monarchy. The arrival of a British Resident (Major Colin Macaulay in 1800), the suppression of various insurrections (particularly that of 1804), and the new treaty of 1805 dismantled the Travancore state politically, militarily and financially. Independent Travancore became a quasi-independent 'princely state'. Though theoretically the Raja governed, the Dewan executed and the British Resident advised, in actual practice it was mostly the reverse. Especially during the first half of the nineteenth century, the British representative took the initiative, the Dewan responded, and the Raja was almost completely left out of political life; he merely sanctioned government rules.

The political changes induced a totally disordered export economy: 'Western' organizational forms did not fit into the south-west Indian context and 'indigenous' modes of production and trade systems conflicted with the ideas and policies of the British. Turning to the 'foreign' mode of production and its agents, there is no doubt that most British planters kept away from Travancore. Despite invitations from the British Residents (Colonel Newall in particular), they preferred to open up coffee estates in a politically more friendly environment, i.e. in directly-ruled areas like Ceylon or the districts of Coorg, Mysore, Nilgiris and Wynad situated in the Madras Presidency. W. Huxham, the only planter who settled in Travancore and opened up Camapalle Estate in 1824, had to deal with many obstacles in the running of his undertakings. Though soil and climate were extremely suitable for the cultivation of coffee, it can be assumed that he met with great difficulties in recruiting and controlling the labour force, besides encountering years of delay in obtaining a legal justification for his seizing possession of lands for cultivation purposes, i.e. a *written* agreement stipulating the conditions of the tenure, from the Dewan.

If the British representatives were opposed by the Dewans in their efforts to encourage European planters, the ministers and Residents joined hands in promoting the emergence of an indigenous coffee sector. This co-operation resulted in a clear-cut policy: the proclamation of July 1824 offered attractive land and tax concessions to beginners, free coffee seeds were distributed among peasants, and the Europeans Winson and Lowe were appointed to implement the programme (1838). The plan, however, failed. The coffee peasants obtained only low prices for the crop, particularly in the period 1844–9. Potentially much more damaging was the draining away of the strength and determination of the Travancore administration to carry out its plans. It is no exaggeration to say that the early nineteenth century governments were much less interested in coffee cultivation than their predecessors, Marthanda and Rama Varma, were concerned with pepper production. And logically so, for nineteenth century coffee, in contrast to eighteenth century pepper, was not synonymous with military power and political prominence. The British government favoured free trade (the pepper monopoly was abolished in 1860) and for the Dewans the impossibility of introducing a coffee monopoly symbolized Travancore's lost autonomy and bred political indifference.

From the late 1850s, it was the British planters themselves who set the political agenda. They requested proclamations and regulations which facilitated the recruitment and disciplining of labour, the acquisition of landed property, and transportation to and from the estate regions. By their actions, the planters took the initiative away from the British Residents, who never flagged in their support for the emergence and development of a plantation sector. The attitude of the first ministers radically changed with the appointment of Dewan T. Madhava Rao in 1858: they too encouraged the rise of plantations, attempting to ameliorate Travancore's political, financial and economic situation. The convergence of all these variables resulted in a very favourable government policy for British agro-entrepreneurs. Following the abolition of slavery in 1855, the Criminal Breach of Contract Act was enacted in 1865, enabling the planters to transform *unfree*, attached (ex-)slaves first into *free* wage labourers and then into *unfree*, indentured plantation coolies. Turning to the status of land, the Memorandum of 8 March 1862, the Rules for the Sale of Waste Land on the Travancore Hills (April 1865), the Rules for the Grant of Grass Land to Proprietors of Coffee Estates (July 1877) and the Rules for the Sale of Waste Land on the Travancore Hills for Coffee and Tea Cultivation (July 1898) were of crucial importance. In their stipulations, these regulations followed Western notions of private property: 'the lands applied for shall be granted in perpetuity as heritable or transferable property'. Accessibility was improved as the Public Works Department spent large amounts of money to construct and maintain roads connecting the plantation districts with the coastal regions. No wonder, then, that European planters flocked to Travancore, instead of avoiding the region as had been the case during the first part of the nineteenth century. The government policy also altered the attitude of the British officials towards Travancore. Reflecting their appreciation, Madhava Rao was knighted, the Raja received the honourary title of Maharaja and Travancore was frequently labelled as 'the Model State'.

The disquisition above in no way suggests that the European planters 'Westernized' south-west Indian agriculture either immediately or at any time fully. In fact, British plantation development itself was to a very large extent based on Indian concepts and realities. This assumption is supported by the European estate managers' use of the widespread advance system, employed earlier in the case of pepper production, to recruit and control the workers. More importantly, the British planters manipulated the caste system to impose stricter discipline on the labourers. As has been discussed in more detail above, high-caste workers especially were employed as staff members and low-caste workers as field workers. Well-acquainted with local customs, the European managements were fully aware that the 'pure' would never side with the 'impure'. In these particular senses, the British plantations were 'Indianized'. It could also be argued that in these respects, the British reinforced the 'Indianness' of India.

In addition, Indian agriculture only seriously 'Europeanized' in the early twentieth century, thereby further 'Indianizing' the nature of plantation development. Then, enterprising Syrian Christians seized the initiative and opened up plantations, copying the British estates' cultivation systems (planting rubber trees in 'scientifically'-determined rows), labour regimes (employing a strict hierarchical order of management, staff and field workers), processing techniques (including the acquisition of European equipment) and marketing methods (enlisting the services of managing agencies). It is no exaggeration to say that indigenous planters used the foreign estates as blueprints for their own enterprises. In their efforts, many local agro-entrepreneurs outperformed the British, especially when achievement is measured in terms of production costs. Paradoxically, the indigenous planters were successful because they strengthened the 'Indianization' of plantation development. Apart from providing advances and adopting caste principles to recruit and discipline workers (like European managers), because of their profound knowledge of south-west Indian languages, labour relations, and society at large, local planters were more in touch with the situation on their estates, and because of their 'Indianness' did not have to fear colonial interference or the nationalists' protests, which meant that worse labour conditions prevailed on their estates compared to those on European estates. The by and large smaller size of their estates meant that these planters had the added advantage of being more flexible, especially in responding to labour difficulties or price fluctuations, than it was possible for their British counterparts to be.

It is fascinating to realize that the Syrian Christian planters' path to political prominence was also characterized by adaptation, making 'foreign' political structures suitable for 'indigenous' use, thereby fundamentally changing the nature of politics. To explain this most clearly it is essential to understand that two distinct, though closely-linked, developments occurred during the first part of the twentieth century. Firstly, the local planting community, like other prominent indigenous interest groups, became familiar with 'Western' notions of the state, including the concepts of citizenship, responsible administration and self-government. Shrewdly, the indigenous planters, like other (proto-)nationalists, used these *Western* concepts to defeat *Western* dominance. Local planters, as well as their partners in the struggle against European plantation development, *appropriated* Travancore. Indigenous agro-entrepreneurs, journalists from newspapers like the *Malayala Manorama* and Syrian Christian members of the constitutional bodies all spoke the same language: Travancore should no longer serve *foreigners;* Travancore represents *our* government, *our* farmers and *our* land. The second interlinked development was the tactics that the Syrian Christian planters used to gain political power, which contributed to the strengthening of a 'non-European' phenomenon in south-west Indian politics: communalism based on caste to be more precise. To protest against the Brooke Bond scheme (1926–7), the implementation of the International Rubber Regulation Agreement (from the mid-1930s onwards) and the Kerala Agrarian Relations Bill (1957–9), the Syrian Christian planters mobilized the *Syrian Christian* members of the legislature, *Syrian Christian*-controlled newspapers (like the *Malayala Manorama*) and *Syrian Christian*-dominated political organizations (including the Congress Party).

Roughly from the gaining of Indian independence, the plantation labourers also emerged as a political factor. Especially encouraged by the activists of the Travancore Labour Association which had its base in the non-estate, western areas of south-west India, plantation workers began to organize themselves into labour unions from the late 1930s. Though initially considered invalid and illegal by both planters and state officials, the trade unions and leftist political parties gradually picked up momentum in the plantation regions. Withstanding the planters' other anti-union strategies, the labour activists were finally able to effect the organization of various Tripartite Conferences (from the very early 1950s), the installation of the Minimum Wage Committee for Plantations (August 1952) and the Minimum Wage Advisory Committee for Plantations (November 1953), and the establishment of the Plantation Labour Committee (from the early 1950s).

Though they gradually gained more clout in south-west India, the political emancipation of the plantation workers was of a different order than the political rise of the British planters in the late 1850s and of the Syrian Christian agro-entrepreneurs in the early twentieth century. In contrast to both categories of planters, the labourers have never become influential at the all-India level. Though gradually represented prominently in the Legislative Assembly of the Travancore, Travancore–Cochin and Kerala states (finally leading to the installation of the communist ministry of E.M.S. Namboodiripad in 1957), at no time did those fully devoting themselves to the (plantation) workers' fortunes obtain a majority in the Lok Sabha, the Indian parliament. Here, the Congress Party remained almost omnipotent, and, despite its often leftist rhetoric, gave priority to the interests of successively the Indian (agro-)industrialists, foreign investors and – lastly – the workers.

This again had repercussions on the situation on the estates. This resulted in a tug-of-war in which the trade unions demanded higher wages, the direct and full implementation of the (central) Indian Plantation Labour Act of November 1951 which was meant 'to provide for the welfare of labour', and the immediate abolition of the exploitative kangany system. The planters, counteracting labour union influence at estate level, taking advantage of the state government's dilemmas, and backed by the central Indian administration, succeeded in slowing down the negotiation processes. This is the reason that in south-west India the putting into effect of the Plantation Labour Act was regulated only in 1959 and labour recruitment by kanganies was formally abolished only in 1962. Even then, many planters were reluctant to implement the decisions finally taken. This had obvious effects: the conditions of the plantation labourers improved only slightly. Though continuing to work long hours, the majority of the workers were not able to reach a satisfactory level of calorie consumption, still lived in overcrowded and poorly-constructed houses, and remained the most backward inhabitants of south-west India in terms of literacy. It is interesting to note that the same political configuration was largely responsible for the impossibility of nationalizing plantations, including estates in the land reforms, and winning major labour struggles like the Kannan Devan Hills strike of 1958. Indeed, the atmosphere of progress, predominantly the fruit of the rise of the leftist movement in the estate regions and the whole of south-west India, was blocked by an overwhelming political reality, the influential planters' lobby at the Congress-controlled central level, shattering the labourers' dreams of fundamentally improving the quality of their lives within a short period of time.

7.2 Nothing is Dissimilar – Everything is Unique

7.2.1 Macro-Developments

Though certainly not claiming specialized knowledge about other Asian regions, I find it difficult not to place the south-west Indian developments in a wider Asian framework. This seems the logical thing to do: at least to a certain extent changes in south-west India were a reflection of (and influenced) broader historical developments, particularly those occurring in the Indian Ocean region.[1]

To take one cogent example, the prominence of Marthanda and Rama Varma, based on their largely successful efforts to control the trade in and production of pepper in eighteenth century Travancore, was not a unique historical phenomenon. At least in two other pre-colonial societies bordering the Indian Ocean that same mechanism was at work. In the early seventeenth century, the rulers of northern Sumatra also based their wealth, prestige and power on pepper. Realizing the great interest shown in pepper by European traders, the Sultan of Kedah opened up pepper 'plantations', on which prisoners, probably slaves, laboured. Sultan Iskandar Muda of Aceh (1607–36) also applied fairly ruthless methods to enhance his status. In order to monopolize the trade in pepper in the Sumatran region, Iskandar Muda destroyed the plantations of his rival, the above-mentioned Sultan of Kedah, and forced the pepper cultivators of Tiku and Pasaman to sell their produce on the market of Aceh.[2]

[1] Some of the more interesting works considering the Indian Ocean as one single unit of analysis include: Chaudhuri 1986; Das Gupta and Pearson 1987; Reid 1993a; Reid 1993b.

[2] Lombard 1967: 66–7, 99; Ricklefs 1981: 31–2. Showing a striking resemblance

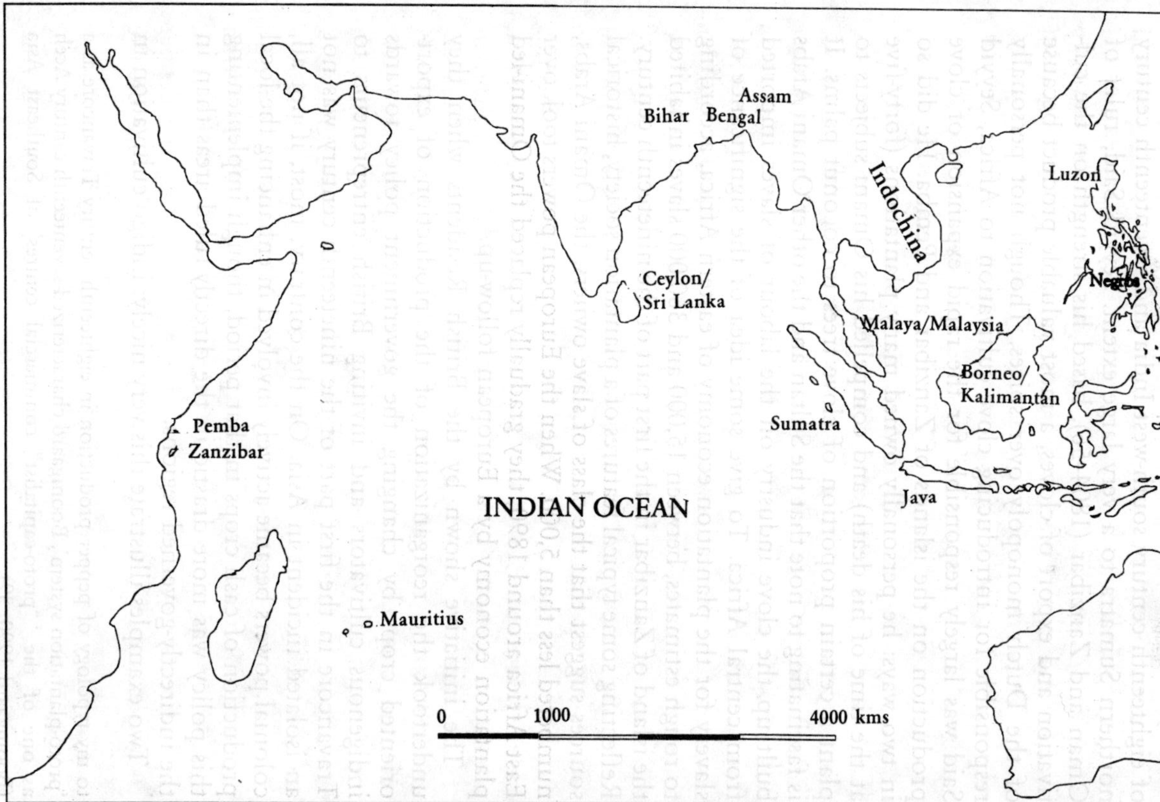

Map 7.1 The Indian Ocean Region

The rise of Sultan Seyyid Said in early and mid-nineteenth century East Africa also resembled the trajectories of the rulers of eighteenth century south-west India and seventeenth century northern Sumatra to a very large extent. Seyyid Said, ruler of Oman and Zanzibar (1804–56), based his strength on the cultivation and export of cloves, a most valuable product because of the Dutch monopoly over spices. Though not personally responsible for introducing clove cultivation to Africa, Seyyid Said was largely responsible for the rapid expansion of clove production on the islands of Zanzibar and Pemba. He did so in two ways: he personally owned many plantations (forty-five at the time of his death) and compelled his Omani subjects to plant a certain proportion of clove trees to coconut palms. It is fascinating to note that the Sultan and the other Omani Arabs built up the clove industry on the labour of slaves imported from central Africa. To give some idea of the significance of slavery for the plantation economy of eastern Africa, according to rough estimates, between 15,000 and 360,000 slaves inhabited the island of Zanzibar in the first part of the nineteenth century. Reflecting some typical features of a plantation society, historical sources suggest that the class of slave owners, the Omani Arabs, numbered less than 5,000. When the European powers took over East Africa around 1890, they gradually replaced the Omani-led plantation economy by a European follow-up.[3]

The initiative shown by the British Residents when they undertook the reorganization of the production of export-oriented crops by changing the government policy towards indigenous cultivators and inviting British entrepreneurs to Travancore in the first part of the nineteenth century was not an isolated incident in Asia. On the contrary, most, if not all, colonial powers became actively involved in enhancing the local production of cash crops in that period, though implementing this policy was more drastic in the directly-ruled areas than in the indirectly-governed regions.

Two examples illustrate this very nicely: indigo cultivation in

to my typology of pepper production in eighteenth century Travancore as a 'proto-plantation system', Boomgaard characterized seventeenth century Aceh as one of the ' "proto-capitalist" commercial centres' of Southeast Asia (Boomgaard 1990: 39).

[3] Cooper 1977: Ch. 2; Austen 1987: 71–3; Sheriff 1987: 1–7, 48–69.

north India and coffee, sugar and indigo production in Java. In both cases, the indigenous cultivators produced (export-oriented) cash crops under Western supervision, toiling under harsh and oppressive labour conditions. In north India, in the late eighteenth and early nineteenth century the English East India Company took the initiative in bringing indigo cultivation under British control. The directors of the company brought over experienced planters from the West Indies and provided capital and licences to all of their countrymen who intended to set up plantations in north India, particularly in Bengal and Bihar. The kinds of European plantations can be roughly divided into two groups. One of these was *nij* or *zerat* cultivation, in which the planter produced on land in his possession, supervising the wage labourers. The other was much more prevalent till around 1860 and under it planters contracted and advanced money to tenants *(ryots)* who subsequently cultivated the blue dye (in *ryotti* or *assamiwar* cultivation). The latter arrangement was especially advantageous to the planters. Though firmly in control of the most pertinent factors influencing the cultivation of the crop, the peasants' land and labour, the planters did not run the risk of crop failure, an ever-present threat in this industry because indigo was dependent on good and well-distributed rainfall. The tenants found themselves faced with other problems: the price paid by the planters was mostly totally unremunerative, causing serious indebtedness; indigo had to be grown on very fertile lands which were also suitable for the cultivation of rice, the staple food crop; and the planters resorted to various forms of compulsion and coercion, including debt servitude and outright violence.[4]

Shifting our focus to Java we find that in 1830 Governor-General J. van den Bosch introduced the *Cultuurstelsel* (the Cultivation System), stipulating that indigenous cultivators were required to set aside part of their land and labour for the production of prescribed export crops, particularly coffee (especially on previously uncultivated lands, mostly situated in remote upland regions) and indigo and sugar (both planted on irrigated fields, initially used for rice cultivation). These

[4] Mittal 1978: Ch. 1; Akanda 1980: 2–13; Das Gupta 1992: 174–9; Rao and Rao 1992.

impositions, which went hand in hand with particular malpractices, not to say abuses, by Dutch bureaucrats, traders, and industrialists, as well as indigenous aristocrats, officials and village chiefs, often meant great hardship for the Javanese peasantry. Generally speaking, the payments peasants received were in no way commensurate with the extra time and effort demanded of them, far below the market value of the crops, and in the case of export crops grown on village land, much less than the value of rice which would otherwise have been planted. The list of problems with which the peasants had to deal is long. Among the gravest of these hardships were ones such as the labour cycles of rice and export agriculture not being complementary; crops like indigo and sugar devouring large areas of the best and most valuable rice land, usually for periods up to eighteen months, thereby exhausting the soil; the export crops having first use of irrigation water; the government not compensating for crop failure which was the outcome of experimentation with new crops and new environments; the peasants being obliged to construct public works especially meant to facilitate the transportation of the crops produced; and various powerholders withholding payments and using peasant labour for strictly personal purposes.[5]

The British planters' leading role in the development of a European plantation industry in Travancore after the mid-nineteenth century was paralleled in other Asian regions as well. This is probably the best-documented period in the history of plantation development in Asia (see Section 1.2). European planters opened up thousands of estates, usually in sparsely-populated mountainous forest areas, especially from the mid-nineteenth century. Aided by the colonial governments in their efforts to acquire land, to recruit and discipline labour, and to transport their products abroad, agro-entrepreneurs rapidly opened up plantations, specializing particularly in coffee (Ceylon, Java, Malaya), tea (Assam, Ceylon and Java) and rubber (Indochina, Malaya, Java and Sumatra) as well as in tobacco (Sumatra) and sugar (Fiji, Malaya and Mauritius).[6]

[5] Elson 1990: 26–31; cf. Furnivall 1939: Ch. 5; Allen and Donnithorne 1954: 79–81; Akanda 1980: 17–20; Wolf 1982: 334; Schutte 1983: 301; O'Malley 1990: 138, 151, 156, 158; Wertheim 1990: 5–6; Knight 1992: 68–71.

[6] Furnivall 1939: Ch. 6 and 7; Jackson 1968: part 2; Tinker 1974; Pelzer 1978;

It should be noted, however, that the change in the production of export crops did not come as a bolt from the blue nor did it signify a total break with the past. In the European mother countries evangelical, humanitarian and mercantile interest groups gradually turned hostile to the continuation of forced peasant cultivation in the colonies. The former groups sympathized openly with the cultivators' lot, stimulated in their resolve by literary works such as Dinobondhu Mitra's drama *Nil Darpan* (1860, translated from Bengali into English as *The Indigo Planting Mirror* in 1861) and Eduard Douwes Dekker's novel *Max Havelaar* (1860, published under the pseudonym of *Multatuli*). For their part, businessmen promoted a new economic order, supporting liberal principles such as free trade and *laissez-faire*. They achieved notable results: in Bengal forced indigo cultivation began to decline in the 1860s and in Java the Cultivation System was fundamentally abandoned with the passing of the Agrarian Law of 1870, providing for the expansion of private enterprise in the Netherlands East Indies. Not all in the garden was rosy, however. Despite their supposed *laissez-faire* stance, the colonial powers' active role in institutionalizing unfree labour remained unchanged, though it shifted from *taking the initiative* in creating an unfree *peasantry* (north India and Java) to *supporting the planters' efforts* to restrict the freedom of *wage workers*, especially through the introduction of penal contracts, i.e. the British Indian Workman's Breach of Contract Act (Act XIII of 1859) and Sumatra's Coolie Ordinance (1880).[7]

In south-west India, the European planting community was increasingly threatened by two related developments during the first half of the twentieth century. The planters lost the (hitherto) unlimited support from the (hitherto) powerful representatives of the colonial state, while an emerging indigenous planting segment quickly gained economic and political significance.

The first feature was one which could also be noted in other

Wolf 1982: Ch. 11; Bandarage 1983; Stoler 1985; Breman 1990a; Barlow and Drabble 1990; O'Malley 1990; Behal and Mohapatra 1992; Carter 1992; Das Gupta 1992; Kelly 1992; Murray 1992; Ramaswamy 1992.

[7] Furnivall 1939: Ch. 5–7; Mittal 1978: Ch. 1; Akanda 1980; Schutte 1983: 301; O'Malley 1990: 165.

Asian regions. For example, the above-mentioned Workman's Breach of Contract Act and Coolie Ordinance were gradually dismantled from the early twentieth century. During the same time, colonial governments were held increasingly responsible for the welfare of the plantation workers; political pressure from Europe in particular resulted in the appointment of *ad hoc* labour enquiry committees and the introduction of more permanent labour inspectorates in the different estate regions of Asia.[8]

The second feature – the growing significance of indigenous producers cultivating what had initially been considered typically 'European plantation products' (particularly tea and rubber) – corresponded to similar developments in other Asian regions as well. For instance, in Java smallholders began cultivating tea, a process which had already got underway in the late nineteenth century.[9] Of greater historical importance was the rise of local rubber segments in Malaya and the Netherlands East Indies (particularly in Sumatra and Borneo/Kalimantan) from the early 1900s, gradually constituting serious threats to the European planting communities in terms of the number of holdings, total acreage, and total production. In colonial Malaya both Malays and Chinese were the main participants, whereas in the Netherlands East Indies it was predominantly *pribumi* (indigenous people) who were involved in rubber cultivation, the Chinese being prohibited from owning land under the Agrarian Law of 1870. Again showing parallel developments with south-west India, in the other Asian regions the local rubber cultivators were also disadvantaged by the International Rubber Regulation Agreements of the 1930s and 1940s, which placed them fairly and squarely in the position of underdog to the European planting interests. Though not much research has been done on this particular topic, it would not surprise me at all if the thwarted development of the indigenous rubber segments encouraged anti-Western sentiments and contributed directly to the political successes of the nationalist movements in Asia.[10]

The rise of an indigenous planter class, owning estates and

[8] Furnivall 1939: Ch. 8; Stoler 1985: 37; Behal and Mohapatra 1992: 167.

[9] Doorn 1994: 202; cf. Furnivall 1939: 320; O'Malley 1990: 168.

[10] Furnivall 1939: 322-3; Jackson 1968; Thomas and Panglaykin 1976: 139–49; Pelzer 1978; Wolf 1982: 330; Clemens 1989; Barlow and Drabble 1990.

not smallholdings, was a much more unique phenomenon in colonial times. Perhaps the growing significance of the Syrian Christian entrepreneurs in south-west India from the early 1900s was matched only by a similar development in the Philippines. Here too an indigenous planting elite emerged under colonial rule, culminating in a wealthy, prestigious and influential modern elite. Making use of indigenous capital, from the mid-nineteenth century Filipino planters opened up sugar *haciendas* in Negros Occidental and Central Luzon. They profited from colonialism in two ways: their alliances with successive colonial powers, respectively Spain and the United States of America (after 1898), safeguarded their landed properties, provided them with troops to suppress labour unrest and aided them in many other ways (in terms of capital, technology and markets) while the anti-colonial nationalists exempted their plantations from critical examination because of their Filipino ownership. After independence (1946), except for the years of the Marcos dictatorship (1972–86), the sugar planters remained very powerful, successfully lobbying for trade concessions, financial assistance and even military support for labour control.[11]

The various developments which took place in south-west India after independence were also mirrored in other areas. To start with, the indigenous planting segments continued to gain ground at the expense of their European counterparts in all kinds of fields, notably that of political influence. In view of

[11] McCoy 1992: cf. Rutten 1994. It is true that in the Dooars (north Bengal) the rise of a Bengali-dominated Indian plantation segment, which began in the mid-nineteenth century, was fairly spectacular. For example, there were 163 Indian-owned tea companies in north-east India in 1933, 88 of them controlled by Bengalis operating from the Jalpaiguri district. In contrast to both Syrian Christian planters and Filipino sugar entrepreneurs, however, the development of Bengali entrepreneurship in tea was not an uninterrupted process. From the late 1930s, and especially after Indian independence, ownership and management were gradually transferred to non-Bengalis (Mukherjee 1976: 487–9). The same development can be traced in the Assam tea sector. Here, Assamese were successful but lost their independence to Marwari trading capital in due course (Guha 1977: 59–61). A sociologically interesting feature is that – generally speaking – both the Assamese and the Bengali entrepreneurs had similar backgrounds, being member of the higher castes and of landowning and/or service-holders' and lawyers' families. See besides Mukherjee 1976 and Guha 1977; Das Gupta 1992a: 59–65.

incipient problematical relations with the newly-emerging Asian nation-states, Western companies in these former colonies opted either for other, more congenial places for investment, particularly colonial eastern Africa, or were compelled to stop their planting activities because of the nationalization of plantations, as was the case in Indonesia (1957–8, Dutch estates) and Sri Lanka (1975, foreign estates).[12]

Throughout this period plantation development over the whole of Asia was being strongly influenced by the rise of the labour movement, 'silently' starting in the late colonial period and overtly thriving in the early post-colonial years. In Sumatra there are indications that trade union activism probably started around 1920, finally culminating in what was considered the single largest and most militant workers' union in the Archipelago during the 1950s, functioning under the aegis of the Indonesian labour organization SARBUPRI. Though trade union activists entered the tea districts of Ceylon in the 1920s and 1930s, the growing influence of the plantation labour movement in the island became clearly visible only after the 1950s: the Ceylon Workers' Congress achieved the so-called Thirteen Points Agreement in 1950, the Employees' Provident Fund in 1958, and a first national collective agreement in 1967.[13]

Quite apart from the planters themselves, both the Indonesian and Sri Lankan governments tried to restrain growing trade union support, radicalism and influence. In Sumatra the Indonesian government, supported by regional military commanders, began to intervene in the labour movement in the early 1950s: communist literature was confiscated, leftist activists were arrested, work-stoppages were prohibited and compulsory arbitration was introduced. With the transfer from Dutch-owned to government-run estates, the military commanders acquired influential posts at both estate (as managers) and supra-estate (in industry-wide bodies) levels, greatly facilitating the ability of state authorities to intervene in union activities and labour disputes.[14] When compared with Sumatra, the Sri Lankan trajectory of plantation development shows many similarities. In the

[12] Pelzer 1982: Ch. 7; Stoler 1985: 141, 147; Rote 1986: 104, 127, 227.
[13] Stoler 1985: 7–8, 53–5, Ch. 5; Rote 1986: 79–80.
[14] Stoler 1985: 126, 139, 141–3, 148.

pre-nationalization years the government opposed the formation of a strong and effective labour movement, in this respect aiding the management of the foreign-owned estates. As a direct result, the labour movements' 'victories' of 1950, 1958 and 1967 were above all cosmetic successes: according to many scholars the labour conditions on the plantations did not improve fundamentally in that period; perhaps they even deteriorated. In the years following nationalization, the workers' situation remained a difficult one: the Singhalese-dominated governments realized they had no interests whatsoever in improving the lot of the predominantly Tamil, mostly stateless, estate labourers. There was not much to gain from such measures, not even political support (most workers were not allowed to vote because of their Indian citizenship or statelessness), while the risks run, including a decrease in the profitability of the plantations, were considerable.[15] Indeed, both in Sumatra and Sri Lanka government authorities opposed the development of a powerful, independent labour movement, defending the plantation industry on the grounds of its financial, economic and political importance for the newly-established nations. At the same time, these comparisons suggest that the attitude and policies adopted by the central Indian government towards the rise of the south-west Indian labour movement were not really exceptional.

7.2.2 *Personal Histories*

Though there are many resemblances between plantation development in south-west India and changes in other Asian regions, particularly at a macro-level, historical events also have their unique side, especially when studied at an individual, personal level. I could also phrase this statement differently. Though the lives of all those involved in Asian plantation development were influenced by particular historical trends (like the decline of European planting dominance, the emergence of indigenous planters and/or the rise of labour unions during the twentieth century), all individuals acted and reacted differently, depending on personal features and circumstances such as character, family background, wealth, prestige, power and health

[15] Rote 1986: 80–2, 202–6, 223, 228, 281.

(thereby again influencing broader historical processes). I will illustrate this statement by following the lives of a European planter, a Syrian Christian agro-entrepreneur, and a Tamil worker, all revolving around plantation agriculture in twentieth century south-west India. As their personal histories will show, their lives were both so typical yet so unique.

The life of W.S.S. Mackay was strongly influenced by the gradual fall of the British planter class in south-west India. He witnessed the rise of an Indian planting segment ('they were proving themselves to be the most successful planters') as well as the growing influence of the labour union movement ('the days of milk and honey are over'). Intriguingly enough, Mackay's career was the exact opposite of the path of British political and economic power during the twentieth century. In short, Walter Smith Sutherland Mackay was born in Scotland in 1904 during the heyday of the British Empire; became an assistant manager in the Kannan Devan Hills in 1924, opening up the last great valley in that region for tea cultivation; was appointed manager in 1932, fearing the possible implications of the rising nationalist and leftist movements; and witnessed Indian independence, the gradual Indianization of the management, and the rapid unionization of the estate workers, as the assistant general manager of the Kannan Devan Hills Produce Company (1946–57). He later continued his career as a controller-cum-advisor in the estate regions of Kenya, Uganda, Tanganyika and Rhodesia. As interesting as Mackay public life was his personal life, including his inner, deeply personal feelings and convictions. Very much at ease in Indian society, perhaps even falling in love with India at large, Mackay married an Eurasian, invited Indian friends to his house, and even became a Hindu. Having started to read the *Veda* in 1936, he became a convinced believer and remained so until his death, stating 'I believe in reincarnation and in the continuity of the human soul' in a last taped interview in the mid-1980s. His choice was almost indubitably not without repercussions: Mackay's 'Indianness' was against the company's policy and perhaps the reason for his non-appointment as general manager of the Kannan Devan Hills Produce Company during the 1950s.[16]

[16] Mackay 1984.

Jose A. Kallivayalil's biography is at least as fascinating. Being the first-born son of (Chacko) Abraham Kallivayalil, the prominent Indian rubber pioneer, Jose Kallivayalil (born 1913) gradually emerged as *the* spokesman of the Syrian Christian planters during the late colonial and early post-Independence period. To sum up his identity in short, he was an enterprising, Syrian Christian planter as well as a convinced nationalist, Keralite, and anti-communist.

Examing this first feature in more detail we see that he opened up estates in Malabar, at the same time leading some fellow Syrian Christian agriculturalists in their combined struggle against certain Muslim wood traders: the Syrian Christians successfully defended their claim that they had legally obtained both the right to cultivate the land and the right to cut down trees on their possession. There were many other highlights in his planting career: Jose became the Chairman of Darragh Smail and Company, initially the British coastal firm *par excellence*, and of Ruby Rubber Works, the forerunner of the large industrial enterprise Apollo Tyres.

His nationalist ideas found expression at a much earlier age. Attending a predominantly British school in Bangalore, he witnessed the beating up of an Indian youngster by British and Anglo-Indian pupils because the Indian was wearing *khadi*. In direct defiance, Jose A. Kallivayalil successfully convinced all his Indian schoolmates to wear khadi the next day, thereby showing their support to their friend and the nationalist cause. Another anecdote also illustrates his nationalist pride. Opposing the initially strict management regulations regarding dress, it is said that Jose A. Kallivayalil was the first Indian to wear khadi in the prestigious Taj Mahal Hotel in Bombay, which, by the way, also implicitly demonstrates Jose's enormous wealth and influence. There is ample evidence of his political prestige: Jose A. Kallivayalil became the first Indian member of the prestigious Trivandrum Club; Jose became good friends with Jawaharlal Nehru, Indira Gandhi and Lord Mountbatten, the Viceroy even visiting Jose's residence on his last tour through India; Jose aided the *Malayala Manorama* following Dewan C.P. Ramaswami Aiyer's official ban on the newspaper by providing financial assistance and helping to set up new distribution networks after the paper had secretly moved to Cochin; and Jose became the treasurer of the Congress-led organization behind the *Aikya Kerala* Movement, trying to unify the Malayali-inhabited regions of Travancore, Cochin and Malabar into one separate political unit.

Perhaps most telling of all are his anti-communist activities. To start with, Jose A. Kallivayalil himself organized a Congress-controlled labour union in the Mundakayam area, counteracting the communist activists and forestalling unity within the labour movement. During the Namboodiripad government, the planter-cum-political activist wholeheartedly supported the Liberation Movement (1959), organizing the picketing in front of government offices and making trips to Delhi to inform Jawaharlal Nehru and Indira Gandhi about the purposes of the movement. To complete the picture, his lifestyle sometimes gave him problems: once, just before the communists came to power in 1957, his house was attacked by leftist activists. His accomplishments, however, were much more telling, including his earlier-indicated installation as the first Indian President of the Mundakayam Planters' Association in 1954. Jose A. Kallivayalil died in the early 1990s.[17]

Joseph is a bilingual estate worker in his mid-thirties. His family's story shows yet another aspect of plantation development in south-west India and is most interesting in the context of the above discussion about free and unfree labour. His account discloses that labour, despite the abolition of the kangany system in 1962, did not become totally free.

Joseph's grandfather came from Tinnevelly District in the Madras Presidency as a poor landless labourer and was recruited by a kangany to work on Semini Valley Estate in the Peermade/Vandiperiyar region. As time passed, he became a kangany himself, mobilizing workers each time the tea plucking season began.

The life of Joseph's father was completely different. He grew up bilingual, speaking the language of his parents (Tamil) and that of the other, Malayali, workers fluently. In addition, Joseph's father settled permanently in the plantation district, working on another estate (though in the same region and owned by the same company) as a supervisor (who did not

[17] Interviews: Michael A. Kallivayalil; P.K. Michael Tharakan.

recruit new workers). He was converted from Hinduism to Christianity by CSI missionaries[18] and became active in the UTUC/RSP labour union during the early 1950s, which was instrumental in enhancing the workers self-respect and improving their working conditions, particularly from the early 1960s.

Joseph does not see much of a future for himself and his family. Factory work, pruning tea plants, spraying chemicals and assisting the managers' family, particularly in cooking and gardening, does not satisfy him much nor are these tasks well-paid. He feels 'stuck': without any formal education since his thirteenth year he cannot find any other, more promising job elsewhere. He is also hemmed in because his social networks remain confined to the plantation areas and his low-caste status does not help him much either in finding outside employment. Should he indeed seek to earn a livelihood outside the estate he would lose his house, which is company-owned, as well as other regulatory benefits. There is no doubt that the segmented labour-market and lack of geographical mobility severely restrict his freedom to choose another employer. Moreover, his current employer also ascribes a great deal of value to educational accomplishments and family/communal background, thereby making any upward mobility within the estate hierarchy practically impossible. For all this, he is not totally desperate: he and his wife try to save some money for their daughters' education, praying that their children may find employment, preferably as nurses, outside the plantation regions.[19]

[18] CSI refers to the Church of South India. The organization includes churches established by Baptists, Methodists, the LMS, the CMS and Syrian Christian religious leaders, all of which merged in 1950 (cf. Lewandowski 1980: 139).
[19] Interview: Joseph.

Appendices

Appendix 1

Monarchs of Travancore

Ruler	*Reign*
Marthanda Varma	1729–1758
Rama Varma	1758–1798
Balarama Varma	1798–1810
Rani Gouri Lakshmi Bai	1810–1815
Rani Gouri Parvati Bai	1815–1829
Swati Tirunal Varma	1829–1847
Utram Tirunal Marthanda Varma	1847–1860
Ayilyam Tirunal Varma	1860–1880
Rama Varma Visakham Tirunal	1880–1885
Sri Mulam Tirunal Rama Varma	1885–1924
Setu Lakshmi Bai Maha Rani Regent	1924–1931
Sri Chitra Tirunal Balarama Varma	1931–1949

Source: Menon 1991

Appendix 2

British Residents (since 1800); Agents to the Governor General, Madras States (after October 1923); and Residents for the Madras States (after April 1937)

British Representatives	*Commencement of Term of Office*
C. Macaulay	1800
J. Munro	March 1810
S. McDouall	April 1819
D. Newall	February 1821
W. Morison	March 1827
E. Cadogan	December 1829
J.A. Casamajor	April 1834
J.S. Fraser	January 1836
A. Douglas	August 1838
T. Maclean	October 1838
W. Cullen	September 1840
F.N. Maltby	January 1860
W. Fisher	May 1862
H. Newill	April 1864
A. MacGregor	February 1867
H. Newill	May 1867
G.A. Ballard	February 1869
J.I. Minchin	March 1870
G.A. Ballard	June 1871
A.F.F. Bloomfield	July 1874
G.A. Ballard	October 1874
W. Hay	April 1875
A. MacGregor	October 1875
H.E. Sullivan	March 1877
J.C. Hannyngton	February 1878
A. MacGregor	March 1879
J.C. Hannyngton	April 1881
W. Logan	May 1883
R.W. Barlow	February 1884
J.C. Hannyngton	August 1884
H.N.D. Prendergast	July 1887
J.C. Hannyngton	October 1887
H.B. Grigg	June 1890

British Representatives	*Commencement of Term of Office*
J.C. Hannyngton	October 1890
H.B. Grigg	July 1891
J.C. Hannyngton	November 1891
H.B. Grigg	November 1892
J.D. Rees	April 1895
J. Thomson	July 1895
J.D. Rees	August 1896
F.A. Nicholson	December 1896
J.D. Rees	July 1897
F.A. Nicholson	August 1898
G.T. Mackenzie	July 1899
James Andrew	November 1904
R.C.C. Carr	March 1906
L. Davidson	May 1908
R.C.C. Carr	February 1909
A.T. Forbes	October 1910
R.A. Graham	February 1912
A.T. Forbes	March 1913
R.A. Graham	March 1915
A.R. Cumming	November 1916
H.L. Braidwood	February 1917
H.H. Burkitt	December 1920
C.W.E. Cotton	June 1923
H.A.B. Vernon	May 1926
C.W.E. Cotton	November 1926
C.G. Crothwaite	April 1928
A.N. Ley Cater	December 1929
H.R.N. Pritchard	October 1930
D.M. Field	November 1932
W.A.M. Garstin	February 1935
C.P. Skrine	November 1936
G.P. Murphy	1940
H.J. Todd	1944
C.G.N. Edwards	1945
A.A. Russell	1946
C.G.N. Edwards	1947

Source: TAD

Appendix 3

Dewans of Travancore

Dewan	*Commencement of Term of Office*
Velu Tampi	1800
Ommini Tampi	1808
Devan Padmanabhan	1814
Shungoo Annavi	1815
Raman Menon	1816
Vencutrao	1817
Vencata Rao	1822
Soobrow	1830
R. Runga Rao	1837
R.R.R. Venkata Rao	1838
Soobrow	1839
V. Krishna Rao	1842
Vencutrow	1843
Sreenivasa Rao	1845
V. Krishna Rao	1847
T. Madhava Rao	1858
A. Seshiah Sastri	1872
N. Nanoo Pillai	1877
V. Ramiengar	1880
T. Rama Rao	1887
S. Shungarasoobyer	1892
K. Krishnaswami Rao	April 1898
V.P. Madhava Rao	March 1904
S. Gopalachariyar	August 1906
P. Rajagopalachari	October 1907
M. Krishnan Nair	May 1914
T. Raghaviah	July 1920
M.E. Watts	June 1925
V.S. Subrahmanya Aiyer	June 1929
T. Austin	February 1932
M. Habibullah	May 1934
C.P. Ramaswami Aiyer	September 1936

Source: TAD

Appendix 4

Information Relating to the Public Auctions: Notice Number, Lot Number, Taluk, Estimated Area, Area as Surveyed by the Survey Department, and Name of the Indigenous Buyer or Owner

No.	*No. Taluk*	*Estimated Area (in acres)*	*Area as Surveyed by the Survey Department (in acres)*	*Name of the Indigenous Buyer or Owner*
10	1 Calculum	70	96.66	V. Vadanayayam
	2 Calculum	90	121.82	V. Vadanayayam
	4 Calculum	45	51.30	C. Luke
	6 Calculum	6	12.88	M. Joseph
	7 Calculum	50	67.18	N. Paramanandem
	11 Calculum	10	23.37	A.R. Benjamin
11	2 Calculum	65	76.14	R. Arumanayayam
	3 Calculum	20	22.68	S. Swammather
	4 Calculum	40	48.14	S. Swammather
	5 Calculum	12	20.95	C. Luke
	6 Calculum	30	29.50	T. Syedpeer
	9 Calculum	20	22.20	Swamiadian
	12 Calculum	20	28.48	N. Poothayoonathan
	14 Calculum	60	54.52	J.R. Moncur
	15 Calculum	20	21.81	P. Nagandran
	16 Calculum	60	67.95	Poobramanien
	26 ?	30	35	S. Emanuel
	40 Kotaracaray	500	-	R. Row
12	3 Calculum	25	48.28	Bagavathi
	5 Calculum	8	6.13	Nagrandran
	6 Calculum	25	35.35	A.R. Benjamin
	8 Calculum	45	48.42	M. Pillai
	9 Calculum	30	45.88	Paramanandem
	10 Calculum	10	14.88	P. Noah
	11 Calculum	10	15.35	V. Samuel
	12 Calculum	15	25.88	L. Granaprecavam

No.	*No. Taluk*	*Estimated Area (in acres)*	*Area as Surveyed by the Survey Department (in acres)*	*Name of the Indigenous Buyer or Owner*
	17 Calculum	15	19.95	A.R. Benjamin
	19 Calculum	10	11.10	M. Pillai
	20 Calculum	20	33.58	V. Swamiadian
	21 Calculum	20	35.21	V. Swamiadian
	22 Calculum	20	26.35	N. Moodakai
	23 Calculum	20	18.65	V. Arunachellom
	24 Calculum	20	20.30	Michel
	30 Velavencode	25	63.45	J.G. David
	31 Velavencode	7	–	J.G. David
	32 Velavencode	15	–	Daniel et al.
	33 Velavencode	5	5.84	Aurulanendem
	34 Velavencode	5	5.28	P. Daniel
	35 Velavencode	10	16.12	P. Arumougam
	36 Velavencode	7	7.35	P. Moses
	37 Velavencode	5	5.40	P. Moses
	38 Velavencode	3	3.10	P. Moses
	39 Velavencode	20	20.75	Ynanaprecasum
	40 Velavencode	5	6.65	Ynanaprecasum
	41 Velavencode	5	9.35	J. Killioor
	46 Needoovengad	35	52.25	M. Pillai
	53 Needoovengad	200	237.80	H.H. Sait
	58 Kotaracaray	300	345.45	J. Row
	64 Kotaracaray	50	53.90	T. Pillay
	76 Meenachel	70	97	Eappen
	87 Meenachel	60	–	Eappen

Notice number 12 announces the auction of 14 October 1874. Most probably notice numbers 10 and 11 are related to auctions organised in 1872 and 1873 (compare Table 4.21).

Source: CSF No. 1925

Appendix 5

Indigenous Proprietors or Lessees of Estates in South Travancore in 1877

Name of the Estate	*Proprietors or Lessees*	*Total Acreage*	*Cultivated Acreage*
Ananden	Madhava Rao	210	107
Annagherry	J.G. Jacob	50	40
Assamboo	First Prince of Travancore and Madhava Rao	300	200
Crownhill South	A.M. Joseph	65	25
Crownhill Central	A. Dewsagaim	65	25
Crownhill East West	P.P. Joseph and C. Isaac	130	50
Cudamapallam	Meeranjee Meah Saib and Sytheen Meah Saib	329	329
Denasi	N. Paramanandam	40	40
Denis	P. Paul	8	8
Eden	A.R. Benjamin	40	25
Goodhope	A. Joseph and Others	24	24
Honey Rock	M. Luke and Others	52	42
Joaja	P.D. Devasahayam	40	40
Kaligasam	A. Abraham and Chetty	42	42
Love Grove	N. Paramanandam	50	50
Love Hill	P. Paul	20	20
Lower Providence	A.R. Benjamin	20	20
Lower Victoria	P.D. Devasahayam	10	10
Moss Wella	A. Moses and A. Welcome	50	35
Mount Pleasant	Mrs Moses	30	30
Newport	A.R. Benjamin and Others	60	35
Nilloomallay	C. Moses and Others	50	35
North Hope	A.R. Nathaniel	40	40

Name of the Estate	*Proprietors or Lessees*	*Total Acreage*	*Cultivated Acreage*
Prasandan	Prasanthan Company	220	40
Providence	P.P. Joseph and Others	125	100
Rockhill	N. Peter and Others	100	65
Rosehill	N. Paramanandam	40	40
Sagayavalley	J. Dewsagaim	55	35
Seevacanda	Poothoor Pillay	150	90
Thylemallay	P.D. Packianathan and Others	80	70
Tugramallay	John Rose	65	45
Victoria	P.D. Devasahayam, P.P. Joseph and C. Isaac	52	52
Wishworth	S. Daniel	35	25

Source: Ferguson 1879: 159-62.

APPENDIX 6

Indigenous Proprietors of Privately-Owned Rubber Estates in Travancore in 1925, 1934 and 1948 (property in acres)

Community	*Year*	*Proprietors*
Syrian	1925	[Mundakayam:] Cherian (20); Curien (150); Devassier (60 + 40); Ittiavira (121); Ittyerah (63); Kunjackȯ (75); Kunjoovarkey (200); Kunjoovarkey (48); Kurian (116); Kuruvilla (100); Mathai (89); Mathai Thomas (30); Maliakel (420); Michael Thomas (45); Thomman (191); Thommi (60); Verkey (30)
	1934	[Mundakayam:] Auseph Auseph (100); Chandy (138); Chandy (85 + 80); Cherian (75); Chacko (40); Devasia (150); Eapen Ittycheria (84); Joseph Thommen (28); Joseph Ittyavirah (150); Joseph Mathew (20); Korutha (10); Koshy (250); Kunjacko Kadamapuzha (148); Kunjacko Kunjacko (200); Kunjommen Thommen (25); Kunjuvarkey (90); Kunjuvarkey (550); Kurian (105); Kurien (150); Kurien (200); Kuruvilla (275); Mrs Kuruvilla Mariam (10); Kuruvilla (120); Mathai (220); Mathai (61); Sakaria (15); Thomas Eapen (100); Thomas, Punnen Kurian and Jacob Thomas (480); Thomas and Ittyera (92); Thomas Jacob (110 + 52 + 40); Thomas Mathai (65); Thomas Michael (25 + 40); Thomas Palampadam (175 + 103); Thommen Dominic (119 + 75); Thommen Dominic and Dominic Joseph (30); Thommey (240 + 60); Thommi Kunjacko (50); Varghese Teressamme (20); Varkey (110 + 40); [north Travancore:] Anthony Pappoo (12); Chacko and Chandy (140); Chacko Pallathu Lookka (10); Kuriakose Pothamko (12); Kurien Kora (81); Mathai George (100); Mathai (65); Mathu Ulahannan Kandirikal (40 + 23); Ouseph Kuripe (25); Pappan Chempakathinkal (60); Poulose (60); Tharakan (100); Thomman (84); Varghese Iype (70); Varkey (25 + 14); [south Travancore:]

Abraham Uthuppan (60); Alexander (50); Anthraper (200 + 30 + 150); Chacko Chandy (12); Chacko Kochukura (200); Chacko Mathulla (56); Chacko (44); Chandy Uthuppu (35); Cheriyan (18); Easo Easo (50); Geevarghese (20); George Kochukudy (70 + 27); Jacob Poonen (80); Joseph Thommen (72); Kuruvilla Pothen (35); Mathews John (50); Matthen (174); Matthen (200); Oonnithan Kuruvilla (20); Pothen Joseph and Thomma Oolakannen (125); Thomas and Chandy (150); Thomas Kochu Koraka (45); Thomma Ulahannan (25); Thommen Zacheriah (30); Uthuppan (65); Varkey Yohanan (20); Varughese Mathai (74); Varughese, Oommen and Abraham (180)

1948 Abraham Cecilia (53); Abraham Kalthoor (26); Abraham Thomas (27); Abraham Varkey (28); Annamma Varughese (30); Augustin Kandathil (42); Augustin Ponnathu (50); Chacko (86); Chacko (44); Chacko (351); Chacko (55); Chacko (22); Chacko (36); Chacko (45); Chandy (20 + 35); Chandy Verghese (35); Cherian (39); Cherian Koshy (198); Devasia Chacko (40); Devasia (27); Devasia (52); Easow (29); Eipes (105); Eipe Georges (50); George and Mrs Varkey (42); George Joseph (35); George (65); George Kochukudy (50); Idiculla (49); Ipe Itoop (38); Ittiavira (36); Ittiavira Joseph (166); Ittiavira (54); Rev. Jacob Chettoor (33); Rev. James Kalacherry (30); Joseph Chandy (38); Joseph Ittiyavirah (425); Joseph Ittiyavirah (158); Joseph Thommen (100); Joseph Mathai (28); Joseph Joseph (35); Joseph George (48); Joseph Panjikaran (87); Mrs Joseph Paulo (80); Joseph Mathew (47); Joseph Teresia (239); Kattakayam Kurian (120); Kochuthuppan (37); Korah and George (28); Korah (141); Koshy (132); Kungithommen Thariath (153); Kunjacko (200); Kunjannamma Chacko (37); Kunju Varkey (93); Kuriaippu (26); Kuriakose (147); Kurian (86); Kurian (41); Kurian Thomas (150); Kurian Onittan (80); Kuriape Ouseph (45); Kuriatthu Chakku

		(25); Kurien (97); Kurien Thommen (40); Kurien Davasia (56); Kurien (40); Rev. Kuruvilla (135); Kuruvilla (106); Kuruvilla Thommen (50); Kuruvilla Itty (43); Kuruvilla Joseph (35); Kuryan (15); Mammen Mappillai (50); Mani Ouseph (32 + 25); Mani Ouseph (45); Mariamma Oonnithen (25); Rev. Mar Theodosius (40); Mathai (37); Mathai (236); Mathai (80); Mathen Kuriappe (38); Mathen Ouseph (30); Matheri Amma (25); Mathew Cherian (39); Mathew Dominic (60); Mathew Joseph (64); Mrs Mathew Rosamma (30); Mathews (28); Mrs Mattom (75); Michael Ittiyavirah (120); Michael Thomas (32 + 10 + 15); Nidiri (70); Ninan Kunjamma (29); Ommen John (43); Oommen George (110); Oommen (33); Ouseph Chacko (44); Ouseph Mapilla and Ulahannan Mapilla (27); Ouseph Ouseph (105); Ouseph Poonnolly (28); Pallivathukal (175); Paulose Annamma (99); Paulose Varkey (25); Philip Cheriyan (185); Philipose (27); Pilo (28); Pothen Itty (33); Punnen (30 + 16); Punnose Joseph (55); Scaria Gnavallil Kurien (38); Scaria (39); Tharakan Hormose (220); Thomas and Mathai (87); Thomas George (110 + 325); Thomas Jacob (38); Thomas Kunjacko (59); Thommen (77); Thommen Dominic (185); Thommen (204); Ulahannan Thoma (46); Uthup Gnavallil Kurien (42); Uthup Varkey (161); Rev. Valliarampeth (29); Varghese Oommen (36 + 8); Varkey Joseph (64); Varkey (309); Varukey (388); Varkey (219); Varkey (141); Varkey Mathoo (30); Varkey (65 + 43); Varughese (59); Varghese (28); Varughese (37); Varughese Rao Bahadur (33); Varughese and George Philip (42); Varughese (33); Varughese Thomas (80); Varughese Varkey (30); Verghese (44); Verghese (50); Verghese (110); Zacheriah Thommen (30)
Chr.	1925	[Mundakayam:] Abraham (98); Mrs Joseph (20); Thomas (15); Thomas (676); [north Travancore:] Luiz (800); Zackariah (60); Zackariah (70); [south Travancore:] Abraham (100)

	1934	[Mundakayam:] Abraham (120); Dominic (20); Joseph (17); Joseph (110); Mathew (25); Nicholas (10 + 25 + 20); Thomas (40); Thomas (150); [north Travancore:] Abraham (55); Joseph (60); Luiz (476 + 300); Zacheriah (177 + 70); Zacheriah (70); [south Travancore:] Abraham (20); Jacob (101); Jacob (41); Mathew (10); Noel (50)
	1948	Abraham (31); Abraham (32); Abraham (413); Abraham (26); Abraham (50); Abraham (30); Abraham (435); Abraham (401 + 221); Abraham (40); Anthony Ammal (25); Daniel (152); Fernandez (78); Isaac (20); Jacob (44); Jacob (71); Mrs Jacob (114); Jacob Perumal (43); Joseph (43); Joseph (48); Joseph (55); Joseph (217); Joseph (210); Joseph (331); Joseph (26); Luiz (313); Mathew (151); Mathew (25); Mathew (58); Mathew (35); Mathew (47); Mathew (175); Sebastian (51); Thomas (44); Thomas (40); Thomas (97); Thomas (68); Thomas (33); Zacheriah (100); Zacheriah (152)
Brahmin	1925	[south Travancore:] Ananda Row (130)
	1934	[Mundakayam:] Krishna Iyer (225)
	1948	Amrithanatha Aiyer (67); Ananda Row (100); Balasubramaniam (47); Krishna Iyer (179); Kumaraswamy (200); Rajankam Aiyer (54); Subramania Iyer (28); Venkatachala Sarma (102 + 42 + 31)
Nayar	1925	[south Travancore:] Oonnittan (165)
	1934	[Mundakayam:] Nilakandha Pillay (105); Ramakrishna Paniker (25); [south Travancore:] Ooonnithan (90)
	1948	Gopala Panicker (80 + 25); Gopalan Oonnithan (35); Karunakara Nair (80); Kesavan Unnithan (36); Kochkunju Pillai (34 + 46); Krishna Pillai (26); Narayana Panicker (41); Narayana Pillai (31); Narayana Pillai (65); Parameswaran Pillai (26); Parameswaran Pillai (58); Parameswaran Pillai (100); Parameswaran Pillai (69); Parameswaran Pillai (42); Ramakrishna Panicker (35); Ramakrishna Pillay (27); Raman Nair (43); Sankaran Pandalai (31)

Muslim	1934	[Kannan Devan Hills:] Syedamean Sahib (2); [Mundakayam:] Mirannan Syed Mohamad (51); Nagoormeeran Rowther (25)
	1948	Abdul Kadir Musaliar (113); Ebrahim Ali (29); Haja Mohideen Rowther (35); Hameed Sayed Mohamed Rowther (25); Hassan Pilla Rowther (102); Koyan Mackar Pillai (36); Mohamed Jaffar Hassan Sait (74); Mohamed Khan (72); Mohamed Meeran Rowther (43); Moideen Kannu Rowther (50); Muhamed Rowther (88); Peer Mohamed Khan (25); Sait Mohamed Rowther (40); Saith Mohamed Rowther (53); Saith Mohamed Rowther (28); Syed Mohamed Khan (40); Syed Mohamed Rowther (25); Thangal Kunju Musaliar (220); Usamia Khan (25)
O,U,U	1934	[Mundakayam:] Raman (120); [north Travancore:] Puravathu (100); [south Travancore:] Perumal Pillai (110)
	1948	Chellamma (33); Kukillya (200); Mahadevan Pillai (52); Pillai (28); Nicholvose (37); Perumal Pillai (96); P.V.K. (290 + 125); Rabindranath (60); Raman (200 + 118); Sankaran (9); Selvanayakam (30); Subramanian Chettiar (450); Valli Ammal (48)

Syrian = Syrian Christian; Chr. = Christian (which can either be a Syrian Christian or a 'new', low-caste Christian); O,U,U = Others, Uncertain, Unknown.

Source: *UPASI Planting Directory* 1925: 88–114; *UPASI Planting Directory* 1934: 126–76; *UPASI Planting Directory* 1948: 107–42 and 144.

APPENDIX 7

Indigenous Proprietors of Privately-Owned Tea Estates in Travancore in 1925, 1934 and 1948 (property in acres)

Community	*Year*	*Proprietors*
Syrian	1925	[central Travancore:] Chacko Pothan (123); Cherian (45); P.J. John (316); P. John (320)
	1934	[central Travancore:] Chacko Pothan (138); Cochumman (293); Kunchira (102); Mathoo (140); John Zachariah (51); [Mundakayam:] Devasia (50); Kunjuvarkey (80); Kurien (149); Mrs Kuruvilla Mariam (15); Mathai (120); Sakaria (10); Thomas Michael (15 + 10); Thomas Pathrose (14); Thommen, Dominic and Dominic Joseph (106); Varkey (60); [north Travancore:] Kuriakose Pothamkot (2)); [south Travancore:] Anthraper (50 + 25); Chacko Kochukura Mundakottakal (30); Chandy Uthuppu (30); Kuriyan (71 + 61); John Mathews (10); Matthen (83 + 115); Joseph Pothen and Thomma Oolakannen (25); Thomas Kayalachakam (50); Thomas Kochu Koraha (2); Thommen Zachariah (120); Varughese, Oommen and Abraham (100)
	1948	Ajoo John Anthony (20 + 4); Davasia (27); Davasia Eapen (15); John Kuruvilla and Kuruvilla David (14); John Kurian (255); Joseph Dominic (106); Joseph Ittiyavirah (80); Kayalachakom Thomas (50); Kochummen (281); Korah (7); Kurian (29); Kurian Kurian (16); Kuryan (72); Mathai (120); Michael Thomas (10 + 15); Ninan (167); Ommen George (100); Philip Cheriyan (14); Pothen (136); Thomas Kunjacko (41); Ulahannan Thoma (16); Varkey (13); Varkey (25); Varkey (278); Varkey (185); Zacheriah Thommen (150)
Chr.	1925	[Mundakayam:] Joseph (200)
	1934	[central Travancore:] Fernandez (25); Thomas (170); Thomas (140); [Mundakayam:] Joseph (2); Joseph (100); Mathew (16); Thomas (25); Thomas (159); Zachariah (128 + 249 + 72); [south Travancore:] Joseph (25)

	1948	Abraham (5); Abraham (87); Abraham (111); Abraham (75 + 174); Dominic (10); Fernandez (83); Fernandez (19); Joseph (14); Mathew (6); Mathew (18); Mathew (15); Mathew (150); Philip (150); Thomas (49); Thomas (32); Thomas (15); Thomas (224); Thomas (66); Thomas (112); Zacheriah (7); Zacheriah (227)
Brahmin	1925	[central Travancore:] Krishna Aiyer (240)
	1934	[central Travancore:] Krishna Iyer (419 + 209 + 161); Nilakanta Iyer (400); Subramoney Aiyer (305); [south Travancore:] Amrithanatha Iyer (50)
	1948	Amrithanatha Aiyer (36); Krishna Iyer (52 + 52 + 430); Krishna Iyer (628); Raman Nampoothiri (29); Subramoney Iyer (209)
Nayar	1948	Raman Nair (9)
Muslim	1934	[central Travancore:] Hussain Rowther (107)
	1948	Haja Mohideen Rowther (322); Hassan Rowther (150)
O,U,U,	1925	[south Travancore:] Devasahayam (48)
	1934	[central Travancore:] Chidambaram Pillai (95); [Kannan Devan Hills:] Ramanathan Chettiar (210); [Mundakayam:] Raman (243); [south Travancore:] Mutunayagam Rao Bahadur (79); Perumal Pillai (65); [south Travancore:] Devasahayam (73)
	1948	Krishnaswami Reddiar (142); Mutunayagam Rao Bahadur (50); Perumal Pillai (155); Raman (244); Ramanathan Chettiar (150)

Syrian = Syrian Christian; Chr. = Christian (which can either be a Syrian Christian or a 'new', low-caste Christian); O,U,U = Others, Uncertain, Unknown.

Source: *UPASI Planting Directory* 1925: 88–114; *UPASI Planting Directory* 1934: 126–76; *UPASI Planting Directory* 1948: 107–42 and 144.

Appendix 8

Indigenous Proprietors of Privately-Owned Rubber and Tea Estates in Kerala in 1960 (property in hectares)

Community	*Crop*	*Proprietors*
Syrian	rubber	[central Travancore and Mundakayam:] Abraham Kallivayalil (46); Mrs Annamma Thomas (42); Chacko Kallivayalil (82); Chacko (46); Chacko (60); Chacko Kunjuvarkey (60); Cyriac (49); Dominic Joseph (66); Eapen (42); Jose Kallivayalil (84); Joseph Chacko (42); Joseph Kallivayalil (69); Joseph Pottenkolam (101); Joseph Thomas (79); Karimpanal (98); Mrs Kochuponnamma (42); Kurian (50 + 43 + 45); Kuruvilla (66 + 75); Mathew Joseph (63); Mathew Ouseph (108); Michael Kallivayalil (54); Pothen (163); Tharakan George (154); Therasiamma Thomas (60); Therasiamma Mathew (45); Mrs Therasiamma Thomas (55); Thomas Chacko Kayalakkam (311); Thomas George and Kuruvilla Thomas (41); Thomas George (121 + 45); Uthup (47); Varghese (64); Varkey Korah (45); Varkey Itten (42); [Cochin:] Joseph Syriac (45); Kochuvareed (520 + 70); [Malabar:] Alikutty George (80); Chacko and Abraham (113); Cherian and others (729); George Kallivayalil (69); Jose Kallivayalil (78); Kurian Uthup (45); Kuruvilla (61); Mathai Eapen (51); Pothen Joseph (171); Thomas Cherian (61); Varkey Devasia (49); [Malabar-Wynaad:] Joseph Jacob (67); Joseph Michael and others (44); Kunhathumma (81); [south-Kerala:] Mrs Annamma Eipe (42); Claramma Anthraper (43); Joseph Panjikaran and Mrs Rosamma Joseph (49); Karimpanal (219); Kochu Thomman Thomas (49); Kunhathumma (86 + 142); Kurian (45); Kuruvanna Kunhavara (58); Mathews John (76); Ninan (71); Oommen (113); Rev. Mar Theodesius (42); Thommen Ittiavirah (61); Varghese and others (41); Varughese (71)

	tea	[central Travancore and Mundakayam:] George Philip, David and others (65); Joseph Ittyavirah (73); Michael Kallivayalil (75 + 176); Mrs Therasiamma Thomas (60); [south-Kerala:] Karimpanal (71); Ninan (68 + 157 + 130)
Chr.	rubber	[central Travancore and Mundakayam:] Abraham (98); Abraham (71); Abraham (43 + 45); John (53); Joseph (210); Joseph (88); Joseph (69); Luiz (129); Rev. Mathew (41); Mathew (92); Peter (42); Philip (81); Thomas (54); Thomas (57); Zacheriah (42); [Cochin:] Joseph (70); Joseph (61); [Malabar:] Abraham (121); D'Cruz (200); Jacob (62); Jacob (405); Joseph (76); Mary Samuel (142); Philip (47); Sebastian (41); [Malabar-Wynaad:] Jacob (53); Joseph (75); Joseph (45); Manuel (75); [south-Kerala:] Abraham (84); Philip (75)
	tea	[central Travancore and Mundakayam:] Abraham (50); Mathew and Joseph (114); Mathew (69); Philip (120); Mrs Salome (65); Thomas (117); Thomas (84); Thomas (42); Zacheriah (142); [south-Kerala:] Abraham (101); Francis Fernandez (46); Zacheria (61)
Brahmin	rubber	[Cochin:] Anatha Raman (59); [south-Kerala:] Ananda Rau (53); Balasubramanier (45)
	tea	[central Travancore and Mundakayam:] Krishna Iyer (187 + 96)
Nayar	rubber	[Cochin:] Appukuttan Nair (49); [south-Kerala:] Gopala Paniker (42); Parameswaran Pillai (45)
Muslim	rubber	[central Travancore and Mundakayam:] Hassan Pillai Rowther (41); [Malabar:] Abdul Khader Musaliar (121); Ahmed Kutty Hajee Koyappathody (121); Hamsa (80); Imbichi Moideen, Haji and others (91); Mohithu (229); Sayed Mohamed Rowther (142); [Malabar-Wynaad:] Amina Umma (53); Kader Kutty Saheb (81); Moideen Kutti Haji (61); Muhammad Kutty and others (188); [south-Kerala:] Akbar Hyder Musaliar and others (100); Mohamed Khan Bava Sahib (59)
	tea	[central Travancore and Mundakayam:] Hassan

		Rowther (102); [south-Kerala:] Haja Mohideen Rowther (130)
O,U,U	rubber	[central Travancore and Mundakayam:] Sankaran (76); [Cochin:] Hari (44); Krishnan (41); [Malabar:] Ammukutty Nethiar (47); Ashiyara Green (121); Chandukutty (71); Kalladi Kumappa (101); Kamappa and Ramu Chettiar (101); Malia Kunhunni Raja (61); Shivakantappa (45); [Malabar-Wynaad:] Kunhambu (65); Kunhi Komman Natakal (105); [south-Kerala:] Alagappa Chettiar (115); Alagappa Chettiar (94); Chellamma (52); Kasi Chettiar (121); Padmanabha Kukillaya and Lakshminarayana Kukillaya (42); Parameswaran (47); Perumal Pillai (54)
	tea	[central Travancore and Mundakayam:] Krishnaswami Reddiar (70); Sahadevan (66); Sankaran (110); [Kannan Devan Hills:] Ramanathan Chettiar (61)

Syrian = Syrian Christian; Chr. = Christian (which can either be a Syrian Christian or a 'new', low-caste Christian); O,U,U = Others, Uncertain, Unknown.

Source: *UPASI Planting Directory* 1960: 214–387.

Bibliography

Though realizing that my choice is an exceptional one, I opt for the subsequent classification of the bibliography. First, the written materials which originate from the government, the planting community and the labour movement are presented. The government sources concern the materials of south-west Indian administrations (Travancore, Travancore-Cochin and Kerala) and 'supra-south-west Indian' states (the Madras Presidency, British India and Independent India), concluding with some individual accounts of government representatives, all of these categories being organized in chronological order (see Section A). The planters' materials involve the sources of planting associations at district, state and supra-state levels, followed by the accounts of planting companies and individual planters, again presented chronologically (see Section B). The sources written by leftist leaders and intellectuals active in the late colonial and early post-Independence period are provided in Section C. In Section D the works of 'independent' scholars and institutions are presented, this time in alphabetical order. In the last section (E), the oral sources are presented, though in some cases – to prevent any disciplinary action being taken against the informants – their names have been replaced by pseudonyms.

I prefer this classification to others, including those based on the division of archival/non-archival and manuscript/printed materials. There is a clear reason for this: a categorization into 'government', 'planting' and 'labour' sources makes explicit the socio-political context of these writings. As underlined particularly in the section regarding the strike in the Kannan Devan Hills in 1958, all individuals and interest groups involved in plantation development write their 'own' history, emphasizing the 'positive' features and avoiding or changing the 'negative' aspects of their acts. This arrangement therefore offers clarity to the readers. At the same time, I have tried to meet the eventual objections to the classification used. All published papers, reports and manuscripts are presented in italics. Moreover, with regard to the sources dealt with in the Sections A, B and C the location of the material, or parts thereof, is given between square brackets.

A. Government Sources

Government of Travancore, Files Dewan's office, in: Cover System Files (CSF) Nos 1681, 1727, 1739, 1760, 1763, 1768-1769, 1772, 1775, 1777-1779, 1791, 1794, 1797, 1815, 1826-1827, 1830-1831, 1858, 1873, 1892, 1925, 2368, 2416, 2441, 3444, 3468, 3928, 3964, 6547, 6586, 6598, 7253, 7408, 7567, 10101, 10104, 10106-10107, 10112-10115, 10127-10129, 10133, 10512, 10557, 10561, 10654, 10657, 10661-10662, 10667, 11219-11220, 11222, 11225-11226, 11231, 11244, 11247, 11249, 11486, 11579, 11624-11626, 11629, 11631, 11648, 11663, 11665-11666, 11671-11672, 11675-11676, 11679, 11684-11685, 11687, 11689-11690, 11716, 11727, 11729-11730, 11735, 11739, 11745-11746, 11798, 11800, 12041, 12064-12065, 12165, 12423, 12446, 13607 and 13700; various departments' files: Bundle Nos 7:1101/21, 7:1174/21, 14:653/21, 29:1244/75/23, 38:1944/1158, 43:2148/2031, 224:5531/1358, 240:6454/1620, 77:3997/944 [KSA].

—, Papers regarding the Kannan Devan Hills [CDS: RF.546, KSA, ULKU].

—, Travancore Legislative Council Proceedings, 1888-1932 [LL].

—, Travancore Sri Mulam Popular Assembly Proceedings, 1904-32 [HDKU, LL, ULKU].

—, Travancore Sri Chitra State Council Proceedings, 1933-1946/7 [HDKU, LL, ULKU].

—, Travancore Sri Mulam Assembly Proceedings, 1933-1946/7 [HDKU, LL, ULKU].

—, Travancore Representative Body Proceedings, 1948 [LL, ULKU].

—, Travancore Legislative Assembly Proceedings, 1948/9 [LL].

—, Weekly Secret Bulletin/Travancore Police Abstracts of Intelligence, nos 1-7, 1934-40 [KSA].

—, *Acts and Proclamations of Travancore, 1010-1086 ME*, Trivandrum: Government Press [CDS, HDKU, ULKU].

—, *Census of Travancore*, 1875, 1881, 1891, 1901, 1911, 1921, 1931 and 1941, Trivandrum: Government Press [CDS, HDUK, IOL&R, LL].

—, *The Regulations and Proclamations of Travancore*, vols 1-14, 1835-1945, Trivandrum: Government Press [CDS, SOAS, ULKU].

Government of Travancore, Statistics of Travancore, Trivandrum: Government Press [CDS, LL, LSE, ULKU].

—, *Travancore Administration Reports*, since 1859/60, Trivandrum: Government Press [HDKU, IOL&R, LL, LSE, ULKU].

—, *Travancore Almanac and Directory*, 1871, 1872, 1876, 1879, 1895, 1903 and 1912, Trivandrum: Government Press [HDKU, IOL&R, ULKU].

—, *Travancore Gazette*, Trivandrum: Government Press [LL].

—, *Travancore Land Revenue Manual*, vol. 2: Departmental Rules and Standing Orders (1915 and supplements in 1921, 1929, 1935, 1943 and 1950), vol. 5: Selections from the Old Records (1916), Trivandrum: Government Press [CDS, HDKU, ULKU].

—, 1874, Vishakham Tirunal Varma, *Our Industrial Status, A Lecture Delivered by the First Prince of Travancore, Fellow of the Madras University, to the Trivandrum Debating Society, 26 September 1874*, Trivandrum: Government Press [ULKU].

—, 1883, *Addresses Presented to, and Replies Given by, H.H. the Maharajah of Travancore, During his Tour to Madras, 1883*, Trivandrum: Government Press [ULKU].

—, 1893, Bourdillon, E.F., *Report on the Forests of Travancore, 1892*, Trivandrum: Government Press [CDS].

—, 1906, Nagam Aiya, V., *The Travancore State Manual*, vols 1–3, Trivandrum: Government Press [CDS, FUA, HDKU, IOL&R].

—, 1914, *Addresses to the Dewan of Travancore with the replies of the Dewan, March-May 1914*, Trivandrum: Government Press [ULKU].

—, 1914, Mahadeva Iyer, *Travancore Revenue Manual*, Trivandrum: Government Press [HDKU].

—, 1920, *Addresses to Dewan Bahadur M. Krishnan Nair, Dewan of Travancore*, Trivandrum: Government Press [ULKU].

—, 1921, S.G. Barker, *Report of the Industrial Survey of Travancore*, Trivandrum: Government Press [LL].

—, 1923, S. Ramanath Aiyer, *Progressive Travancore*, Trivandrum: Government Press [ULKU].

—, 1925, R. Veeraraghava Iyengar, *Report of the Travancore Central Flood Relief Committee*, Trivandrum: Government Press [LL].

—, 1926, *Papers Regarding the Ten Square Miles Concession*, Trivandrum: Government Press [ULKU].

—, 1928, Unemployment Enquiry Committee, *Report*, Trivandrum: Government Press [LSE].

—, 1928, Travancore Indian States Committee, *Report*, Trivandrum: Government Press [ULKU].

Government of Travancore, 1930, Banking Enquiry Committee, *Report*, Trivandrum: Government Press [CDS, LSE].

—, 1931, Economic Depression Enquiry Committee, *Report*, Trivandrum: Government Press [CDS].

—, 1938, Bala Varma, *Speeches and Messages of His Highness Sri Bala Varma, Maharajah of Travancore*, Trivandrum: Government Press [ULKU].

—, 1940, T.K. Velu Pillai, *The Travancore State Manual*, vols 1-3, Trivandrum: Government Press [CDS, HDKU, IOL&R].

—, 1947, *Travancore Rubber Production and Marketing Act*, Trivandrum: Government Press [LL].

Government of Travancore-Cochin, Travancore-Cochin Legislative Assembly Proceedings, 1949-56 [HDKU, LL].

—, *Five Year Plan*, 1951-6 (1st) [HDKU, LL].

—, *Travancore-Cochin Administration Reports* [LL, LSE, ULKU].

—, *Travancore-Cochin Gazette* [LL].

—, 1949, The Agrarian Problem Enquiry Committee, *Report*, Ernakulam: Government Press [KSA].

—, 1951, *Census of Travancore-Cochin*, Trivandrum: Government Press [CDS, LL, LSE].

—, 1951, The Land Policy Committee, *Report*, Trivandrum: Government Press [LL].

—, 1953, The Minimum Wages Committee for Plantations (Travancore-Cochin), *Report 1952*, Trivandrum: Government Press [CDS].

—, 1954, The Minimum Wages Advisory Committee for Plantations, *Report 1953*, Trivandrum: Government Press [CDS].

—, 1954, *Opinions Received by Government Regarding the Bills Relating to Land Reforms*, Trivandrum: Government Press [LL].

—, 1955, The Minimum Wages Committee for Plantations, *Report*, Trivandrum: Government Press [LL].

Government of Kerala, Kerala Legislative Assembly Proceedings 1956-60 [HDKU, LL].

—, *Five Year Plan*, 1956-61 (Second) and 1961-6 (Third) [CDS, HDKU, LL].

—, *Kerala Administration Report* [LL, ULKU].

—, *Kerala Gazette* [LL].

—, 1957, *Kerala Agrarian Relations Bill*, Trivandrum: Government Press [KSA, LL].

—, 1957, Travancore-Cochin Land Policy Committee, *Report 1949-1950*, Trivandrum: Government Press [LL].

—, 1958, *Opinions on the Kerala Agrarian Relations Bill, 1957*, vols 1 and 2, Trivandrum: Government Press [LL].

Government of Kerala, 1959, Department of Statistics, *Report on the Nutritional Diet Survey of Workers in Tea Plantations*, Ernakulam: Government Press [CDS, LL].

—, 1959, S. Govinda Menon, *Report on the Enquiry into Matters Arising from the Labour Strike in Munnar*, Trivandrum: Government Press [LL].

—, 1959, *The Kerala Plantation Labour Rules*, Trivandrum: Government Press [LL].

—, 1960, *Kerala Plantation (Additional Tax) Act, Act XVII of 1960*, Trivandrum: Government Press [LL].

—, 1960, Mammen Tharakan, P.K., *Report of the Committee for Plantation Industry, Wages and Bonus, 1957-59*, Trivandrum: Government Press [CDS, LL].

—, 1961, *Census of Kerala*, Trivandrum: Government Press [CDS, LL, LSE].

—, 1961, *Kerala Agrarian Relations Act, 1960 (Act IV of 1961)*, Trivandrum: Government Press [LL].

—, 1962, A.S. Menon, *Kerala District Gazetteers, Kozhikode*, Trivandrum: Government Press [LL].

—, 1964, *Kerala Land Reforms Act, 1963 (Act I of 1964)*, Trivandrum: Government Press [LL].

—, 1968, Bureau of Economics and Statistics, *Land Reforms Survey in Kerala, 1966-1967*, Trivandrum: Government Press [LL].

—, 1970, K.T. Jacob, *Kerala Foreign Owned Plantations Enquiry Commission Report*, Trivandrum: Government Press [CDS].

—, 1976, Department of Public Relations, *Kerala through the Ages*, Trivandrum: Government Press [FUA].

—, 1978, Bureau of Economics and Statistics, *Kerala in Maps*, Trivandrum: Government Press [UA].

—, 1990, *100 Years of Legislative Bodies in Kerala; 1888-1988; A Centenary Souvenir*, Trivandrum: Government Press [LL].

Government of the Madras Presidency, Records of Fort St George, Letters to Tellicherry, vol. 1, 1726-8 [ULKU].

—, *Madras Almanac* [IOL&R].

—, 1896, The South of India Planters' Enquiry Committee, *Report*, Madras: Government Press [MA].

—, 1903, *Madras Act I of 1903*, Madras: Government Press [UPASI].

—, 1908, R.L. Proudlock, *Report on the Rubber Trees at Nilambur and at Calicut, South Malabar*, Madras: Government Press [RTI].

Government of British India, Crown Representative's Records, Political Department, R/1, nos 1/1644, 1/1889, 1/2065, 1/2238, 1/3641, 1/3786, 1/3927, 1/4063, 1/4217 [IOL&R].

Government of British India, Crown Representative's Records, Indian States, R/2, nos 879/8, 879/10, 879/14, 879/18, 879/19, 879/22, 879/25, 879/30, 879/32, 879/39, 879/41, 880/69, 880/83, 880/84, 880/86, 880/89, 882/108, 882/125, 884/147, 884/155, 884/163, 887/187, 888/207, 889/373, 897/351, 900/402, 900/409, 901/415, 901/416, 903/441 [IOL&R].

—, 1931, Royal Commission on Labour, *Report*, Calcutta: Government Press [LL].

Government of Independent India, Lok Sabha Debates [LL].

—, 1948, Labour Bureau, *Report on an Enquiry into the Cost and Standard of Living of Plantation Workers in South India*, Calcutta: Government Press [LL].

—, 1950, D.V. Rege, *Report on an Enquiry into Conditions of Labour in Plantations in India*, Delhi/Simla: Government Press [CDS].

—, 1951, *Plantation Labour Act*, Delhi: Government Press [LL].

—, 1956, P. Madhava Menon, *Plantation Enquiry Commission; Reports regarding Tea and Rubber*, Delhi: Government Press [CDS, HDKU, IOL&R].

—, 1964, Labour Bureau, *A Study on Labour Conditions in Rubber Plantations, 1961-62*, Delhi: Government Press [RRII].

—, 1968, Ministry of Commerce, *Report of the Rubber Small Holdings Economics Enquiry Committee*, Kottayam: Rubber Board [RRII].

—, 1968, National Commission on Labour, *Report of the Study Group for Plantations (Coffee/Rubber)*, New Delhi: Government Press [RRII].

—, 1980, Labour Bureau, Ministry of Labour, *Socio-Economic Conditions of Women Workers in Plantations*, Government Press [LL].

Individual Accounts, Tottenham, Assistant Resident of Travancore and Cochin (1919-20), Memoirs [CSAS].

—, 1930, Watts, M.E., Travancore, in: *Asiatic Review*, October [RTI].

B. Planters' Materials

Central Travancore Planters' Association, Minutes Central Travancore Planters' Association [CTPA].

—, 1928, *The Journal of the Central Travancore Planters' Association* [UPASI].

—, 1974, *Centenary Souvenir: 1874-1974*, Kottayam: Santa Press [CTPA].

Kannan Devan Planters' Association, Minutes Kannan Devan Planters' Association [KDPA].

Kannan Devan Planters' Association, 1958, *The High Range Strike; October 4th to October 30th, 1958*, Kottayam: CMS Press [IOL&R: MSS Eur F 154].

—, 1988, *Hundred Years of KDPA; 1888-1988*, KDPA [KDPA].

Mundakayam Planters' Association, Minutes Mundakayam Rubber Planters' Association [MPA].

—, Minutes Mundakayam Planters' Association [MPA].

The Association of Planters of Kerala, 1989, *50 Years of Service (1939-1989)*, Cochin: APK [APK].

United Planters' Association of Southern India, UPASI proceedings, since 1893 [UPASI].

—, *The Planting Opinion*, later renamed *The Planters' Chronicle*, since 14 September 1895 [RRII, UPASI].

—, *UPASI Planting Directory*, 1896 (first edition), 1925 (fourth edition), 1934 (ninth edition), 1948 (twelfth edition), 1960 (sixteenth edition) [UPASI: second edition (1902) and third edition (1924) not available].

—, 1953, Wilson Mayne (ed.), *UPASI 1893-1953*, Coonoor: UPASI [UPASI].

—, 1981, *Plantations in South India: Past, Present and Future*, Coonoor: UPASI [UPASI].

—, 1985, *A Note on Plantation Labour*, Coonoor: UPASI [UPASI].

—, 1993, S. Muthiah, *The First 100 Years of the UPASI; 1893-1993*, Delhi: Affiliated East West Press [UPASI].

Indian Tea Association, Collection, 1879-1982 [IOL&R: MSS Eur F 154].

Company Sources, Archives and Papers of The Kannan Devan Hills Produce Company, 1908-64, and of its predecessor, the North Travancore Land Planting and Agricultural Society [IOL&R: MSS Eur D 1204].

—, *Pamphlet*, dated *c.* 1960, issued by J. Finlay & Co., to describe the tea estates owned by the Company in the Kannan Devan Hills [IOL&R: Photo Eur 256].

—, 1961, *The Malankara Rubber and Produce Company, 1910-1961; A Short History*, Kottayam [RRII].

—, 1962, W.K.M. Langley (ed.), *Century in Malabar; The History of Pierce Leslie and Co. Ltd, 1862-1962*, Madras: The Madras Advertising Company [CDS].

—, 1978, Anita Baig and William Henderson, *A Centenary of Planting in the Kannan Devan Hills Concession; 1878-1978*, Munnar: Tata-Finlay Limited [TT].

Company Sources, 1978, K.M. Ganapathy (ed.), *Malayalam Plantations; An Era of Planting in South India*, Cochin [CDS].

Individual Accounts, Including Family Histories, 1879, A.M. Ferguson, *The Planting Directory of India and Ceylon*, Colombo [IOL&R].

—, 1883, S. Mateer, *Native Life in Travancore*, London: W.H. Allen & Co. [IOL&R].

—, 1931, Aylmer Ffulke Martin, *History of the High Range Planting District, North Travancore*, Bangalore [IOL&R].

—, 1967, W.S.S. Mackay, The Back People, in: *The James Finlay House Magazine*, Winter, pp. 6-8 [TT].

—, 1969, Chacko A. Kallivayalil, in: *First All India Agricultural Industrial Science Exhibition, Souvenir 1968-69*, Mundakayam [informant's private collection].

—, 1969, K.V. Thomas, in: *First All India Agricultural Industrial Science Exhibition, Souvenir 1968-69*, Mundakayam [informant's private collection].

—, 1972, Heather Lovatt, A Short History of the Peermade/Vandiperiyar District, Vandiperiyar; unpublished manuscript [CDS].

—, 1974, K. George Mathew (ed.), Kandathil Kudumba Charithra Committee, *Kandathil Kudumban* [The Kandathil Family], Kottayam: A & A Printers [informant's private collection].

—, 1974, W.W. Mayne, United Planters' Association of Southern India, in: *The James Finlay House Magazine*, Autumn, pp. 20-1 [TT].

—, 1976, J.B. Soutar, The High Range; Ten Years of Transition in Planting, in: *The James Finlay House Magazine*, Autumn, pp. 22-4 [TT].

—, 1980, George Alakkapally, *Kottukappally* [The Kottukapally Family], Trivandrum: St Joseph's Press [informant's private collection].

—, 1983, Editorial Board, Palakunnel Kudumba Charitha, *Palakunnel Kudumbavum Kerala Kristhavarum* [Palakunnel Family and Christians in Kerala], Kottayam: Catholic Mission Press [informant's private collection].

—, 1984, W.S.S. Mackay, Memoir describing his career as a tea plantation manager in the High Range of Travancore [IOL&R: MSS Eur D 1063].

—, 1984, R.R. Tewson, Memoir: The Carefree Days of Planting [IOL&R: Photo Eur 253].

—, 1984, Mrs M. Clough, Interview on tape, 20 January [IOL&R: MSS Eur R 148].

—, 1984, W.S.S. Mackay, Interview on tape, 16 June [IOL&R: MSS Eur R 166].

C. Leftist Literature

Leftist Literature, New Age, monthly journal of the CPI since 1933 [CPI].

—, 1958, E.M.S. Namboodiripad, *What is Behind the Congress Attack on Kerala?; Chief Minister Replies Congress Government*, Trivandrum: Department of Public Relations [LL].

—, 1959, A.K. Gopalan, *Kerala; Past and Present*, London: Lawrence & Wishart [CDS].

—, 1959, E.M.S. Namboodiripad, *The Agitation in Kerala, Its Real Nature and Reactions of the National Press*, Trivandrum: Department of Public Relations [LL].

—, 1968, E.M.S. Namboodiripad, *Kerala; Yesterday, Today and Tomorrow*, Calcutta: National Book Agency [CDS].

D. Other Readings

Abraham, N.M., 1971, Journalism in Kerala, in: *National Christian Council Assembly; Souvenir*, Kottayam: SG Press.

Akanda, Latifa, 1980, A Study of the Pattern of Plantation under Colonial Rule in Bengal and Java in the Nineteenth Century: An Analogy, paper presented at the Eight Conference International Association of Historians of Asia, Kuala Lumpur, 25–29 August.

Alexander, K.C., 1989, Caste Mobilization and Class Consciousness: The Emergence of Agrarian Movements in Kerala and Tamil Nadu, in: F.R. Frankel and M.S.A. Rao (eds), *Dominance and State Power in Modern India; Decline of a Social Order*, vol. 1, Delhi: Oxford University Press, pp. 362–413.

Allen, G.C. and A.G. Donnithorne, 1954, *Western Enterprise in Indonesia and Malaya; A Case Study in Economic Development*, London: George Allen and Unwin.

Arnold, David, 1986, *Police Power and Colonial Rule; Madras 1859–1947*, Delhi: Oxford University Press.

Austen, Ralph A., 1987, *African Economic History; Internal Development and External Dependency*, London: James Currey.

Baak, Paul E., 1990, Het Overheidsbeleid van Travancore ten aanzien van de Plantagesector in de Tweede Helft van de Negentiende Eeuw; Achtergronden, Maatregelen en Consequenties [The Policy of the Travancore Government towards Plantations during the Second Half of the Nineteenth Century: Its Historical Background, Measures and Implications], M.A. Thesis Free University Amsterdam.

Baak, Paul E., 1991, The Conference on Capitalist Plantations in Colonial Asia, September 26-29, 1990, Amsterdam, in: *Archipel*, no. 42, pp. 3-10, and in: *South Asia Research*, vol. 11, no. 1, pp. 89-95.

—, 1992, Planters' Lobby in Late 19th Century; Implications for Travancore, in: *Economic and Political Weekly*, vol. 27, no. 33, 15 August, pp. 1747-53, and in: *Journal of Kerala Studies*, vol. 13, pp. 23-43.

—, 1993, Out of the Shadow, Into the Spotlights: The Indigenous Aspects of Plantation Development; The Case of the Emergence of Syrian Christian Planters in Princely Travancore in the First Half of the 20th Century, in: *Itinerario*, vol. 17, no. 2, pp. 74-96.

—, 1994, The Successful Planters' Lobby in a Leftist 'Problem' State; Plantation Development in Southwest India in the Early Post-Colonial Period, in: *Abstracts of the International Congress on Kerala Studies*, vol. 3, pp. 67-8.

—, 1995, About Enslaved Ex-Slaves, Uncaptured Contract Coolies and Unfreed Freedmen: Some Notes about 'Free' and 'Unfree' Labour in the Estate Regions of Southwest India, 1824-1962, in: P.E. Baak (ed.), *CASA Nova; Aspects of Asian Societies 1*, Amsterdam: Thesis Publishers, pp. 157-83.

Bandarage, Asoka, 1983, *Colonialism in Sri Lanka; The Political Economy of the Kandyan Highlands, 1833-1886*, Berlin: Hildebrand.

Barlow, C. and J. Drabble, 1990, Government and the Emerging Rubber Industries in Indonesia and Malaya, 1900-40, in: A. Booth, W.J. O'Malley and A. Weidemann (eds), *Indonesian Economic History in the Dutch Colonial Era*, New Haven: Yale University Southeast Asia Studies, pp. 187-209.

Bastianpillai, B., 1964, From Coffee to Tea in Ceylon. The Vicissitudes of a Colony's Plantation Economy, in: *The Ceylon Journal of Historical and Social Studies*, vol. 7, no. 1, pp. 43-66.

Bayly, Susan, 1984, Hindu Kingship and the Origin of Community; Religion, State and Society in Kerala, 1750-1850, in: *Modern Asian Studies*, vol. 18, no. 2, pp. 177-213.

Beckford, George L., 1972, *Persistent Poverty; Underdevelopment in Plantation Economies in the Third World*, New York: Oxford University Press.

Behal, Rana P. and Prabhu P. Mohapatra, 1992, Tea and Money versus Human Life; The Rise and Fall of the Indenture System in the Assam Tea Plantations 1840-1908, in: *The Journal of Peasant Studies*, vol. 19, nos 3 and 4, pp. 142-72.

Bhaskaran, N.K., 1981, The Malayali Memorial, in: *Journal of Kerala Studies*, vol. 8, parts 1-4, pp. 225-48.

Bhowmik, Sharit, 1981, *Class Formation in the Plantation System*, New Delhi: People's Publishing House.

Binsbergen, W. Van and P. Geschiere, 1982, *Oude Produktiewijzen en Binnendringend Kapitalisme; Antropologische Verkenningen in Afrika* [Traditional Modes of Production and the Rise of Capitalism; Anthropological Explorations in Africa], Amsterdam: Free University Press.

Boeke, J.H., 1940, *Indische Economie* [The Economy of the Netherlands East Indies], Haarlem: Tjeenk Willink.

Boomgaard, Peter, 1990, Why Work for Wages? Free Labour in Java, 1600–1900, in: *Economic and Social History in the Netherlands*, vol. 2, pp. 37–56.

Breman, Jan, 1990a, *Taming the Coolie Beast; Plantation Society and The Colonial Order in Southeast Asia*, Delhi: Oxford University Press.

—, 1990b, *Labour Migration and Rural Transformation in Colonial Asia*, Amsterdam: Comparative Asian Studies No. 5.

—, 1992, Controversial Views on Writing Colonial History, in: *Itinerario*, vol. 16, no. 2, pp. 39–60.

Brocheux, P., 1990, The Plantations Labor and the Labor Movement in Vietnam, paper presented at the Conference on Capitalist Plantations in Colonial Asia, Centre for Asian Studies Amsterdam, 26–29 September.

Carter, Marina, 1992, Strategies of Labour Mobilisation in Colonial India: The Recruitment of Indentured Workers for Mauritius, in: *The Journal of Peasant Studies*, vol. 19, nos 3 and 4, pp. 229–45.

Chaput, Pascale, 1992, Les Allégeances Croisées d'un Saigneur Malayali; Monography d'une Plantation d'Hévéas [The Conflicting Loyalties of a Malayali Rubber Tapper: The Monography of a Rubber Estate], in: *Sociologie du Travail*, no. 3, pp. 331–52.

Chatterjee, Partha, 1993, *The Nation and its Fragments; Colonial and Postcolonial Histories*, Princeton: Princeton University Press.

Chattopadhyay, Ramkrishna, 1987, Social Perspective of Labour Legislation in India 1859–1932: As Applied to the Tea Plantations, Unpublished Ph.D. Thesis Calcutta University.

Chaudhuri, K.N., 1986, *Trade and Civilization in the Indian Ocean; An Economic History from the Rise of Islam to 1750*, Cambridge: Cambridge University Press.

Chaudhury, Saumitra, 1984, Indian Bourgeoisie and Foreign Capital: A Study of Congress Policy towards Foreign Capital, 1931–61, in: *Social Scientist*, vol. 12, no. 132, pp. 3–22.

Chiriyankandath, James L., 1985, Social Change and the Development of 'Modern' Politics in Travancore: From the Late 19th Century to 1938, Ph.D. Thesis University of London.

Clarence-Smith, W.G., 1990, Planters and Peasants in Portuguese Timor in the 19th and 20th Centuries; A Preliminary Note, paper presented at the Conference on Capitalist Plantations in Colonial Asia, Centre for Asian Studies Amsterdam, 26-29 September.

Clay, R., 1911, *Rubber in Southern India*, London.

Clemens, A.H.P., 1989, De Inheemse Rubbercultuur in Jambi en Palembang Tijdens het Interbellum, in: A.H.P. Clemens and J.Th. Lindblad (eds), *Economische Expansie en Koloniale Staatsvorming in de Buitengewesten van Nederlands-Indië, 1870-1942* [Economic Expansion and State Formation in the Outer Islands of the Netherlands East Indies, 1870-1942], Amsterdam: Neha, pp. 213-37.

Clerkx, Lilly E. and Wim F. Wertheim, 1991, *Living in Deli; Its Society as Imagined in Colonial Fiction*, Amsterdam: Comparative Asian Studies No. 6.

The Cochin Chamber of Commerce and Industry, 1982, *1857-1982: Post Centenary Silver Jubilee Commemoration Volume.*

Cooper, Frederick, 1977, *Plantation Slavery on the East Coast of Africa*, New Haven: Yale University Press.

Courtenay, P.P., 1971, *Plantation Agriculture*, London: G. Bell & Sons.

Dalton, Eira, 1963, *The Baker Family in India*, Kottayam: CMS Press.

Darwin, John, 1991, *The End of the British Empire; The Historical Debate*, Oxford: Basil Blackwell.

Das Gupta, A., 1976, *Malabar in Asian Trade; 1740-1800*, Cambridge: University Press.

Das Gupta, A. and M.N. Pearson (eds), 1987, *India and the Indian Ocean, 1500-1800*, Oxford: Oxford University Press.

Das Gupta, Ranajit, 1986, From Peasants and Tribesmen to Plantation Workers; Colonial Capitalism, Reproduction of Labour Power and Proletarianisation in North East India, 1850s to 1947, in: *Economic and Political Weekly*, vol. 21, no. 4, 25 January, pp. PE2-PE10.

—, 1992a, *Economy, Society and Politics in Bengal: Jalpaiguri 1869-1947*, Delhi: Oxford University Press.

—, 1992b, Plantation Labour in Colonial India, in: *The Journal of Peasant Studies*, vol. 19, nos 3 and 4, pp. 173-98.

Desai, S.S.M., 1980, *Economic History of India*, Bombay: Himalaya Publishing House.

Dharmarajan, S., 1991, Aspinwall; Seaside Saga, in: *Contemporary Tea Time*, vol. 2, no. 9, September, pp. 27-8.

Doorn, J.A.A. Van, 1994, *De Laatste Eeuw van Indië; Ontwikkeling en Ondergang van een Koloniaal Project* [The last Century of the

Netherlands East Indies; The Rise and Fall of a Colonial Project, Amsterdam: Bert Bakker.

Doshi and Liebl, M., 1988, *Kerala; India; Week by Week*, Bangkok, Media Transasia Limited.

Edwin, P.G., 1978, British Impact on Kerala, in: *Journal of Karala Studies*, vol. 5, pp. 271-86.

Elson, R.E., 1990, Peasant Poverty and Prosperity Under the Cultivation System in Java, in: A. Booth, W.J. O'Malley and A. Weidemann (eds), *Indonesian Economic History in the Dutch Colonial Era*, New Haven: Yale University Southeast Asia Studies, p.p. 24-48.

Emmer, P.C., 1986, The Meek Hindu: The Recruitment of Indian Indentured Labourers for Service Overseas, 1870-1916, in: P.C. Emmer (ed.), *Colonialism and Migration; Indentured Labour Before and After Slavery*, Dordrecht: Martinus Nijhoff, pp. 187-207.

Franke, W. Richard and Barbara H. Chasin, 1992, *Kerala: Development through Radical Reform*, New Delhi: Promilla.

Füller, C.J., 1976, Kerala Christians and the Caste System, in: *Man*, vol. 11, no. 1, pp. 53-70.

Furnivall, J.S., 1939, *Netherlands India; A Study of Plural Economy*, Cambridge: Cambridge University Press.

Ganesh, K.N., 1990, The Process of State Formation in Travancore, in: *Studies in History*, vol. 6, no. 1, pp. 15-33.

—, 1991, Ownership and Control of Land in Medieval Kerala: Janmam-Kanam Relations during the 16th-18th Centuries, in: *Indian Economic and Social History Review*, vol. 28, no. 3, pp. 299-323.

George, Alex, 1987, Social and Economic Aspects of Attached Labourers in Kuttanad Agriculture, in: *Economic and Political Weekly*, vol. 22, no. 52, 26 December, pp. A141-A150.

Ghosh, H.H., 1928, *The Realm of Rubber, Being a Record of Existing Conditions in the Rubber Industry*, Calcutta: J.B. Raymond.

Graham, Edgar and Ingrid Hoering, 1984, *The Modern Plantation in the Third World*, London: Croom Helm.

Griffiths, P., 1967, *The History of the Indian Tea Industry*, London: Weidenfeld & Nicolson.

Guha, Amalendu, 1968, A Big Push without a Take-off: A Case Study of Assam, in: *The Indian Economic and Social History Review*, vol. 5, no. 3, pp. 199-221.

—, 1977, *Planter-Raj to Swaraj; Freedom Struggle and Electoral Politics in Assam 1826-1947*, New Delhi: People's Publishing House.

Guha, Ranajit and Gayatri Chakravorty Spivak (eds), 1988, *Selected Subaltern Studies*, New York: Oxford University Press.

Haridasan, V., 1967, *Family Budget and Social Security Benefits of Rubber Plantation Workers in India*, Kottayam: Rubber Board.

—, 1975, Early History of Rubber Planting in India, in: *Rubber Board Bulletin*, vol. 12, no. 3, pp. 118-21.

—, 1978, Rubber Planting Industry in India: First World War to Independence, in: *Rubber Board Bulletin*, vol. 15, nos 3 and 4, pp. 63-6.

—, 1980, Small Rubber Growers in India, in: *The Planters' Chronicle*, April, pp. 173-5.

—, 1986, *Dr John Matthai, 1886-1959; A Biography*, Cranganore: Sunil.

—, 1992, *Management of Plantations; A Study of Rubber*, Changanacherry: Nalanda Printers.

Harler, C.R., 1956, *The Culture and Marketing of Tea*, London: Oxford University Press.

Hatch, Emily Gilchrist, 1939, *Travancore*, London: Oxford University Press.

Herring, R.J., 1980, Abolition of Landlordism in Kerala: A Redistribution of Privilege, in: *Economic and Political Weekly*, vol. 15, no. 26, June, pp. A59-A69.

Hettne, Bjorn, 1978, *The Political Economy of Indirect Rule; Mysore 1881-1947*, London: Curzon Press.

The Hindu

Hitchcock, R.H., 1983, *Peasant Revolt in Malabar; A History of the Malabar Rebellion 1921*, New Delhi: Usha Jain (first published in 1925).

Hollup, Oddvar, 1990, Caste, Cultural Continuity and Social Change among Tamil Plantation Workers in Sri Lanka, paper presented at the Conference on Capitalist Plantations in Colonial Asia, Centre for Asian Studies Amsterdam, 26-29 September.

Houben, Vincent J.H., 1987, Native States in India and Indonesia: the Nineteenth Century, in: *Itinerario*, vol. 11, no. 1, pp. 107-34.

—, 1993, 'Colonial History' Revisited; A Response to Breman, in: *Itinerario*, vol. 17, no. 1, pp. 93-7.

Hurd, John, 1972, The Economic Consequences of Indirect Rule in India, in: *The Indian Economic and Social History Review*, vol. 12, no. 2, pp. 169-81.

Hüsken, Frans, 1990, The Making of a Plantation Society: The Residency of Jepara during the Cultivation System, paper presented at the Conference on Capitalist Plantations in Colonial Asia, Centre for Asian Studies Amsterdam, 26-29 September.

The Indian Trade Journal

International Labour Office, 1966, *Plantation Workers: Conditions of Work and Standards of Living*, Geneva: ILO.

—, 1989, *Socio-Economic Conditions in Plantations in India: Proceedings of a National Tripartite Workshop*, Coonoor, 27-30 November, Geneva: ILO.

Jackson, James C., 1968, *Planters and Speculators; Chinese and European Agricultural Enterprise in Malaya, 1786-1921*, Kuala Lumpur: University of Malaya Press.

Jacob, George, 1974, *Labour Conditions in Rubber Estates*, Kottayam: Rubber Board.

Jain, Ramindra Kumar, 1990, Colonial Ambience and Immigrant Tamil Labour: A View from Malayan Plantations, paper presented at the Conference on Capitalist Plantations in Colonial Asia, Centre for Asian Studies, Amsterdam, 26-29 September.

Janson, Christien, 1990, Might of the Managing Agencies; Control of the Coastal Firms; A study on Controlled Economic Development; Travancore 1921-31, paper presented at the 11th European Conference on Modern South Asian Studies, Amsterdam, 2-5 July.

Javernesa Begom, M.J., 1983, Impact of the Land Reforms of 1958 and 1963 on Kerala, in: *Journal of Kerala Studies*, vol. 10, parts 1-4, pp. 139-45.

Jeffery, Roger, 1982, Merchant Capital and the End of Empire; James Finlay, Merchant Adventures, *Economic and Political Weekly*, vol. 17, no. 7, 13 February, pp. 241-8.

Jeffrey, Robin, 1975, The Politics of 'Indirect Rule', Types of Relationship among Rulers, Ministers and Residents in a 'Native State', in: *Journal of Commonwealth and Comparative Politics*, vol. 13, no. 3, pp. 261-81.

—, 1976, *The Decline of Nayar Dominance: Society and Politics in Travancore, 1847-1908*, New Delhi: Vikas Publishing House (reprinted by Manohar Press 1994).

—, 1978 (ed.), *People, Princes and Paramount Power: Society and Politics in the Indian Princely States*, New Delhi: Oxford University Press.

—, 1979, Peasant Movements and the Communist Party in Kerala 1937-57, in: D.B. Miller (ed.), *Peasant and Politics: Grass Roots Reactions to Change in Asia*, London: Edward Arnold, pp. 131-48.

—, 1980, What the Statues Tell: The Politics of Choosing Symbols in Trivandrum, in: *Pacific Affairs*, vol. 53, no. 3, pp. 484-502.

John, K.C., 1991, *Kerala: The Melting Pot; A Political History*, New Delhi: Nunes (first published in 1975).

Jones, William O., 1968, Plantations, in: *International Encyclopaedia of the Social Sciences*, vol. 12, pp. 154-9.

Jose, A.V., 1980, Agricultural Labour Force in Kerala: An Historical cum Statistical Analysis, Thesis Centre for Development Studies.

Joseph, K.V., 1988, *Migration and Economic Development of Kerala*, Delhi: Mittal Publications.

Kannan, K.P., 1988, *Of Rural Proletarian Struggles: Mobilization and Organization of Rural Workers in Southwest India*, Delhi: Oxford University Press.

Karnik, V.B., 1966, *Indian Trade Unions; A Survey*, Bombay: Manaktalas.

Kaushal, G., 1979, *Economic History of India, 1757-1966*, New Delhi: Kalyani Publishers.

Kelly, John D., 1992, 'Cooly' as a Labour Commodity: Race, Sex, and European Dignity in Colonial Fiji, in: *Journal of Peasant Studies*, vol. 19, nos 3 and 4, pp. 246-68.

Kerala at a Glance, 1988, Trivandrum.

Kieniewicz, Jan, 1969, The Portuguese Factory and Trade in Pepper in Malabar During the 16th Century, in: *The Indian Economic and Social History Review*, vol. 6, no. 1, pp. 61-84.

Kirk, Colin, 1987, *People in Plantations, A Literature Review and Annotated Bibliography*, IDS Research Reports No. 18, Brighton: Institute of Development Studies.

Knight, G.R., 1992, The Java Sugar Industry as a Capitalist Plantation: A Reappraisal, in: *The Journal of Peasant Studies*, vol. 19, nos. 3 and 4, pp. 68-86.

Kooiman, Dick, 1984, The Gospel of Coffee: Mission, Education and Employment in 19th Century Travancore, in: Kooiman, Van den Muyzenberg and Van der Veer (eds), *Conversion, Competition and Conflict: Essays on the Role of Religion in Asia*, Amsterdam: Free University Press, pp. 185-215.

—, 1989, *Conversion and Social Equality in India; The London Missionary Society in South-Travancore in the 19th Century*, Delhi: Manohar Press.

—, 1991, Conversion from Slavery to Plantation Labour; Christian Mission in South India, 19th Century, in: *Social Scientist*, vol. 19, nos. 8 and 9, August-September, pp. 57-71.

—, 1992a, State Formation in Travancore; Problems of Revenue, Trade and Armament, in: A.W. Van den Hoek, D.H.A. Kolff and M.S. Oort (eds), *Ritual, State and History in South Asia; Essays in Honour of J.C. Heesterman*, Leiden: E.J. Brill, pp. 587-609.

—, 1992b, Plantations in Southern Asia: Indigenous Plants and Foreign Implantations, in: *South Asia*, vol. 15, no. 1, pp. 53-79.

Kooiman, Dick, 1993, Political Rivalry Among Religious Communities; A Case Study of Communal Reservations in India, in:

Economic and Political Weekly, vol. 28, no. 7, 13 February, pp. 287–94.

Kooiman, Dick, 1994, Separate Electorates: Separate Identities? Experiences from Colonial India, in: Joachim Heidrich (ed.), *Changing Identities: The Transformation of Asian and African Societies Under Colonialism*, Berlin: Das Arabische Buch, pp. 255–73.

—, 1995, *Communities and Electorates; A Comparative Discussion of Communalism in Colonial India*, Comparative Asian Studies No. 16, Amsterdam: VU University Press.

Koshy, M.J., 1972, *Constitutionalism in Travancore and Cochin*, Trivandrum: Kerala Historical Society.

—, 1973, *Last Days of Monarchy in Kerala*, Trivandrum: Kerala Historical Society.

Koshy, P.A., 1977, Raja Sir T. Madhava Rao and the Modernization of the Travancore Administration, Ph.D. Thesis Kerala University.

Krishnan Kutty, P.N. and V. Haridasan, 1976, *Family Budget of Rubber Plantation Workers in Kozhikode District; Report of a Survey*, Kottayam: Rubber Board.

Kumar, Dharma, 1961, Caste and Landlessness in South India, in: *Comparative Studies in Society and History*, vol. 4, pp. 337–63.

—, 1965, *Land and Caste in South India; Agricultural Labour in the Madras Presidency During the Nineteenth Century*, Cambridge: Cambridge University Press.

Kumar, Dharma and Tapan Raychaudhuri (eds), 1991, *The Cambridge Economic History of India, Vol. 2: c.1757-c.1970*, New Delhi: Orient Longman (first published in 1982).

Kurian, Rachel, 1990, The Plantation Labour Regime under Colonialism; The Case Study of South India, paper presented at the Conference on Capitalist Plantations in Colonial Asia, Centre for Asian Studies Amsterdam, 26–29 September.

Kurup, K.K.N., 1984, The Colonial Investment and the Abolition of Slavery in South India: A Case Study of Kerala, in: *Journal of Kerala Studies*, vol. 11, parts 1–4, pp. 187–99.

—, 1988, *Modern Kerala; Studies in Social and Agrarian Relations*, Delhi: Mittal Publications.

—, 1989, *Agrarian Struggles in Kerala*, Trivandrum: CBH Publications.

Kusuman, K.K., 1973, Travancore's Status in Relation to British Paramountcy; A Case Study, in: *Journal of Kerala Studies*, vol. 1, pp. 299–309.

Kusuman, K.K., 1987, *A History of Trade and Commerce in Travancore (1600-1805)*, Delhi: Mittal Publications.

Lannoy, Mark De, 1986, A 'Dutchman' in the Service of the Raja of Travancore: Eustache Benoît de Lannoy (1715-77), in: *Journal of Kerala Studies*, vol. 13, parts 1-4, pp. 1-16.

Leur, J.C. Van, 1983, *Indonesian Trade and Society: Essays in Asian Social and Economic History*, Dordrecht: Foris Publications (first published in 1955).

Lewandowski, S., 1980, *Migration and Ethnicity in Urban India; Kerala Migrants in the City of Madras, 1870-1970*, New Delhi: Manohar.

Lieten, Georges Kristoffel, 1975a, Nature of Travancore's Economy Between the Two World Wars, in: *Journal of Kerala Studies*, vol. 2, part 1, pp. 105-23.

—, 1975b, The Economic Structure of Kerala in the Mid-Fifties, in: *Journal of Kerala Studies*, vol. 2, part 4, pp. 449-78.

—, 1979, Progressive State Governments: An Assessment of First Communist Ministry in Kerala, in: *Economic and Political Weekly*, vol. 14, no. 1, 6 January, pp. 29-39.

Lombard, Denys, 1967, *Le Sultanat d'Atjéh au Temps d'Iskandar Muda, 1607-1636* [The Sultanate of Aceh During Iskandar Muda, 1607-1636], Paris: École Française d'Extrême-Orient.

Lucassen, Jan, 1993, Free and Unfree Labour Before the Twentieth Century: A Brief Overview, in: Tom Brass et al. (eds), *Free and Unfree Labour*, Amsterdam: International Institute for Social History.

Lukose, Ani, 1991, *Labour Movements and Agrarian Relations*, Delhi: Rawat Publications.

Mahadevan, Raman, 1991, Industrial Entrepreneurship in Princely Travancore, 1930-47, in: S. Bhattacharya (ed.), *The South Indian Economy: Agrarian Change, Industrial Structure and State Policy, c. 1914-1947*, New Delhi: Oxford University Press, pp. 159-207.

Mahadevan Nadar, S., 1980, Commercialization of Agricultural Products and the New Economic Order in Travancore, in: *Journal of Kerala Studies*, vol. 7, parts 1-4, pp. 215-23.

Mammen, P.M., 1981, *Communalism vs Communism; A Study of the Socio-religious Communities and Political Parties in Kerala, 1892-1970*, Delhi: South Asia Books.

Mani, Ajit, 1984, Between Labour and Management: Estate Staff in Plantations, in: *Economic and Political Weekly*, vol. 19, no. 19, 12 May, pp. 791-2.

Mathew, P.M., 1984, Estate Staff in Plantations; Some Misconceptions, in: *Economic and Political Weekly*, vol. 19, no. 38, 22 September, pp. 1677-80.

Matthen, C.P., 1951, *I Have Borne Much*, Madras (first published in 1943).

McCoy, Alfred W., 1992, Sugar Barons: Formation of a Native Planter Class in the Colonial Philippines, in: *The Journal of Peasant Studies*, vol. 19, nos. 3 and 4, pp. 106–41.

Menon, A.S., 1991, *A Survey of Kerala History*, Madras: S. Viswanathan Publishers.

Menon, K.P.S., 1981, *Many Worlds Revisited; An Autobiography*, Bombay: Bharatiya Vidya Bhavan.

Menon, P.S., 1978, *A History of Travancore: From the Earliest Times*, New Delhi: Sterling Publishers.

Meyer, Eric, 1992, 'Enclave' Plantations, 'Hemmed-In' Villages and Dualistic Representations in Colonial Ceylon, in: *Journal of Peasant Studies*, vol. 19, nos. 3 and 4, pp. 199–228.

Miller, P., 1988, Kerala; Jewel of India's Malabar Coast, in: *National Geographic*, vol. 173, no. 5, May, pp. 592–617.

Mishra, G., 1966, Indigo Plantation and the Agrarian Relations during the 19th Century, in: *The Indian Economic and Social History Review*, vol. 3, pp. 332–55.

Mittal, S.K., 1978, *Peasant Uprisings and Mahatma Gandhi in North Bihar*, Meerut: Anu Prakashan.

Moldrich, Donovan, 1989, *Bitter Berry Bondage; The Nineteenth Century Coffee Workers of Sri Lanka*, Kandy: Co-ordinating Secretariat for Plantation Areas.

Moore, J. Daniel, 1981, Plantation Labour in Mysore 1871–1941: An Historical Approach to Migration Analysis, in: N.G. Barrier (ed.), *The Census in British India: New Perspectives*, New Delhi: Manohar, pp. 1–40.

Mukherjee, S., 1976, Emergence of Bengalee Entrepreneurship in Tea Plantation in a Bengal district, 1879–1933, in: *The Indian Economic and Social History Review*, vol. 13, no. 4, pp. 487–512.

Murray, M.J., 1992, 'White Gold' or 'White Blood'?: The Rubber Plantations of Colonial Indochina, 1910–40, in: *The Journal of Peasant Studies*, vol. 19, nos. 3 and 4, pp. 41–67.

Nayar, P.K.B. (ed.), 1972, *Development of Kerala; Problems and Promises*, Trivandrum: Department of Sociology, Kerala University.

Nossiter, T.J., 1982, *Communism in Kerala; A Study in Political Adaptation*, Delhi: Oxford University Press.

—, 1988, *Marxist State Governments in India; Politics, Economics and Society*, London: Pinter Publishers.

O'Malley, William J., 1990, Plantations 1830–1940: An Overview, in: A. Booth, W.J. O'Malley and A. Weidemann (eds), *Indonesian*

Economic History in the Dutch Colonial Era, New Haven: Yale University, Southeast Asia Studies, pp. 136–70.

Oommen, M.A., 1971, *Land Reforms and Socio-Economic Change in Kerala*, Madras: The Christian Literature Society.

—, 1977, Rise and Growth of Banking in Kerala, in: *Social Scientist*, vol. 5, pp. 24-46.

Ouwerkerk, Louise, 1994, *No Elephants for the Maharajah; Social and Political Change in the Princely State of Travancore, 1921-1947* (Edited with an Introduction by Dick Kooiman), New Delhi: Manohar.

Pandian, M.S.S., 1990, *The Political Economy of Agrarian Change, Nanchilnadu, 1880-1939*, New Delhi: Sage.

Panicker, P.G.K., T.N. Krishnan and N. Krishnaji, 1978, *Population Growth and Agricultural Development*, Rome: FAO.

Pannikar, K.N., 1977, Land Control, Ideology and Reform: A Study of the Changes in Family Organisation and Marriage System in Kerala, in: *The Indian Historical Review*, vol. 4, no. 1, pp. 30–46.

Pelzer, Karl J., 1978, *Planter and Peasant; Colonial Policy and the Agrarian Struggle in East Sumatra, 1863-1947*, Den Haag: Martinus Nijhoff.

—, 1982, *Planter Against Peasant; The Agrarian Struggle in East Sumatra, 1947-1958*, Den Haag: Martinus Nijhoff.

Perlin, Frank, 1983, Proto-Industrialization and Pre-Colonial South Asia, in: *Past and Present*, no. 98, pp. 30–95.

Pillai, K.P., 1959, *The Red Interlude in Kerala*, Trivandrum: Kerala Pradesh Congress Committee.

Prakash, B.A., 1982, Evolution of Land Tenure in Kerala, in: P.P. Pillai (ed.), *Agricultural Development in Kerala*, New Delhi: Agricole Publishing Academy.

—, 1987, Agrarian Development of Kerala from 1800 AD to 1980 AD; A Survey of Studies, Ulloor: Centre for Development Studies Working Paper No. 220.

Prakash, Gyan, 1990, *Bonded Histories: Genealogies of Labor Servitude in Colonial India*, South Asian Studies No. 44, Cambridge: Cambridge University Press.

Pylee, M.V. 1984, *An Introduction to the Constitution of India*, Calicut: Mathrubhumi Printing and Publishing Company.

Raj, K.N. and P.K. Michael Tharakan, 1983, Agrarian Reform in Kerala and its Impact on the Rural Economy; A Preliminary Assessment, in: Ajit Kumar Ghose (ed.), *Agrarian Reform in Contemporary Developing Countries*, New York: Croom Helm, pp. 31–90.

Ramachandran Nair, K., 1973, *Industrial Relations in Kerala*, New Delhi: Jullundur Sterling Publishers.

Ramachandran Nair, S., 1979, British Relations with Travancore 1840-60, M. Phil. Thesis, University of Calicut.

Ramakrishna Pillai, K., 1974, *Response of Small Growers of Rubber to the Aid Schemes of the Rubber Board*, Kottayam: Rubber Board.

Ramasamy, P., 1992, Labour Control and Labour Resistance in the Plantations of Colonial Malaya, in: *The Journal of Peasant Studies*, vol. 19, nos. 3 and 4, pp. 87-105.

Rangaswami, Vanaja, 1981, *The Story of Integration; A New Interpretation in the Context of the Democratic Movements in the Princely States of Mysore, Travancore and Cochin, 1900-1947*, New Delhi: Manohar.

Rao, Amiya and B.G. Rao, 1992, *The Blue Devil: Indigo and Colonial Bengal*, Delhi: Oxford University Press.

Ravi Raman, K., 1986, Plantation Labour: Revisit Required, in: *Economic and Political Weekly*, vol. 21, no. 22, 31 May, pp. 960-2.

—, 1990, Global Capital and Peripheral Labour: Tea in South India, *c.* 1860-1951, paper presented at the Conference on Capitalist Plantations in Colonial Asia, Centre for Asian Studies Amsterdam, 26-29 September.

—, 1991, Labour under Imperial Hegemony: The Case of Tea Plantations in South India, *c.* 1914-46, in: S. Bhattacharya (ed.), *The South Indian Economy: Agrarian Change, Industrial Structure and State Policy, c. 1914-1947*, New Delhi: Oxford University Press, pp. 243-67.

Reid, Anthony, 1993a, *Southeast Asia in the Age of Commerce, 1450-1680; Volume Two: Expansion and Crisis*, New Haven: Yale University Press.

—, 1993b (ed.), *Southeast Asia in the Early Modern Era; Trade, Power and Belief*, Ithaca: Cornell University Press.

Ricklefs, M.C., 1981, *A History of Modern Indonesia; c. 1300 to the Present*, London: Macmillan Press.

Rote, Ronald, 1986, *A Taste of Bitterness; The Political Economy of Tea Plantations in Sri Lanka*, Amsterdam: Free University Press.

Rothermund, Dietmar, 1988, *An Economic History of India; From Pre-Colonial Times to 1986*, New York: Croom Helm.

—, 1911, *Rubber in Southern India*, London: R. Clay & Sons.

Rubber Board, 1978, *Rubber Growers Companion*, Kottayam: Rubber Board.

Rutten, Rosanne, 1994, Popular Support for the Philippine Revolutionary Organization CPP-NPA: Perspectives from a Hacienda, paper presented at the Proseminar on Political Mobilization and Conflict, New School for Social Research, 8 February.

Salleh, H., 1990, Labour Control in Colonial Plantations in Malaya, paper presented at the Conference on Capitalist Plantations in Colonial Asia, Centre for Asian Studies Amsterdam, 26-29 September.

Sankaranarayanan, K.C. and V. Karunakaran, 1985, *Kerala Economy*, Delhi: Oxford University Press.

Sankarankutty Nair, T.P., 1977, *A Tragic Decade in Kerala History*, Trivandrum: Kerala Historical Society.

Saradamoni, K., 1980, *Emergence of a Slave Caste: Pulayas of Kerala*, New Delhi: People's Publishing House.

Sarojini, P., 1994, Genesis of Social Change Among the Nadar Caste in Travancore, in: *International Congress on Kerala Studies, 27-29 August 1994*, Supplementary Volume Abstracts, pp. 23-4.

Schenk, Hans, 1986, Views on Alleppey; Socio-Historical and Socio-Spatial Perspectives on an Industrial Port Town in Kerala, South India, Ph.D. Thesis, University of Amsterdam.

Schutte, G.J., 1983, De Koloniale Geschiedschrijving [Historiography Regarding Colonialism], in: W.W. Mijnhardt (ed.), *Kantelend Geschiedbeeld; Nederlandse Historiografie sinds 1945* [Shifting Perspectives; Dutch Historiography Since 1945], Utrecht: Het Spectrum, pp. 289-310.

Sharma, G.K., 1963, *Labour Movement in India; Its Past and Present*, Jullundur: University Publishers.

Shea, T.W., 1959, Barriers to Economic Development in Traditional Societies: Malabar, A Case Study, in: *Journal of Economic History*, vol. 19, no. 4, pp. 504-27.

Sheriff, Abdul, 1987, *Slaves, Spices and Ivory in Zanzibar; Integration of an East African Commercial Empire into the World Economy, 1770-1873*, London: James Currey.

Singh, Raghubir, 1986, *Kerala: The Spice Coast of India*, London: Thames and Hudson.

Sivanandan, P., D. Darayana and K. Narayanan Nair, 1985, Land Hunger and Deforestation: A Case Study of the Cardamom Hills in Kerala, Ulloor: Centre for Development Studies Working Paper No. 212.

's Jacob, Hugo, 1976, De VOC en de Malabarkust in de 17e Eeuw [The Dutch East India Company and the Malabar Coast in the 17th Century], in: Meilink-Roelofsz (ed.), *De VOC in Azië* [The Dutch East India Company in Asia], Bassum: Fibula-Van Dishoeck.

—, 1986, Dutch Memoirs and Instructions on Kerala, in: *Journal of Kerala Studies*, vol. 13, parts 1-4, pp. 45-58.

Shlomowitz, Ralph, 1981, Market for Indentured and Time-Expired Melanesian Labour in Queensland, 1863-1906, in: *The Journal of Pacific History*, vol. 16, nos. 1-2, pp. 70-91.

Smail, John R.W., 1961, On the Possibility of an Autonomous History of Southeast Asia, *Journal of Southeast Asian History*, vol. 2, pp. 72-102.

Somarajan, C.N., 1983, A Survey of Land Reforms in Kerala upto 1956, in: *Journal of Kerala Studies*, no. 10, parts 1-4, pp. 131-7.

Stavrianos, L.S., 1981, *Global Rift; The Third World Comes of Age*, New York: Quill.

Stoler, Ann Laura, 1985, *Capitalism and Confrontation in Sumatra's Plantation Belt, 1870-1979*, New Haven: Yale University Press.

—, 1989, The Company's Women: Labour Conditions in Sumatran Agribusiness, in: K. Young (ed.), *Serving Two Masters; Third World Women in Development*, New Delhi: Macmillan India, pp. 49-66.

—, 1990, Making Empire Respectable: The Politics of Race and Sexual Morality in 20th-Century Colonial Cultures, in: J. Breman (ed.), *Imperial Monkey Business; Racial Supremacy in Social Darwinist Theory and Colonial Practice*, CASA Monographs No. 3, Amsterdam: VU University Press, pp. 35-70.

—, 1992, 'In Cold Blood': Hierarchies of Credibility and the Politics of Colonial Narratives, in: *Representations*, vol. 37, Winter, pp. 151-89.

Struben, Paulette Y., 1992, State Formation of Travancore in the Eighteenth Century, paper presented at the Seminar on European Expansion: Trade and the World Economy, Lisbon, 28 September-4 October.

Sugeetha, G., 1981, The Constitutional Progress in Travancore in the 19th and 20th Centuries, in: *Journal of Kerala Studies*, no. 8, parts 1-4, pp. 7-27.

Tharakan, P.K. Michael and Sunil Mani, 1984, Competitiveness in the Indian Tyre Industry: 1936-84, Ulloor: Centre for Development Studies Working Paper No. 203.

Tharakan, P.K. Michael, 1976, Migration of Farmers from Travancore to Malabar, from 1930 to 1960: An Analysis of its Economic Causes, M. Phil. Thesis, Jawaharlal Nehru University, Delhi.

—, 1984a, Socio-Economic Factors in Educational Development; Case of 19th Century Travancore (Part 1), in: *Economic and Political Weekly*, vol. 19, no. 45, 10 November, pp. 1913-28.

—, 1984b, Socio-Economic Factors in Educational Development; Case of 19th Century Travancore (Part 2), in: *Economic and Political Weekly*, vol. 19, no. 46, 17 November, pp. 1959-67.

Tharakan, P.K. Michael, 1990, Socio-Religious Reform Movements and Demand for Indicators of Development: Thiruvithamkoor 1860-1930, paper presented at the 11th European Conference on Modern South Asian Studies, Amsterdam, 2-5 July.

Tharamangalam, Joseph, 1984, The Penetration of Capitalism and Agrarian Change in South-West India, 1901 to 1941: A Preliminary Analysis, in: *Bulletin of Concerned Asian Scholars*, vol. 16, no. 1, January-March, pp. 53-62.

Tharian George, K., 1984, The Crises of the South-Indian Tea Industry: Legacy of the Control by British Tea Multinationals, Ulloor: Centre for Development Studies Working Paper No. 191.

Tharian George, K. and P.K. Michael Tharakan, 1986, Penetration of Capital into a Traditional Economy; The Case of Tea Plantations in Kerala, 1880-1950, in: *Studies in History*, 2.2 n.s., pp. 199-229.

Tharian George, K., V. Haridasan and B. Sreekumar, Role of Government and Structural Changes in Rubber Plantation Industry, in: *Economic and Political Weekly*, vol. 23, no. 48, 26 November, pp. m158-m166.

Thomas, E.J., 1985, *Coalition Game Politics in Kerala After Independence*, New Delhi: Intellectual Publishing House.

Thomas, K.D. and J. Panglaykim, 1976, The Chinese in the South Sumatran Rubber Industry; A Case Study in Economic Nationalism, in: J.A.C. Mackie (ed.), *The Chinese in Indonesia: Five Essays*, Melbourne: Thomas Nelson, pp. 139-98.

Thomas Isaac, T.M. and P.K. Michael Tharakan, 1986, An Enquiry into the Historical Roots of Industrial Backwardness of Kerala; A Study of Travancore Region, Ulloor: Centre for Development Studies Working Paper No. 215.

Thompson, Edgar T., 1983, *The Plantation; An International Bibliography*, Boston: C.K. Hall.

Tinker, H., 1974, *A New System of Slavery: The Export of Indian Labour Overseas 1830-1920*, London: Oxford University Press.

Törnquist, Olle, 1991, *What's Wrong with Marxism?; On Peasants and Workers in India and Indonesia*, New Delhi: Manohar Publications.

Turlach, M., 1970, *Kerala: Politisch-Soziale Struktur und Entwicklung eines Indischen Bundeslandes* [Kerala: The Socio-Political Structures and Development of an Indian State], Wiesbaden: Harrassowitz.

Uma Devi, S., 1989, *Plantation Economies of the Third World*, Bombay: Himalaya Publishing House.

Unny, R.G. and George Jacob, 1972, *Rubber Small Holdings in India; Report of a Sample Survey 1969-1970*, Kottayam: Rubber Board.

Varghese, T.C., 1970, *Agrarian Change and Economic Consequences; Land Tenures in Kerala, 1850-1960*, Bombay: Allied Publishers.

Vijayakumar, K., 1981, The Influence of Caste in Kerala Politics: A Historical Perspective, in: *The Journal of Kerala Studies*, vol. 8, parts 1-4, pp. 259-66.

Wallerstein, Immanuel, 1974, *The Modern World System, Part 1, Capitalist Agriculture and the Origins of the European World-Economy in the 16th century*, New York: Academic Press.

—, 1980, *The Modern World System, Part 2, Mercantilism and the Consolidation of the European World-Economy, 1600-1750*, New York: Academic Press.

—, 1986, Incorporation of Indian Subcontinent into Capitalist World-Economy, in: *Economic and Political Weekly*, vol. 21, no. 4, 25 January, pp. PE28-PE39.

—, 1989, *The Modern World System, Part 3, The Second Era of Great Expansion of the Capitalist World-Economy, 1730-1840s*, New York: Academic Press.

Watson Andaya, Barbara, 1993, Cash Cropping and Upstream-Downstream Tensions: The Case of Jambi in the Seventeenth and Eighteenth Centuries, in: A Reid (ed.), *Southeast Asia in the Early Modern Era; Trade, Power and Belief*, Ithaca: Cornell University Press, pp. 91-122.

Werff, Peter Van der, 1992, *Modern Poverty; The Culture of Distribution and Structural Unemployment in the Foothills of Kerala*, New Delhi: Manohar.

Wertheim, Wim F., 1990, On Sugar, Tobacco, Rubber . . . and Coolies; A Comparison between Deli and Java Plantations, paper presented at the Conference on Capitalist Plantations in Colonial Asia, Centre for Asian Studies Amsterdam, 26-29 September.

Wesumperuma, D., 1990, Plantation Labour System in Colonial Sri Lanka, paper presented at the Conference on Capitalist Plantations in Colonial Asia, Centre for Asian Studies Amsterdam, 26-29 September.

Wolf, Eric R., 1982, *Europe and the People Without History*, Berkeley: University of California Press.

Xaxa, Virginius, 1985, Colonial Capitalism and Underdevelopment in North Bengal, in: *Economic and Political Weekly*, vol. 20, no. 39, 28 September, pp. 1659-65.

Yesudas, R.N., 1977, *British Policy in Travancore, 1805-1859*, Trivandrum: Kerala Historical Society.

E. Informants

Informant, Abraham Kurien and his son Abraham Eappen (Palampadom): Owners of large rubber estates.

—, Alex Manuel: Labour Adviser, United Planters' Association of Southern India.

—, N.A. Balagovind: Deputy Marketing Manager, Trivandrum Rubber Works.

—, Estate workers, Gudurale and Pasupara Estates.

—, Hannibal Dominic (Paripparambil): Son of the prominent Indian planting pioneer Dominic Thommen.

—, Dr George Tharian: Scholar, employed by the Economic Division of the Rubber Research Institute of India.

—, Dr V. Haridasan: Deputy Director, Economic Division, Rubber Research Institute of India.

—, J. Itty: Rubber small grower.

—, Joseph: Plantation worker, employed on a large Indian-owned estate.

—, K.J. Joseph: Secretary, Association of Planters of Kerala.

—, V.J. Joseph and V.J. Shaju: Rubber small growers.

—, Michael A. Kallivayalil: Planter and brother of Jose A. Kallivayalil.

—, C.A. Kurien: President, Devikolam Estate Workers; President, National Union of Plantation Staff; State AITUC Secretary; General Secretary of Plantation Workers Federation; ex-member Legislative Assembly.

—, N.E. Kurien and N.K. Elias: Rubber small growers.

—, Marthanda Varma and Karthika Thirunal Maharani: Respectively brother and sister of the last and late Maharaja of Travancore.

—, P. Madhavan: Secretary, National Union of Plantation Staff.

—, M. Mathew: Rubber small grower.

—, M.C. Mathew: Group Manager, Harrisons Malayalam Limited.

—, P.K.V. Nair: Chief Minister of Kerala (October 1978–October 1979) and prominent member, Communist Party of India.

—, E.M.S. Namboodiripad: Chief Minister of Kerala (April 1957–July 1959 and March 1967–October 1969) and prominent member, Communist Party of India (Marxist).

—, M.Y. Ouseph: General Secretary, Devikolam Estate Workers.

—, P. Sethuram: Secretary, Cochin Chamber of Commerce and Industry.

—, P.S. Sundar: Information Manager, United Planters' Association of Southern India.

Informant, Suresh: Assistant Manager, employed on a large Indian-owned estate.

—, P.K. Michael Tharakan: Fellow, Centre for Development Studies and nephew of Jose A. Kallivayalil.

—, K.T. Thomas and T.J. Joseph: Representatives, Central Travancore Agricultural Association.

—, A. Venugopal: Head, Personnel and Legal Department, Tea Estates India Ltd.

—, K.U. Warier: Journalist, *New Age* and *New Kerala.*

Index